Video Scrambling & Descrambling

for Satellite & Cable TV

Rudolf F. Graf & William Sheets

HOWARD W. SAMS & COMPANY
A Division of Macmillan, Inc.
11711 North College, Suite 141, Carmel, IN 46032 USA

PLEASE NOTE
This work is published for the commercial and hobbyist fields to increase the awareness of the many non-video opportunities existing in the satellite communications field. The publisher and the authors do not approve or condone the misuse of the information in this book, in any manner, by any individual, company, or group.

FIRST EDITION
FIFTH PRINTING—1990

International Standard Book Number: 0-672-22499-2
Library of Congress Catalog Card Number: 86-62918

Acquisitions Editor: *Greg Michael*
Editor: *Frank N. Speights*
Designer: *T. R. Emrick*
Illustrator: *Don Clemons*
Cover Artist: *Gregg Butler*
Compositor: *Pine Tree Composition, Inc., Lewiston, Maine*

Printed in the United States of America

Contents

Preface

Scrambling, a technique which has caused confusion, anxiety, rumors, and ''doom and gloom'' in the TVRO (television receive only) industry, is the subject of this book, and, hopefully, a lot of questions on the use of scramblers and the subject of scrambling will be answered.

We started kicking around the idea for this book early in 1985, partly out of curiosity, and partly for the opportunity to write a book, and slowly became aware of the need for such a book. Most of the material that seemed available was bits of information listed in the small, obscure advertisements on the back pages of several video and electronics magazines. Much research had to be done. Naturally, manufacturers of scrambling and descrambling equipment are not too eager to give out this information. Now and then, however, some articles on the subject were published, but they were few and far between. As more information was gathered, it became apparent that there was much duplication of data, and, therefore, increasingly difficult to get something new. Thus, we felt that there was a need for this book.

It has all the information that a videophile, TV technician, hobbyist, or anyone else interested in television scrambling would likely need, in one place, without spending several hundred dollars for all this ''privileged'' information. Most of the currently used methods (June 1986) are discussed, with the necessary technical background included. Since digital scrambling methods will undoubtedly supplant analog methods, we have included information on digital/analog conversion, phase-locked loops, digital techniques, and IC data sheets that the average reader might not be able to find easily without access to all the manufacturers' literature. Satellite scrambling is discussed in detail, and the U.S. patents on three satellite systems have been included for reference. Several representative decoders are discussed in detail, along with the construction details for several ''typical'' circuits.

This book should prove useful for the reader who wants to be ''brought up to speed'' on the subject of scrambling. It provides the basic background information needed to understand the subject, and assumes little or no previous knowledge of scrambling. We stress theory and circuit techniques, not specific decoders. When the reader becomes familiar with the theoretical concepts herein, he or she will have no trouble in understanding specific decoder models and modifications. It is better to know the ''how and why'' rather than just a few detailed specifics that are applicable in only limited cases.

We have also discussed some of the political and legal developments in force at the time of the writing of the book. This information may be somewhat dated by the time the book is published, but it makes for interesting reading. Many issues have yet to be resolved and new ones will surely come up.

The authors do not intend for this book to be a ''video pirates manual.'' There are laws about the unauthorized viewing of pay-TV services and the use of illegal decoders. The information in this book is presented for experimental and educational purposes only. We make no guarantees of any kind whatsoever that the decoders described herein will perform in any way, or to any standard or performance level. The circuits are merely representative of practical decoders. We apologize, however, for any technical inaccuracies, contradictions, or errors. Much of our information has come from scattered sources and some of it had to be pieced together and clarified. A lot of material is presented which had to be rewritten or edited to fit this book.

We would like to thank all the companies and their personnel that provided us with technical assistance, illustrations, and data sheets. Specifically, we thank M/A-COM LINKABIT, Inc., Scientific Atlanta, Oak Orion, Motorola, National Semiconductor, RCA, Fairchild, TRW, Intel, AT&T, and anyone that we may have forgotten.

We also extend our thanks and appreciation to our typist, Mrs. Stella Dillon, whose cooperation and skill at the word processor helped us produce the manuscript for this book in record time.

New developments are coming at a fast pace, and it is not easy to stop writing, and adding, but we have to call it a day somewhere. Thank you and enjoy the book.

RUDOLF F. GRAF AND WILLIAM SHEETS

Dedications

To my wife, Bettina, in appreciation for her patience and understanding.

RUDOLF F. GRAF

To my daughter, Suzanne, who has been anxiously waiting to see her daddy's book in print.

BILL SHEETS

Note to the Reader

As of October 9, 1986, VideoCipher II has reportedly been successfully pirated—both video and audio. While the *absolute* truth of this fact is still somewhat in doubt, there has been enough similarity in the rumors, from several unassociated sources, to substantiate the facts beyond a reasonable doubt. It has reportedly been done by several firms in the United States and Canada. We do not have details at this time, but much concern has been raised in the industry as to the security of VideoCipher II. A large amount of money has been invested by both program suppliers and subscribers in what may turn out to be an obsolescent system.

R.F.G. & W.S.

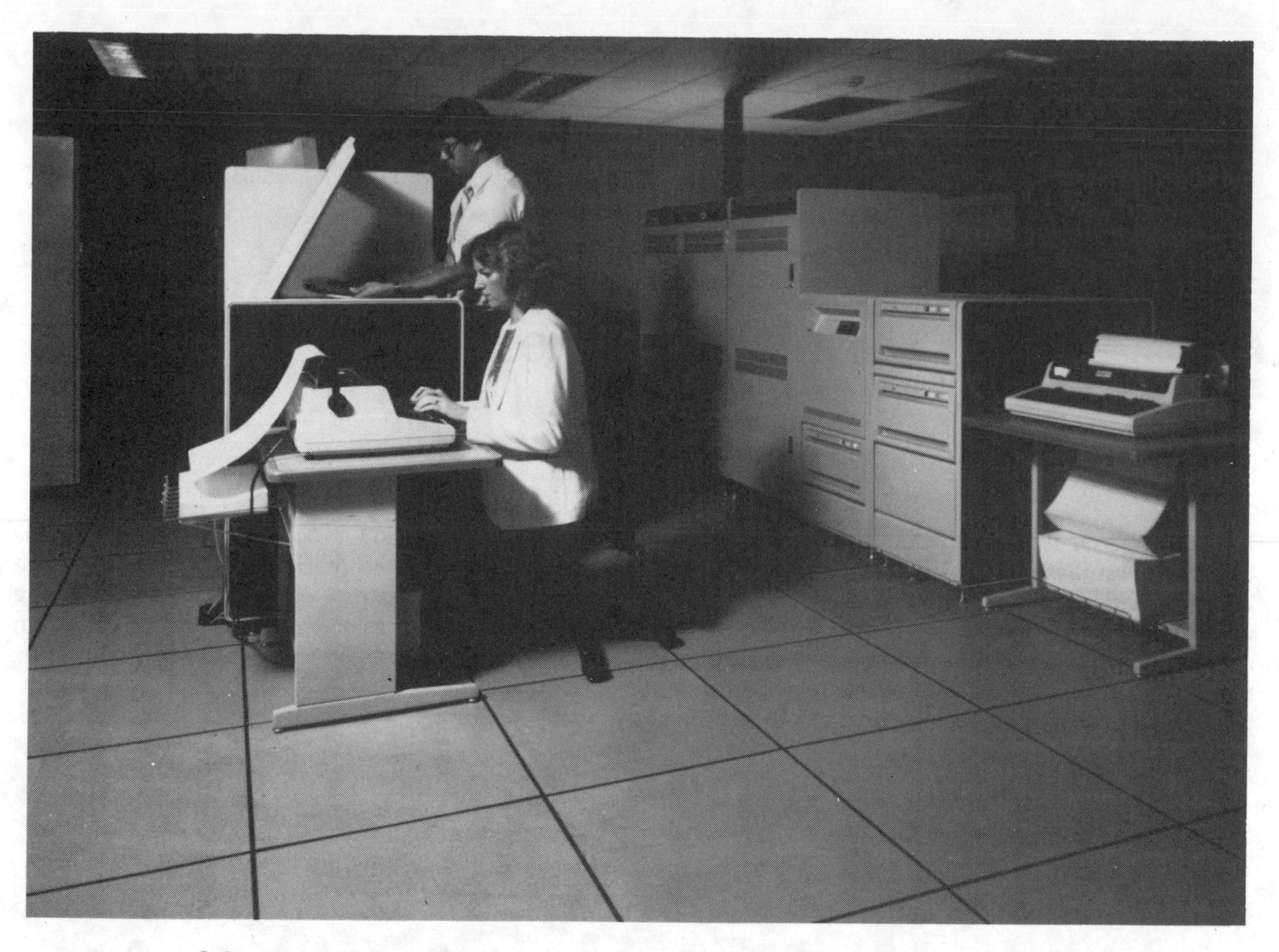

Subcenter authorization center for VCII™ system (Courtesy MA-COM LINKABIT, Inc.).

1

Introduction

The growth of video, and the increase in both the number of homes with cable and the variety of cable channels, has given rise to the concept of ''pay TV'' as an alternative to common ''on the air'' routine programming. The pay-TV concept is to provide channels whereby programming is made available to the viewer, either for a monthly fee or as a program-by-program ''pay as you watch'' arrangement. A problem of the pay-TV company is how to collect the charges for individual programs or premium channels and how to prevent unauthorized reception by nonpaying viewers (or the downright theft of service).

Encoded Signals

Encoding the signal so as to make it unwatchable and/or inaudible to nonsubscribers is commonly referred to as ''scrambling.'' The signal is broadcast on a given channel, either on cable or over the air, and then appears, in its scrambled form, at the video output (TV set) as ''garbage,'' with no visual, and maybe no audible, resemblance to the original program. No adjustment of the television receiver will produce a normal picture and/or sound. To receive the signal, a device called a ''decoder'' must be used, generally between the antenna and the television receiver. It is possible to build one *into* a TV receiver, but unless the receiver is specially designed for the decoder, anyone with less electronics know-how than an expert technician or engineer would have a difficult time doing so.

Scrambling Methods

Many methods have been developed to accomplish scrambling and descrambling. (Scrambling is done at the transmitter or by the cable-TV company while descrambling is done at the individual receiver.) Some scrambling schemes are basically very simple and provide only minimal security from theft of programs (soft scrambling), while others are very sophisticated and provide a high level of security. The digital techniques and low-cost mass-produced IC chips of today make a secure scrambling system economically feasible for home use.

The cost of the decoder must be borne either by the cable company or the viewer, and it must be amortized over a period of time. Thus, to be practical for home use, a scrambling system has to be low cost. Not many people are going to pay more than, say, $20.00 a month for any premium service and, if they do, they will not subscribe to many of these unless they are either very affluent or avid TV addicts. Also, not everyone will be happy having several boxes for separate services in their home, all wired up to their TV receiver. Therefore, some standardization in equipment is necessary, but there is a trade-off between security and cost. And with the many premium services available, cost is really important, because companies are all competing for an audience whose budgets will only accommodate a few pay channels. Therefore, some formidable economic problems can arise.

Descrambling

An approach that uses a standardized descrambler setup would undoubtedly be the best solution. However, sooner or later, someone may defeat the system or even provide bootleg descramblers. This problem would not seriously affect a local cable operator, but it could be disastrous to a nationwide satellite-TV scrambling operation. In the case of satellite TV, the many rural areas in the United States, formerly deprived of TV or severely limited in their choice of the three networks, could be a lucrative market for illegal (private) satellite descramblers. Many rural residents have a financial stake in some type of satellite system and, having spent several thousand dollars, would be rather unhappy to have their new-found entertainment cut off by scrambling, especially since they may be denied access due to selfish CATV (community antenna television/cable television) interests. Satellite-TV scrambling may not work out, due to competition from equally good programming that is unscrambled, free to all viewers, and paid for by advertising revenues. However, if satellite program scrambling does succeed, let us hope that suitable satellite-TV descramblers will be made readily available to everyone at a price that the average working man with a family can afford. Country dwellers, with no cable TV, should have the same access to programs that city dwellers have, and at the same cost.

Coding/Decoding Systems

We will discuss several scrambling-descrambling systems, which are either in use or possible in theory; both system operation and circuitry will be described. Also, we will describe several decoders that can be built by an experimenter for his own use and experience.

However, in order to understand these systems, some knowledge of the structure of a video signal (unscrambled, normal content) is required. We will briefly describe such a signal, emphasizing the material necessary for understanding the encoding and decoding processes. For more detailed information, consult one of the many TV engineering or servicing books that are to be found in your local electronics store or library.

Don't Be a Pirate

We wish to make it absolutely clear that the information given in this book is *not* intended to aid the "video pirate," or anyone else, who might be planning the theft or unauthorized reception of commercial scrambled signals. (At the time of this writing, there may be legal consequences to this.) Therefore, do not attempt to illegally "rip off" your local cable company, MDS System, or pay-TV station. It is better to subscribe, pay the fee, and know that if you do not like the programming, you can get rid of the service, and you will not be stuck with several hundred dollars worth of "purchased" decoder boxes, or all the lost time and effort occurred in building one. The authors have heard of several instances where private individuals and business have been prosecuted by the cable companies, and have received either fines, jail sentences, or judgments against them for the theft of services. So be duly warned about the illegality and possible dire consequences of being a "video pirate."

Standard NTSC Format

We will discuss those systems that use the standard NTSC (National Television System Committee) video signal format (USA, Canadian, etc.). Referring to Fig. 1-1, a simplified diagram of an NTSC signal (the common, ordinary, composite video signal transmitted by your local TV station), several things are evident. The video information, which may have frequency components up to 4 MHz or so, appears between periodic, regularly shaped pulses of large amplitude. These signals are the horizontal blanking (narrow) and vertical blanking (wide) pulses. The horizontal blanking pulse has a narrower pulse "piggybacked" on top of it. This narrower pulse is called the synch pulse. These synch pulses occur during the sweep retrace of the video signal and normally do not produce a visible component on the TV image. They are used to synchronize the TV horizontal sweep circuitry to the video signal. The widest part of the pulse is called the blanking pulse, and has an amplitude such that it reduces the luminance on the CRT (cathode-ray tube) to zero. Right after the horizontal synch pulse trailing edge, but before the blanking pulse trailing edge, is a signal consisting of eight sine-wave cycles at 3.58 MHz (actually 3.579545 MHz). This signal is the color synch (burst) signal; it is used to synchronize the color circuitry in the TV receiver and turn it on for color programs. The color synch signal is not used if the transmitted program is black and white. Thus, in its absence (during a B/W program), the color circuitry in the TV receiver automatically shuts down. This prevents colored snow and streaks from interfering with the black-and-white program.

Scanning

There are 15,734 pulses transmitted per second; this corresponds to 525 lines of 59.94 fields transmitted per second. Each field consists of 262½ lines (interlaced scanning), and two fields make a frame. Alternate lines are scanned on each field. (For example; lines 1, 3, 5, 7, etc., of the first field are scanned, and then, lines 2, 4, 6, 8, etc., of the second field are scanned.) For simplicity, we will round off these numbers to 15,750 horizontal and 30 vertical frames per second. Note, that the vertical pulses occur at a

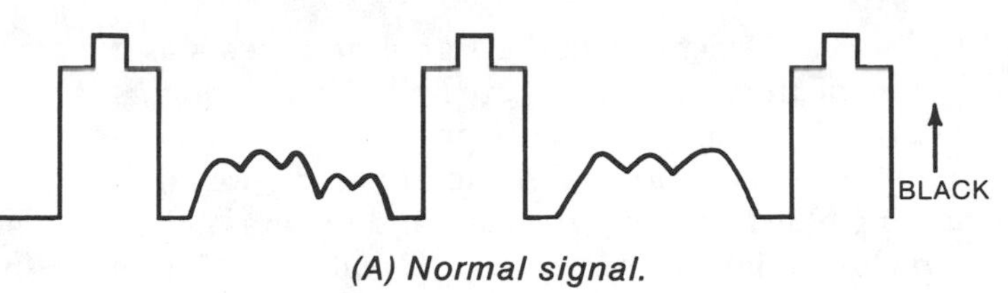

(A) Normal signal.

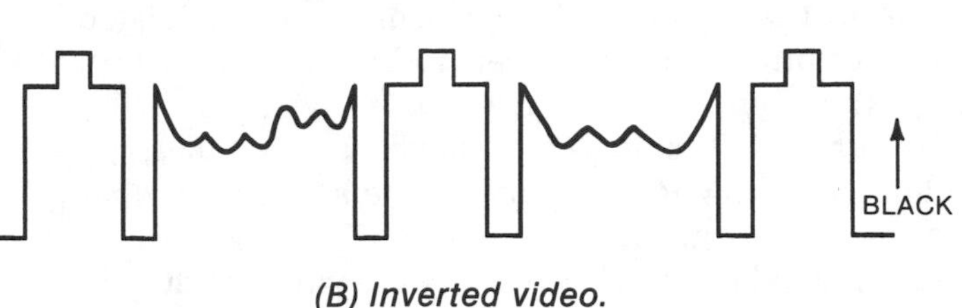

(B) Inverted video.

Fig. 1-1. A simplified diagram of an NTSC signal.

60-Hz rate, but that alternate pulses are slightly different, due to the interlacing. The main point to remember is that these horizontal and vertical pulses—and the color burst signal—are used for the timing and synchronization necessary to reproduce the TV picture. If any of these pulses are absent, distorted, or altered, it becomes difficult or impossible to synchronize the TV picture. The screen will roll, tear, and the colors will be incorrect, rendering the picture unwatchable. Naturally, the video level must not be so great as to exceed the synch pulse level, otherwise the level-sensitive synch circuits will be "confused" and the TV picture will not synchronize properly.

Audio Transmission

Audio information is frequency modulated on a 4.5-MHz signal that is added into the composite video signal, but it may also be modulated on a carrier that is 4.5 MHz higher or lower than the picture carrier frequency. This carrier is like a standard FM (frequency modulated) radio signal except the deviation is limited to ±25 kHz (monaural) instead of the ±75 kHz used for FM broadcasting. In the TV receiver, this 4.5-MHz signal appears at the video detector and is picked off and processed in a 4.5-MHz limiter/detector (or quadrature demodulator) in much the same way as the 10.7-MHz system used in an FM broadcast receiver. If the signal is not 4.5 MHz, the sound will be absent and only random noise (hiss) will be heard.

Chroma Transmission

The color signal is a bit more complex to understand but, basically, it is a 3.58-MHz (actually 3.579545 MHz) signal that is both amplitude modulated and phase modulated. The amplitude of this signal determines the saturation of a color (whether it is, for example, white, light pink, rose, or red), and the phase determines the hue (whether it is red, orange, yellow, etc.). If the burst-signal phase is used as a reference, a signal in phase with it would produce a greenish-yellow hue. It is important to understand that the *amplitude* of the composite color signal is used to determine the *difference* from white; *not the intensity* of a particular color. In this way, a color signal with a zero level would produce a white raster, allowing compatibility between color and black-and-white reception on the same TV receiver. Whether the color is dark red or bright red is determined by the luminance component of the signal.

We will not go any further into TV color theory here but, instead, will refer the reader to any good text on television, since proper treatment of the subject requires a lot more space than this book permits. However, it may now be evident that by controlled alteration, the three key components of the TV signal can be used as a means to encode or "scramble" a TV picture (video signal). This is what will be discussed later. In addition, the video information between the synch pulses may also be modified in a controlled manner. The picture may not necessarily be scanned in the conventional way, and it is possible to continuously modify the scrambling method at various times—maybe even several times per second, or even every frame.

Some Scrambling Techniques

Several simple scrambling schemes are in wide use today on both cable and pay-TV systems. One of the simplest techniques used is "video inversion." In this method, the video information is sent with reversed polarity. The TV set may or may not synch on it (depending whether the synch pulses are inverted also).

If the set can synchronize the signal, it appears as a negative picture—dark areas are light, and highlights are dark. Colors are reversed, appearing complementary to those in the original scene. If you do not mind watching blue faces, brown skies, and red grass, then the picture might be watchable. Usually nothing is done to the sound, although it may be modulated onto an ultrasonic subcarrier (to be discussed later), and the sound carrier may have what is called "barker audio" on it. This is usually a taped message telling you how to subscribe to the service, or what you are missing. The inverted video signal is easily unscrambled by merely inverting the video polarity in a video amplifier. This can easily be done in the TV set. The only precaution is that it must be done at a suitable point in the video system. For example, if the synch pulses are not inverted, video inversion must be done after synch takeoff. Some sets use separate detectors for sound/synch takeoff. Fig. 1-1B shows this method of scrambling.

A More Sophisticated Approach

A more sophisticated method is the inverting of the video signal at several intervals during the frame or every several frames (or during alternate scan lines). However, this results in a picture with a very annoying flicker or a superimposed pattern. The picture would be unwatchable, unless you did not mind severe eyestrain or a headache. Fig. 1-2 illustrates this method. Decoding this type of signal requires an arrangement where the video polarity can be instantaneously changed in synchronization with the scrambling signal. Fig. 1-3 is a block diagram of such a system. The descrambling information can be on the sound carrier, or sent with the picture itself, or obtained from the synch section of the TV set.

If alternate frames are inverted, the eye will tend to add the adjacent frames due to persistance of vision. This would show up as either a weak positive or negative image, or, ideally, a blank raster. However, the video inversion method and its more sophisticated versions are not very "secure," in the sense that descrambling is easily accomplished. Video inversion is best used along with a more complex method that also acts on the synch pulses.

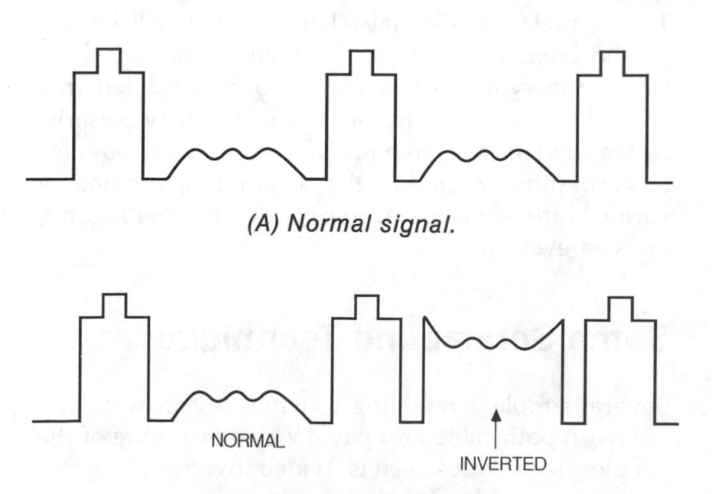

(A) Normal signal.

(B) Changed signal.

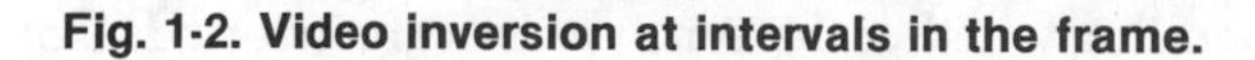

Fig. 1-2. Video inversion at intervals in the frame.

Sine-Wave Scrambling

A scrambling method that is currently popular is the "sine wave" method. This technique uses the addition of a 15.75-kHz (or other frequency) sine wave to the video signal. If the negative peak of the sine wave corresponds to the positive peak of the signal (synch pulse) or vice versa (see Fig. 1-4), the synch signal is suppressed below the peak video level. This "confuses" the synch separator circuits in the TV set and they cease to function properly. The resultant picture "tears" and "rolls," and there is a vertical dark bank or bands through the middle of the picture. Colors are lost or out of synch, since the TV set cannot tell where the synch pulse is. Some video signals that have mostly very light areas (highlights) may momentarily synchronize properly, but most will roll, tear, or distort. The picture is generally unstable and cannot be watched. Unscrambling is done by taking the "scrambled" signal and mixing it with a sine wave of proper phase so as to cancel the added sine wave. This restores the proper synch-tip level and cancels the sine-wave component added to the video. The sine wave is usually obtained from a phase-locked loop (PLL) circuit that is locked to a 15.75-kHz signal encoded on the audio signal. This method of synch recovery operates very much like a PLL that is used for stereo FM pilot-carrier regeneration. In fact, the circuitry is very similar and FM stereo chips can be used to accomplish this in a descrambler circuit. See Fig. 1-5 for a block diagram illustrating the details. This will be discussed in a subsequent chapter.

Synch Alteration Techniques

Another method, similar in that it operates by upsetting the synch-pulse amplitude in relation to the video signal, is shown in Fig. 1-6. This method is called "gated synch" scrambling, and it has the effect of making the picture tear and roll. The picture acts like the synch pulses are missing, and the synch circuitry, which generally picks out the most positive (or

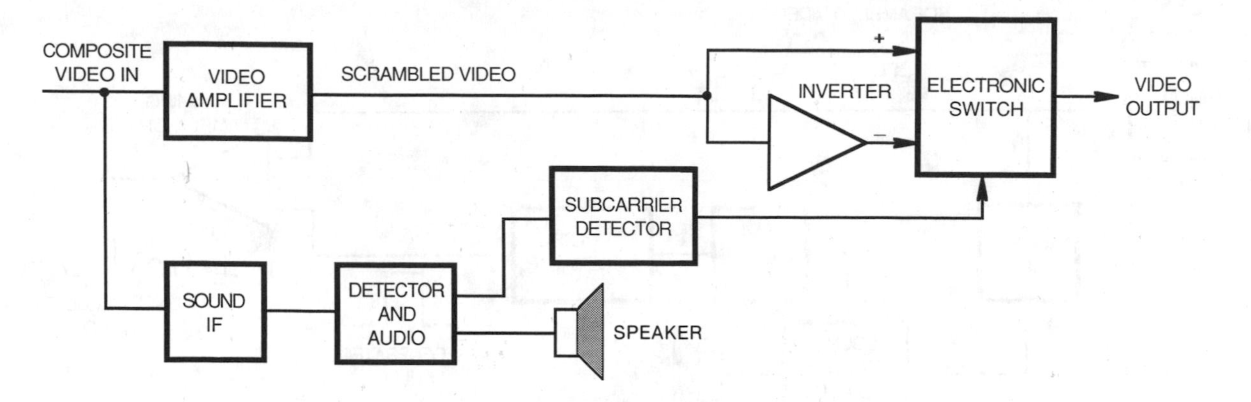

Fig. 1-3. Block diagram of a circuit for changing video polarity.

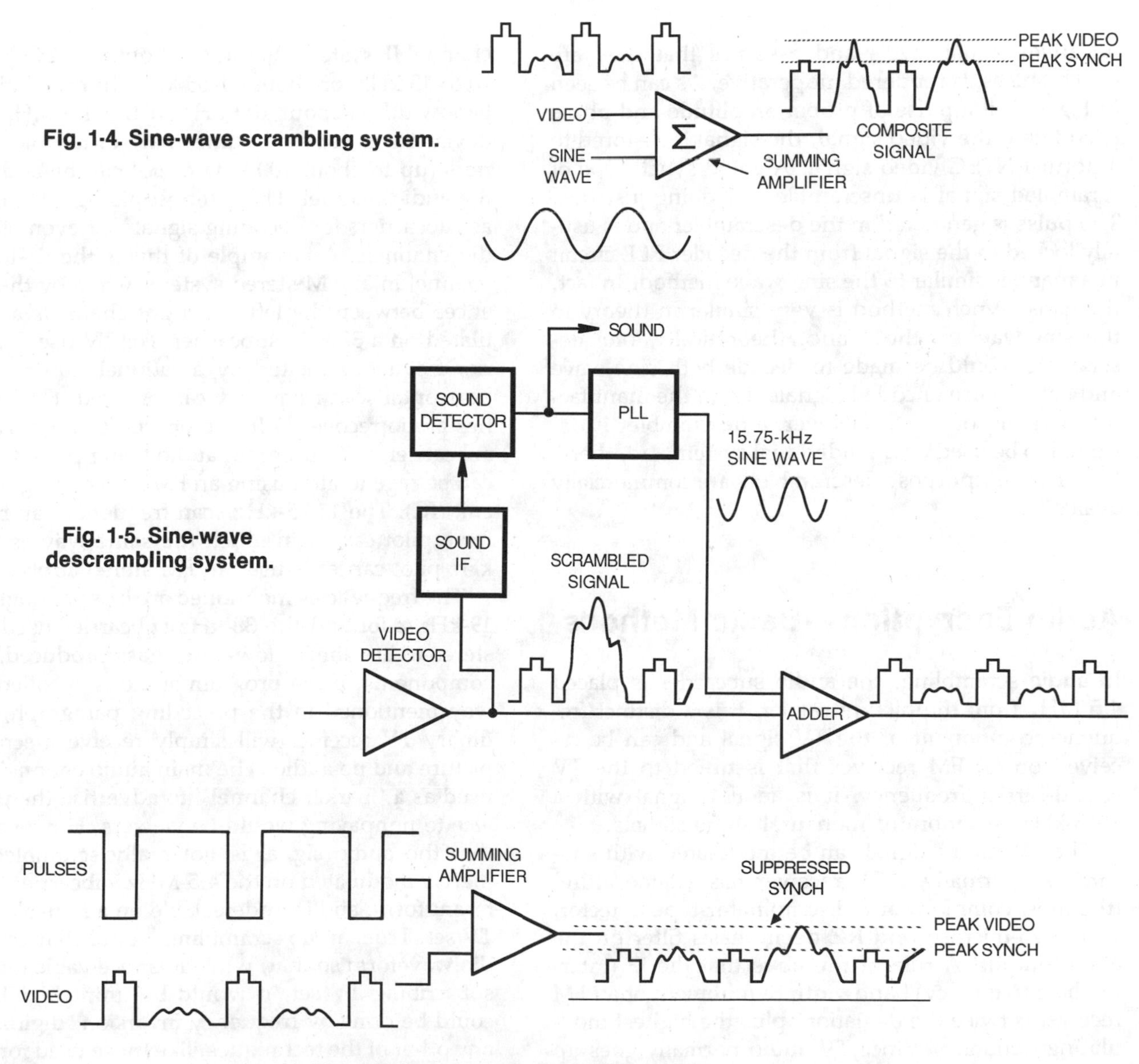

Fig. 1-4. Sine-wave scrambling system.

Fig. 1-5. Sine-wave descrambling system.

Fig. 1-6. Gated pulse scrambling system.

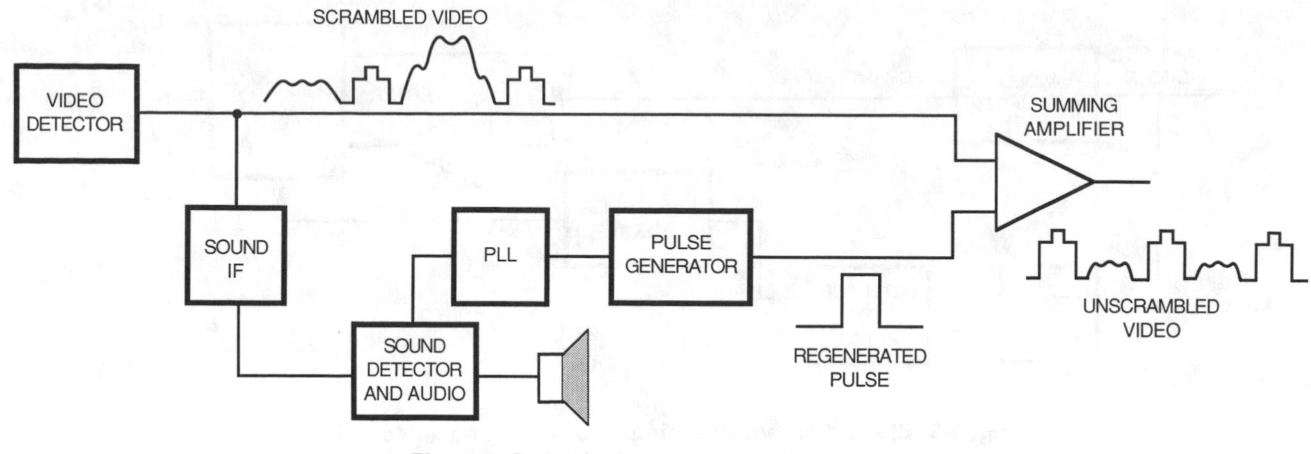

Fig. 1-7. Gated pulse descrambling system.

negative) signal levels and assumes that they are synch pulses, is rendered inoperative. As can be seen in Fig. 1-7, if a pulse of proper amplitude and phase is added to the video signal, the signal is restored to a normal NTSC video signal. A gated synch type of scrambled signal is unscrambled by doing just this. The pulse is generated in the descrambler and is usually locked to the signal from the decoder PLL circuit in a manner similar to the sine-wave method. In fact, the gated synch method is very similar in theory to the sine-wave method, and, theoretically, one descrambler could be made to decode both sine-wave and gated pulse encoded signals. From the manufacturers' point of view, however, a descrambler is designed to be used on an individual system and, therefore, no multipurpose descramblers are commercially available.

Audio Encryption—Basic Methods

In audio scrambling, the audio subcarrier is placed 4.5 MHz from the picture carrier. It is a distinct frequency component of the TV signal and can be received on an FM receiver that is tuned to the TV sound-carrier frequency. It is an FM signal with a ±25-kHz deviation for monaural audio signals.

The TV sound signal can be modulated with subcarriers. Normally, a TV receiver has a quadrature- (the most common) or a discriminator-type detector, with a 75-microsecond RC deemphasis filter on the audio output. A rule of thumb is that the IF (intermediate frequency) bandwidth requirement of an FM receiver is twice the deviation, plus the highest modulating frequency. Since TV audio normally goes up to about 10–15 kHz, the bandwidth of the audio-channel IF system should be about 2 × 25 kHz, plus 10 to 15 kHz, or about 60–65 kHz. In practicality, the bandwidth is about 100 kHz. If the 4.5-MHz carrier deviation is reduced somewhat, ultrasonic subcarriers (up to about 100 kHz or so) can be added into the audio channel. These ultrasonic signals can serve as subcarriers for decoding signals, or even other audio channels. An example of this is the (L–R) audio channel in an FM stereo system, whereby the differences between the left and right channels are modulated on a 38-kHz subcarrier. For TV use, a subcarrier frequency related by a rational number to the horizontal scan frequency can be used. (This simplifies audio recovery.) It is practical to use a 31.5-kHz subcarrier, with program audio being placed on it. It can be regenerated using an FM stereo integrated circuit (IC). The 15.75-kHz scan frequency can be used for a pilot carrier in much the same way as the 19-kHz pilot carrier is used in FM stereo applications.

The frequencies mentioned are close enough to the 19-kHz pilot and the 38-kHz subcarrier used in FM stereo to use similar low-cost, mass-produced, circuit components. If the program audio is encoded in the way mentioned in the preceding paragraph, an ordinary TV receiver will simply receive a scrambled picture and no audio. The main audio channel can be used as a "barker channel" to advertise the pay service to nonpaying would-be viewers. However, note that the audio signal is not really scrambled. It is merely modulated on the 4.5-MHz subcarrier in a different form, and is undetectable on a nonsubscribing TV set. True, audio scrambling would change the audio waveform so that, while it is receivable on a nonsubscribing TV set, it would be unintelligible. This could be done by frequency inversion, digitizing, or any other of the techniques, like those used for secure 2-way radio systems. However, this must be done so

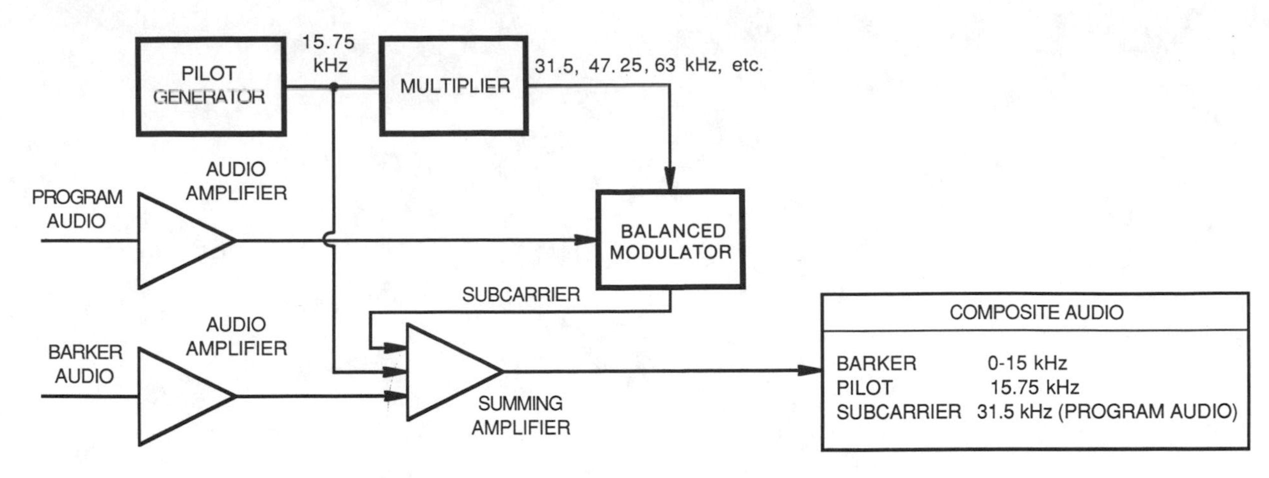

Fig. 1-8. An audio encoding-scheme block diagram.

as not to restrict the usable audio bandwidth. Fig. 1-8 shows an audio encoding-scheme block diagram.

Cable System Approaches

The common-box type of descrambler used in CATV systems generally consists of an RF (radio frequency) converter to convert all cable channels to one fixed channel, usually Channel 3 or 4. This way, the descrambler circuits only have to operate on one channel; this eliminates tuning many circuits simultaneously. In the case of sine-wave or gated synch descramblers, some actually detect the video signal and extract the necessary decoding signals from it, and then use these decoding signals to act upon the CH3 or CH4 IF signals. By using a variable-gain IF amplifier on CH3 or CH4, the decoding signals can modulate the IF gain, thus reestablishing the proper synch signal level and canceling the sine-wave modulation, (or reestablishing the synch pulse level, in the case of gated synch systems). The encoding and decoding signals may also be supplied in CATV systems at frequencies outside the received channels (called Outband Systems). Many different approaches are possible but the gated synch and sine-wave systems and the Zenith SSAVI™ (Synch Suppression and Video Inversion Techniques) system have been popular scrambling schemes for both cable and over-the-air systems.

Another method used on some cable systems is the deliberate insertion of keyed carriers in the video channel used for premium TV shows. An RF carrier can cause severe cross-hatching. Keying the carrier causes an annoying flicker, and the resultant picture looks like a severe case of TVI (television interference) caused by a CB or amateur station. In order to provide service to a subscriber, a high-Q trap is inserted in the transmission line between the CATV signal tap-off and the TV set. This method, while inexpensive, is easy to defeat. A high-Q cavity notch filter can be made out of a coffee can and several feet of ⅛-inch copper tubing. These materials can be used to construct a helical resonator and the tune-up is merely adjusting the cavity for a notch at the interfering frequency. Also, a high-Q stub transmission-line section can do the same thing. We have known cases where a few feet of aluminum foil wrapped around the outside of a length of 300-ohm twin lead (with the aluminum-foil position adjusted for best picture) did a fairly good job of rendering the picture watchable. However, note that this is not true scrambling. It is merely a deliberate, controlled interference.

Next, we will discuss the more sophisticated (hard) scrambling techniques, both actual and possible approaches.

2

More Scrambling Systems and Future Possibilities

In Chapter 1, several simple scrambling techniques were discussed. These all operated on the basis of making the picture unwatchable by either reversing the video polarity or by altering the synch so that the TV receiver could not lock its synch circuitry onto the video signal. There are several other modifications that can be made to these systems. In order to "unscramble" the picture, we had mentioned the possibility of using a "hidden pilot signal" on the audio subcarrier, since the first twenty or so horizontal pulses (which occur immediately after the last frame of the TV picture is scanned) are included in the vertical synchronizing pulse during vertical retrace and represent over one millisecond of time. This retrace interval can be seen if one adjusts the vertical hold control of a TV receiver so that the picture "slips" out of vertical lock; a black bar appears between the bottom of one frame and the top of the next. This interval is another place on the TV signal to place information for use by the descrambler. It can be used to send coding signals to tell the descrambler what to do (for example, normal or invert video). Pilot subcarriers are hidden on the audio and may be any frequency. Table 2-1 gives the most commonly used subcarrier frequencies. There is no reason why these specific subcarrier frequencies are used (there could be others), other than having an integral relationship to the horizontal scan frequency which represents an advantage with regard to system design.

In addition to using the audio subcarrier, it is also possible to modify the sine-wave system by using a double frequency (31,468 kHz) sine wave instead of the 15,734-kHz sine wave. This way, every other sine-wave cycle would cancel the synch pulse and the picture would be vertically streaked twice instead of once. (This is done in one system.) Also, this method would put many "pirate" sine-wave decoders out of business unless the circuitry was redesigned.

Table 2-1. Common Subcarriers Used in Pay-TV Systems

System	Frequency (kHz)
Gated Synch	15.734 31.468 (alternate frequency)
Video inversion and synch suppression	39.335
Sine wave	15.754 63.000
FM stereo (for comparison)	19.000 38.000 94.000
Telease	
VC II	None. Digital data stream appears during horizontal blanking interval (HBI).

The same thing could be done using the gated synch method, by using a complex pulse instead of a simple horizontal pulse with a flat top. A multiple set of pulses per frame would also further scramble the picture and render many "pirate" gated synch decoders inoperative.

All the aforementioned systems have one fault, however. They are "static" in the fact that the scrambling algorithm (method) never changes. Once the method is known, a descrambler can be easily designed using conventional circuit-design techniques. These systems operate in "real time." The video waveform is altered at the instant of generation and transmission. Scrambled or not, the video is still transmitted line by line, frame by frame, in the same order as it was scanned. The audio is either unaltered or hidden on a subcarrier somewhere on the 4.5-MHz sound carrier.

Dynamic Scrambling

If the picture can be stored somehow, it is not necessary to transmit the picture line by line and exactly in sequence. This would comprise a "dynamic" scrambling system in which the method of scrambling is the same but the "algorithm" is constantly changing, possibly from frame to frame. Many of the static scrambling schemes in use on cable and pay-TV systems are usually not difficult to decode for the following reasons:

1. There is a horizontal synch pulse (or "hole") every 63.5 microseconds, and a new field every 16.68 milliseconds.
2. The video signal, at every instant, has a fixed relationship with the particular picture element (pixel) being scanned. It may differ by the superimposed scrambling waveform, but it still has a constant relationship to the video signal at that instant.
3. All necessary decoding or unscrambling information is present in the video signal, or on another frequency or channel (as in a cable pay-TV system).
4. Adjacent frames generally show little or no change in picture content, since in one-sixtieth of a second, very little movement occurs in the average televised scene. (Note that this is not true for abrupt scene changes.)

As an example, if the video signal was sent without synch pulses, the signal would be either zero or some constant value where the synch pulse originally was, and these periodic "holes" in the waveform could be restored to their proper waveform content by a locally generated synch signal that is synchronized to the missing synch pulses. Thus, it should be quite evident that most of the popular scrambling techniques really do not *scramble* the picture. They just make it difficult or impossible to *synchronize* in the conventional manner. The picture is still scanned from left to right and top to bottom.

Scanning Techniques

Suppose we considered a given scene to be televised. As an example, looking out of my window, I see a line of distant hills and the sky, and in the foreground below the line of hills is a hedgerow and a stone wall. If this scene were to be photographed, the camera lens would form an image of the whole scene at once. No one part of the scene registers on the film before or after any other part (assuming the shutter opened on the entire film at once; however, this is not strictly true for short exposures with some cameras). During a long time exposure, the whole scene is registering on the film at once, with the brightest areas affecting the film the most. The hills and stone wall are affecting the film at the same time, with varying luminosities and shades of color.

In a video camera, the top of the picture is scanned first since the circuitry in the TV monitor (or receiver) scans the top line first. Therefore, the sky and distant hills are reproduced slightly earlier in time than the stone wall or hedgerow. This is not normally noticed due to the phenomenon of persistence of vision, but actually it does occur.

In order to convey the picture content, there is no necessity to scan the picture elements in any particular order. The picture is a collection of elements known as *pixels.* These can be likened to the pieces of a jigsaw puzzle. No matter in what order the jigsaw puzzle pieces are put in place, the picture is still the same if they are in the correct position with respect to each other. However, a TV camera scans the scene from left to right, top to bottom. Therefore, the picture elements (pixels) are scanned in a given order. This is assumed in the operation of all TV receivers and monitors, with regard to the synchronization of the picture.

What would happen if we reversed the scan process? A little thought would reveal the fact that the picture would be inverted—top to bottom, left to right. Thus, turning the TV receiver upside down would unscramble the picture (or, with a little more work, reversing the yoke connections on the set).

Nonstandard Scanning Methods

Suppose we began the scan at the midline of the picture. The bottom half would appear above the top half, with things appearing cut in half. It does not take much thought to realize that this process could be continued further. We could cut the picture into four strips and scan them out of order, say, the second strip first, then the fourth (bottom) strip, the third, and then the first (top) strip. Now we would have a nearly unrecognizable picture. Suppose, to go several steps further, we scanned the 262½ lines per field in any random order. Our resulting picture would be unrecognizable—like an unassembled jigsaw puzzle. Or, instead, suppose we broke each line into random segments and scanned them in random sequence every line? All the pixels are still there; however, they would appear on the TV screen in incorrect order. The picture would now appear as colored confetti, like someone tore the picture into bits and pieces, and threw them on the floor.

To do this in real time, with existing TV camera sweep circuitry and deflection components, is possible in theory but impractical. A CRT that used electrostatic deflection and suitable drive circuitry could achieve this type of scanning, but the existing magnetic deflection circuitry and components used in a modern color-TV receiver (or monitor) could not do this. The sweep circuitry would have to generate, on command, very complex high-speed waveforms, with good precision. This is not very practical.

The Digital Approach

Deliberate alterations of the scan process, while theoretically possible, is not a practical approach. However, today's VLSI technology makes another way possible, using digital techniques and memory arrays.

In this method, the picture is scanned in the normal way. Each pixel is digitized, with the luminance and chroma information converted to a digital format. If each pixel could, for example, be represented by 16 different luminance levels and 16 chroma values, there would be 256 possible combinations to describe each pixel—or, one byte (8 bits) per pixel if an 8-bit code (256 states) is used. A 25-inch color-TV screen, with a screen width of about 20 inches and a 3.5- to 4-MHz video bandwidth, would allow details on the screen to be about one-tenth of an inch in size, or about 200–250 pixels per line. A 256 × 256, or 64K-byte, memory could store this amount of information, assuming 8 bits (one byte) per pixel. More memory would be required if higher video definition is desired.

About 50 microseconds are required to scan one line. This is the visible portion of the line (allowing some overscan). Now, in this line-scanning time, about 200–250 bytes of information have to be processed. This is so for our earlier example of 16 levels for both chroma and luminance. As an example, a typical home computer has the memory capability to store one whole digitized frame. One line could be stored in a 256-byte memory.

Actually, good secure scrambling can be done by sending the lines of the picture in order, but scrambling the order of the pixels in each line. However, the video memory has to be accessed at a rate of 64,000 bytes per frame (16.68 milliseconds), or 3.84 megabytes per second, which is rather fast but possible (assuming serial dots). The memory would be accessed in a random order, producing an out-of-sequence pixel arrangement. Scrambling line by line uses less memory, but it has fewer possible combinations, while scrambling frame by frame offers more possible combinations. Doing both at the same time multiplies the number of possible combinations for an even more secure system.

Now that the picture (or line) is stored in memory (one frame or one line), what do we do next? It is a situation similar to a jigsaw puzzle in the box. If we could assign a numerical code to each piece, and if we had a map or chart showing where each piece should be placed, the puzzle could be assembled without wondering where each piece fits or having to figure out what part of the picture (or which pixel) it represents. In the case of our television picture made up of pixels, there are several options.

If lines only are scrambled, the pixels could be arranged in many combinations. Each part of a scan line (consisting of several pixels) could be assigned a number. For example, the leftmost segment could be called 000 (binary), the next segment to the right would be 001, the third 010, and so forth, until the extreme right segment (111) is scanned. (We are assuming eight segments per line.) During the horizontal synch pulse, a digital code word giving the sequence, or the sequence itself, could be inserted as a stream of data. Each line could have a different code. Or, each individual frame could have its line segments chopped up the same way, and the proper sequence given for the entire frame during the vertical synch pulse (or the first one or two horizontal lines could be used for the data). In the case of scrambling by individual lines, rearrangement (shuffling) of the lines could be sent in a prearranged order, with the algorithm (method of arrangement) being sent before each frame. This algorithm could be represented by a data pulse stream that is sent during vertical synchronization or on the first few horizontal lines. This technique is known as *line dicing,* whereby lines are broken into pieces and sent in random sequence. *Line shuffling* is another technique in which individual lines are sent in random order. Both methods offer a high degree of security at the price of increased cost, complexity, and equipment reliability. More complex and sophisticated systems usually have more failure modes; however, ULSI approaches to system implementation would increase reliability.

Possible Digital System Approaches

Fig. 2-1 shows a probable system that can be used to encode signals using the two methods mentioned. Both methods are basically the same, but line dicing uses fewer memory elements. The video is digitized, loaded into memory for one frame, and then retrieved in a different sequence as a scrambled signal.

If we had 8 segments, there are 8! ways (combinations) to arrange them. This figures out to 40,320 possible combinations for each line. The total number of line-segment arrangements with a 525-line frame is astronomical, since each of the lines may be arranged in over 40,000 different ways. If line shuffling is used, there are 525 lines to shuffle around. This figures out to a total of 525!. (Factorial N is a number equal to $N \times (N-1) \times (N-2) \times \ldots$ until the last term is 1. For example, 8! (eight factorial) equals $8 \times 7 \times 6 \times 5 \times 4 \times 3 \times 2 \times 1$, or 40,320.) Imagine 525!. We'd rather not—it will literally blow your mind. (It blew our T157 calculator's.) Anyway, these figures show that without knowing the scrambling algorithm, the chance of unscrambling any given frame is about the same as that of living a million years—no way. Remember, we are only considering *one* frame. There are 60 frames per second (59.94 to be exact), 3,600 per minute, and 108,000 for a half-hour soap opera or quiz show. Even if you knew how the system worked, there are simply too many combinations. The security level is therefore very high.

Referring again to Fig. 2-1, video and audio are digitized and stored in the memory banks. The memory is set up so the sequence of operation is as follows:

1. Read out the previously stored data for the pixel that is being transmitted.
2. Store (write) the pixel from the identical place in the next frame, line, etc., that will be transmitted next.

Note that the scrambled video is delayed one line, field, or frame, since, at any given time, the entire line, field, or frame contents are in the memory. This has no visible effect on the picture since it represents merely a time delay. Steps 1 and 2 (reading and writing) are done in different sequences. This is how the scrambling occurs.

Note that the audio can be digitized and scrambled in the same manner, since it is just another waveform like the video, except that there are no synchronizing pulses in the audio signal. A code generator generates the scrambling algorithm and adds it to the video signal (or transmits it via other means). The totally

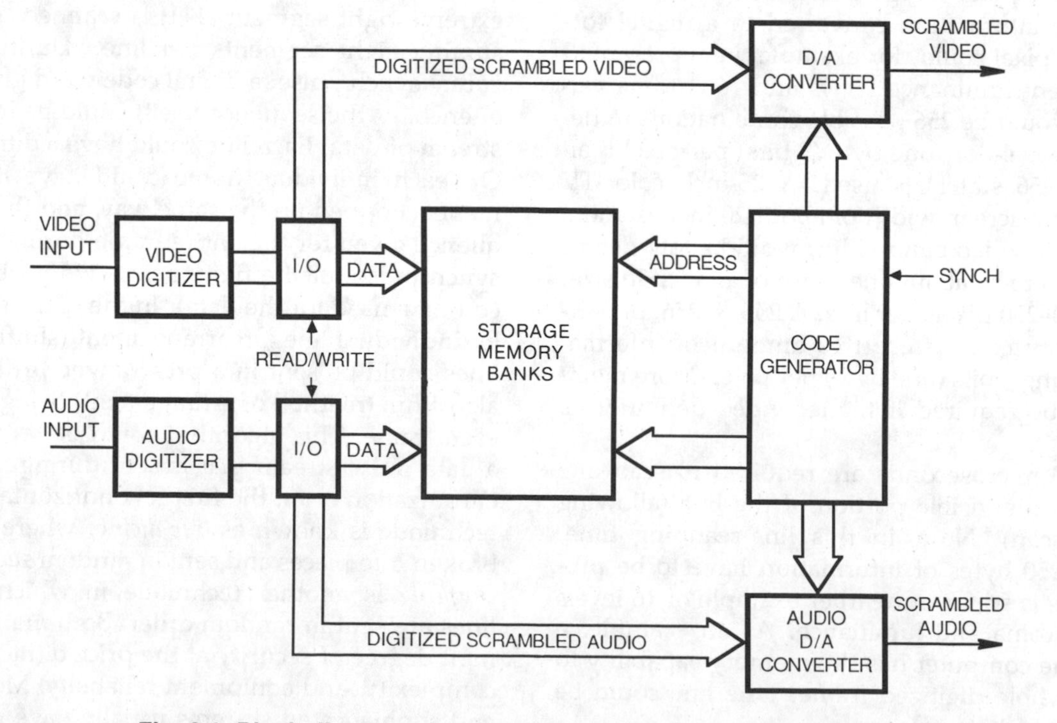

Fig. 2-1. Block diagram of a possible system for encoding signals.

scrambled audio and video are then converted back to analog video (but not absolutely necessary) and can be transmitted by normal methods. It is necessary to modify the synch pulses. A TV receiver without a decoder will simply get garbage. Synchronized or not, garbage is garbage. Actually, the picture and the audio will resemble colored confetti and random noise. This system is literally a video pirate's nightmare. In fact, without the algorithm, the manufacturer of the system would be unable to descramble the picture. See Fig. 2-2 for an illustrative example of this type of scrambling.

Future Possibilities

These systems, while at present somewhat too expensive, could very well be made very cheaply in the future. One only has to look at today's VLSI chips and inexpensive home computers. And, do not underestimate the economies brought about by mass production. There is a potential market for a digital descrambler in the millions of units. In addition, these descramblers could have dedicated computers built into them, with the computer keeping track of the program fees, billings, and pay-TV subscribers; the computer could also control the TV set, a VCR, etc., and could restrict access to programs by those members of the family not authorized to view certain programming. The built-in computer could even provide interactive functioning—one could possibly select programs on an individual basis, at any preferred viewing time.

(A) Lines are in proper sequence in original picture.

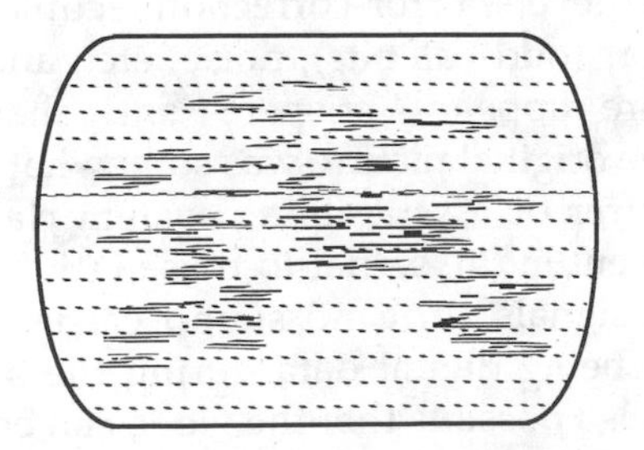

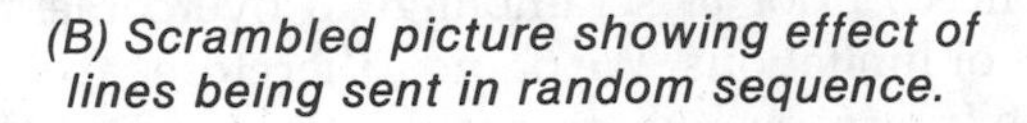

(B) Scrambled picture showing effect of lines being sent in random sequence.

Fig. 2-2. Example of a digitized scrambled signal.

As previously mentioned, audio scrambling can also be done and it will most likely be included in the newer designs. Rock concerts and other musical shows can be enjoyed without the video, so we think that audio scrambling will become commonplace.

At the present time, some pay-TV operators and cable program distributors are using a system that has a relatively primitive video scrambling system together with a sophisticated digital audio scrambling method. Several approaches are under consideration and undergoing testing and evaluation. Currently, the scrambling methods are being used for satellite-TV uplink scrambling to prevent the unauthorized viewing of a popular pay-TV cable service.

Other Approaches to Scrambling

Other static scrambling methods are available that use the older analog approaches. However, these methods still suffer from the defect that all static methods using a fixed algorithm suffer from; i.e., understand the system and you can design and build a descrambler. One scrambling method used worked as follows: Horizontal lines can be displaced with respect to each other, causing a severe case of ghosting and introducing distortion in the picture (Fig. 2-3). This technique makes use of a delay line—a component found in all color-TV receivers. The delay line is electronically switched in and out of the circuit during the display of various horizontal lines, causing line displacement. This method can be combined with video inversion and/or synch suppression techniques, so that even with the sophisticated pirate decoders available on the market today, the picture is still somewhat scrambled and its entertainment value destroyed. This scrambling can also be done on a frame-by-frame basis.

Other scrambling techniques consist of using hidden subcarriers to control the video polarity, the use of deliberately introduced interference, or even the use of nonNTSC television waveforms so that the TV receiver cannot properly process them (for example, using SECAM, etc.).

Practical Considerations

As mentioned earlier, there is a cost/security tradeoff. Simple scrambling techniques, such as the sine

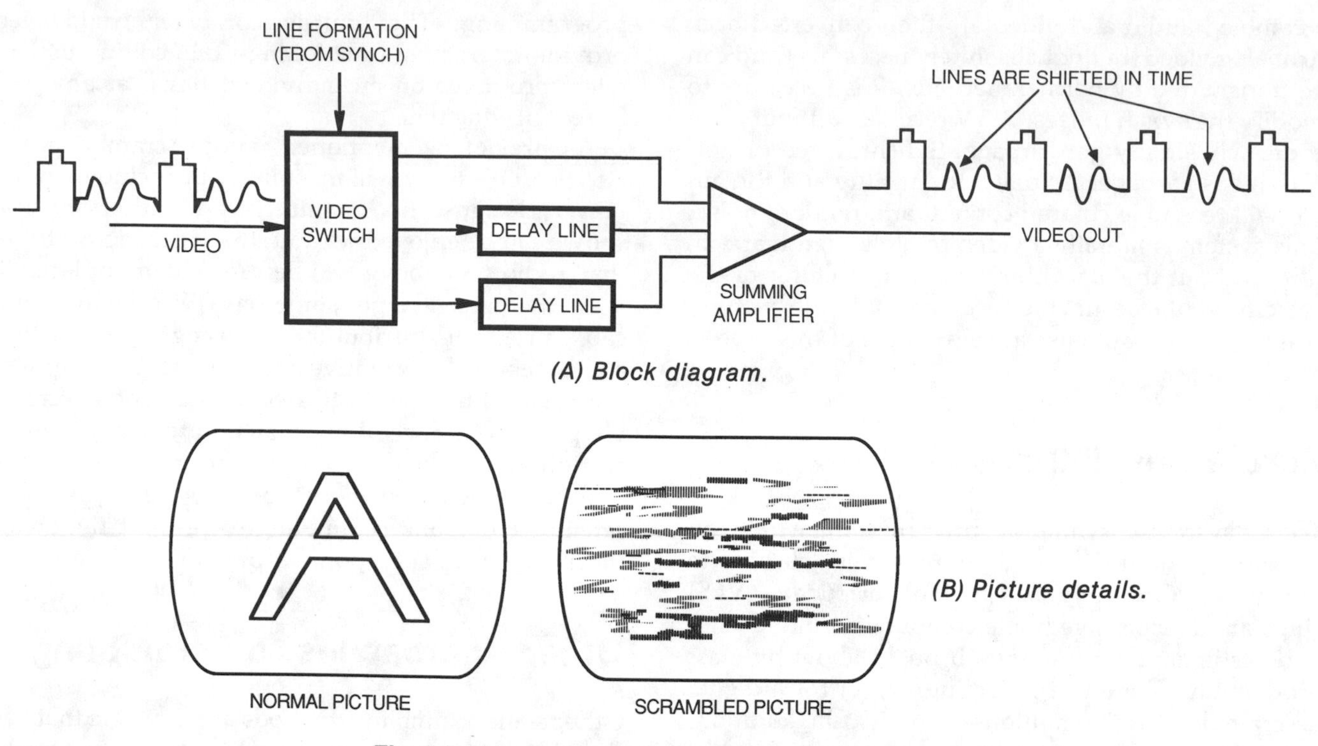

Fig. 2-3. Line displacement method of scrambling.

wave, gated synch, etc., will not permit the average TV viewer to see the program, but will not deter the would-be video pirate. A close examination and some "reading between the lines" in any video/electronics magazine will generally reveal that many sources are making and selling decoders. These decoders can be used for theft of service from pay-TV operators, and they are not expensive. More complex scrambling methods are more expensive to operate, and the pay-TV operator has to weigh the cost of the scrambling (security) versus the lost revenues from the nonpaying "pirate" viewers.

A Word about Picture Quality

Another problem with any scrambling method is the loss of signal quality. Gated-synch scrambling, sine-wave scrambling, and other related methods introduce deliberate interference into the video signal. While decoders do a good job of removing the interference, they are not perfect. There is always some residual video noise or interference left over which could be noticeable on a good large-screen receiver. Also, the audio channel might not be used to its fullest capabilities due to some loss of audio bandwidth. Again, this may not be a problem with an "average" 19-inch portable TV set. However, some of today's monitors are capable of a very high degree of picture quality, and the video degradation and loss of bandwidth that a descrambler may introduce might be noticed. This degradation will naturally occur on premium "scrambled" pay-TV programs where a viewer would be least tolerant of a poor or degraded picture quality caused by a misaligned descrambler. If you have to pay extra for it, you want good TV reception, and should expect and demand both a quality picture and good audio.

Some *digital* scrambling methods could actually *improve* the picture quality. When a signal is digitized and stored in computer memory, it is in a digital form. Each bit is either a zero (0) or a one (1), and there are only so many bits and, therefore, only so many discrete video levels. Error-correction techniques can remove errors (odd-ball bits), noise, etc., and can make the picture appear "crisper" and sharper, even though the original picture was somewhat noisy. One manufacturer of a scrambling system claims a 2-dB improvement in the signal-to-noise ratio of the video baseband signals since no sound carrier is needed, the audio being digital data during the synch intervals. It is also possible that the video can be artificially enhanced, prior to scrambling, to overcome system flaws or limitations. We do not, therefore, at any time in the future, see where degradation of picture quality will be a very good argument against scrambling, when digital methods, including error-correction techniques, can be employed.

What Next?

Only time will tell if and how the scrambling of the free-TV signals of today will occur. More and cheaper sources of video and other entertainment undoubtedly will arise, making the practicality of pay-TV questionable. Why should someone pay for the viewing of a movie when the same movie can be rented as a video cassette (several months later, at worst), and be viewed at a more convenient time than the pay-TV showing, and probably at a lower cost? Today, video cassette recorders (VCRs) can be bought for less money than a sophisticated descrambler, and no cable-TV system or satellite-system installation is required. Unless some form of interactive, view-on-demand, pay-TV service is available, with a wide choice of programming materials to choose from, scrambled pay-TV may never really catch on as a popular entertainment source.

Right now, DBS (satellites) TV has run into some problems of a political nature and may never materialize as originally envisioned. In fact, other than for military and business security, the only application of sophisticated scrambling systems may be in preventing satellite-TV system owners from the unauthorized viewing of such popular cable-TV movie channels as HBO, Cinemax, or the ESPN sports channels. With the wide variety of choices available, who really cares anyway? The cost to the programmers may not be worth it. Small CATV operators may decide to take advantage of the new competitors that could come along, with their unscrambled signals, no special equipment, and the lack of attendant costs and headaches.

Possibly, satellite dishes and equipment will be taxed either periodically or upon purchase (as a one-time charge), and the premium-type pay-TV operators subsidized with this tax revenue. Therefore, ultrasophisticated decoders may never become commonplace for commercial-TV broadcast use.

The next chapters will describe the circuitry used to decode pay-TV signals and will describe several decoders in detail, with construction hints and PC board layouts.

3

Scrambling/Descrambling Techniques—Basic Circuitry

In the last two chapters, various scrambling system approaches were discussed. Although the systems differ somewhat, they all have many circuits in common. For the large part, in this chapter, we will confine our discussions to those systems that use a fixed scrambling algorithm, such as the gated pulse system, the sine-wave system, or the synch suppression and active video-inversion system. By first discussing the common circuit features, a lot of repetitive material is avoided, and a more general view of design approaches can be obtained.

Clamping

One simple circuit used in scrambling/descrambling is the *video clamping circuit.* This is used when a video signal must have a definite DC base level that defines the black or white level. Many video amplifiers and CRT circuits depend on DC levels to establish their operating point. Since there are often only AC-coupled amplifiers, or sources of video that do not have the DC component present along with the video signal, the exact black and/or white reference for the signal is unknown. Fig. 3-1 shows the effect of transmitting a video signal through an AC-coupled circuit.

As is evident, the waveforms look alike but the DC level has shifted down by about 2 volts. The blanking pulses that were at +3 volts are now lower than +3 volts and the synch tips are only up to +3 volts (Fig. 3-1A). A portion of the signal now goes negative down to −2 volts. The result is that, for this particular portion of the signal, the grays would approach white, and the black levels would be gray, resulting in a washed-out picture. Worse still, the very light grays would also appear as white, since anything less than zero is maximum white. The synch tips would approach black, if they were visible. (The picture may also roll because of poor synch, due to the fact that the synch separator may not work properly.)

The cure to this problem is the addition of a diode across the resistor R (Fig. 3-1B). A small portion of the signal is rectified and develops as a DC bias, reestablishing the DC level automatically as required. Assuming a perfect diode is used (Fig. 3-2), the video output can go no further negative than the reference voltage. The video signal may be said to be "clamped" to the reference voltage. (This, of course, can be any required DC level, or ground.) Since the peak-to-peak level of the composite video is constant and equal to the synch pulse level, the video reference level remains quite constant. The RC combination must have a sufficiently large time constant such that the lowest-frequency video components are passed with little (0.5 dB approximately) attenuation. For standard TV signals, this time constant may be 0.05 to 1 second—or more, where little low-frequency losses (tilt) can be tolerated. A transistor switch can be used in place of the diode, and turned on with pulses timed to synchronize with the place on the waveform where clamping is desired (usually the blanking or synch pulse).

Modulation

Another circuit used in many descramblers is the modulator. This is a circuit in which one signal is mixed with another, but we will limit our discussion to the amplitude modulator. This is a circuit used for correcting or reinserting missing components in a video signal.

In the gated synch system, the usual method of scrambling reduces the level of the synch pulse. An-

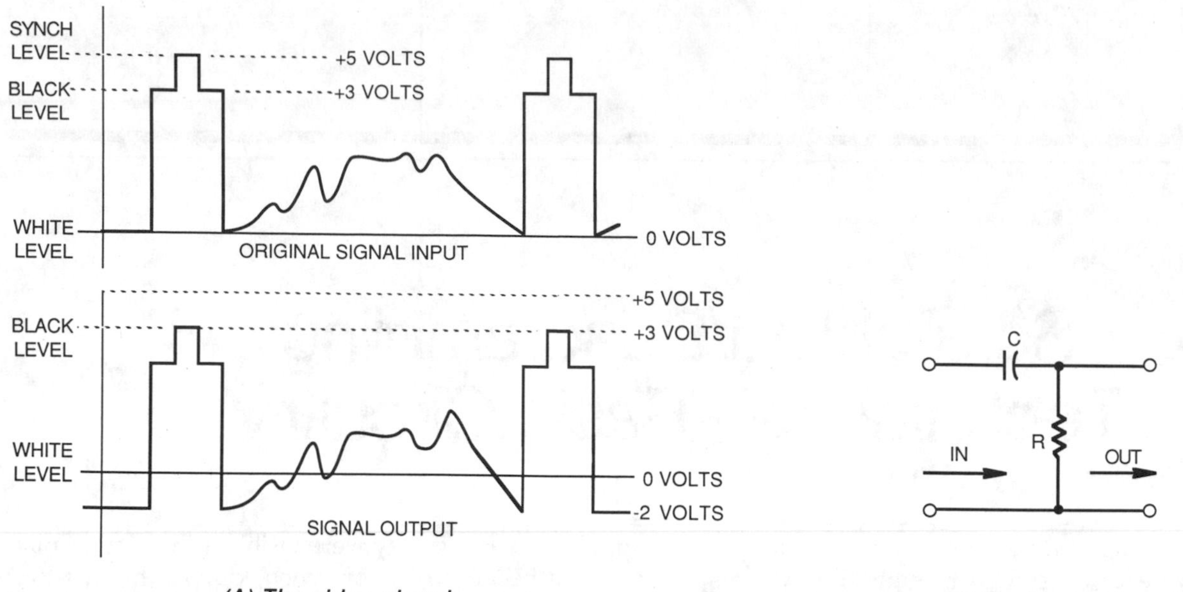

(A) The video signal. *(B) Basic circuit.*

Fig. 3-1. Waveforms of an AC-coupled circuit.

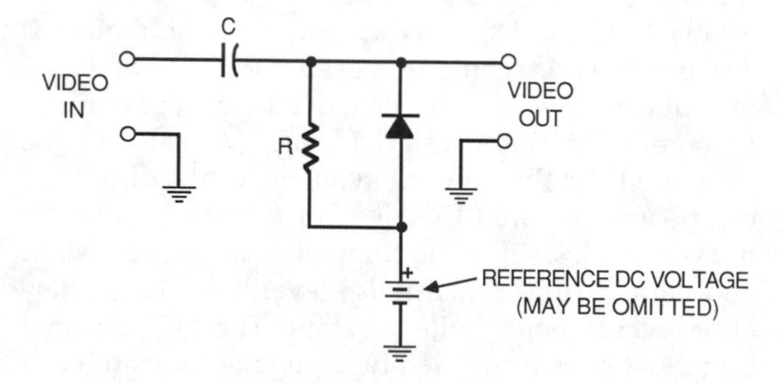

Fig. 3-2. Video clamp circuit.

other way of looking at the situation is simply to take a reverse look at it. The video level is increased to a level much too high for the synch pulse (Fig. 3-3). Either the video is too large (high) or the synch is too small (low). We could correct this situation either by reducing the video gain between synch pulses, or by increasing video gain during the synch pulses. If we had a network of some kind that would have different gain or losses for video and synch, this would neatly solve the problem.

There are two basic approaches—one uses a variable-gain amplifier, the other uses a variable attenuator. Either one will work, and choice of which method to use depends on the designer. Fig. 3-4 shows circuits for a basic fixed attenuator and three commonly used variable attenuators. The first two types of variable attenuators have a continuously variable attenuation characteristic, while the transistor-switched video attenuator has more of an off/on (two-level) characteristic. Any of these three variable attenuators would work well in either the gated pulse or synch suppression and video-inversion systems as system blocks. However, the sine-wave system would require a linear-type circuit (like the first two) since we need some degree of system "linearity." These attenuator schemes could be used either on the RF video signal or the video itself. RF video use, say at VHF/UHF or the IF frequencies (usually 45 MHz), would be somewhat simple because the capacitors

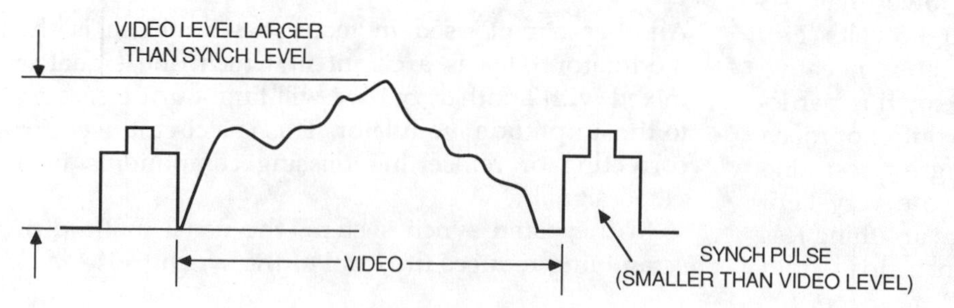

Fig. 3-3. Gate synch circuit levels.

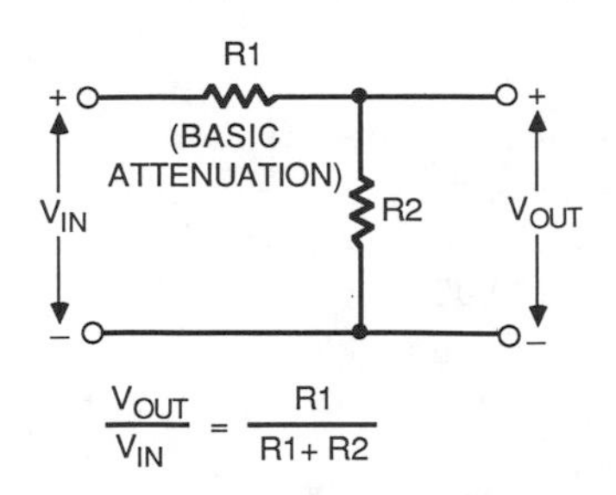

(A) Basic fixed attenuator.

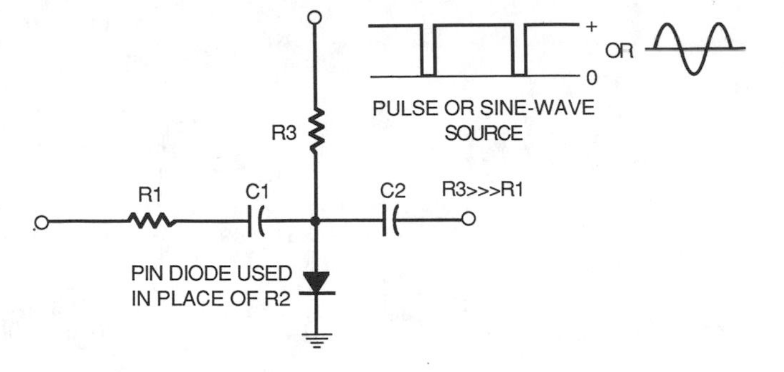

(B) Diode voltage-controlled video attenuator.

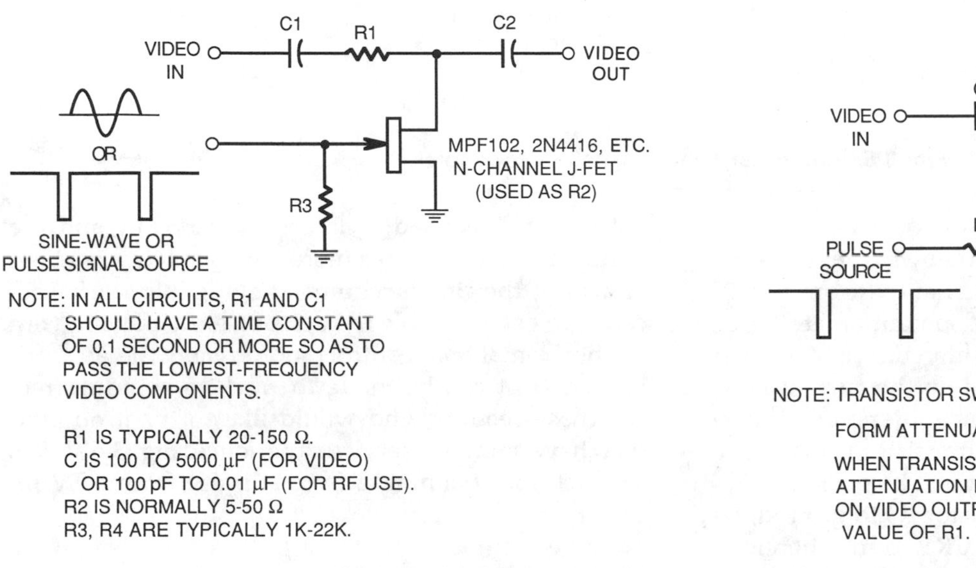

(C) JFET voltage-controlled video attenuator.

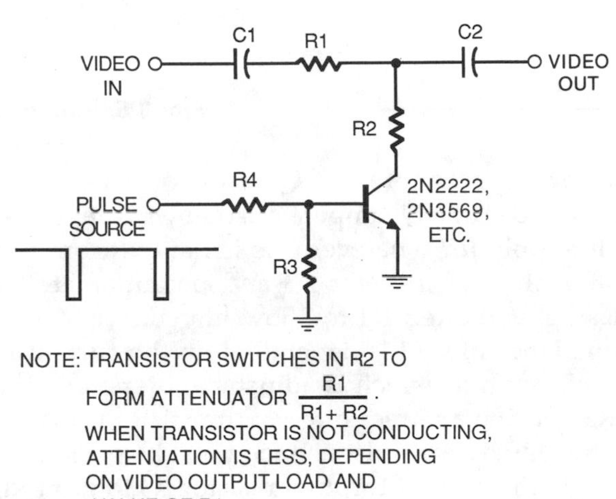

(D) Transistor-switched video attenuator.

Fig. 3-4. Voltage-variable attenuators.

could be much smaller in size. Disc capacitors with values from 100 pF to .01 μF would be adequate for RF use. For use at video frequencies, large electrolytic capacitors would be necessary, with up to 5000 μF in those low-impedance (75-ohm) circuits that have tight specifications for low-frequency tilt. For 1000-ohm impedance levels, much smaller values could be used. In a practical TV receiver circuit, this would be the case.

These variable attenuator circuits require the active elements to be biased *on* (low resistance) during the scan period when the video signal is present. This has the effect of attenuating the video signal during this period. During the blanking interval, the active device is biased *off,* or to a state of high resistance. The blanking, synch, and color burst signals are passed with little attenuation. This increases the relative amplitude of the synch pulses, restoring the encoded video signal to normal (in the case of a gated synch system). In order to do this, the pulse source must provide a positive level that is sufficient in amplitude during the scan period and a zero (or negative) level during synch pulses. In the case of a sine-wave system, a sinusoidal signal would be required. We will see later that, more exactly, a predistorted sine wave may be required in some cases.

The circuits in Fig. 3-4, along with a few other components, can, in theory, and also in practice, decode a gated synch signal directly. A circuit to do this is shown in Fig. 3-5. It will only work on a gated synch system.

A Very Simple "Descrambler"

The circuit in Fig. 3-5 operates by taking a small sample of the flyback pulse from the TV set. The 2N3565 transistor is biased *on* only during retrace intervals.

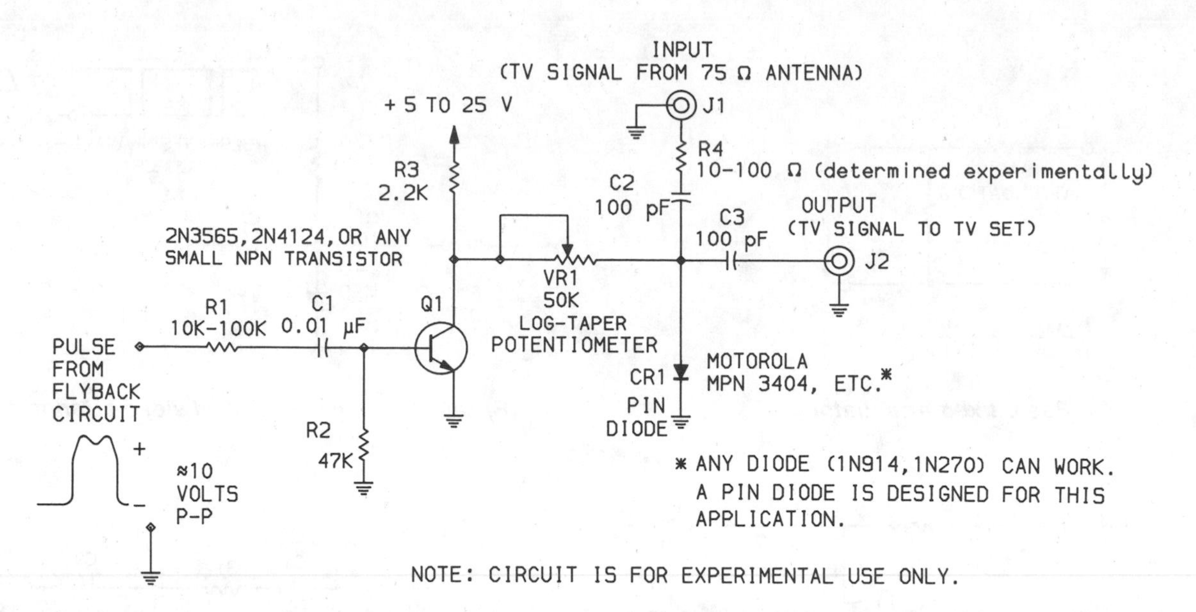

Fig. 3-5. Elementary gated synch descrambler.

During the scan interval, Q1 is cut off and a DC current of about 1 milliampere or so flows from the +12-volt supply (or whatever else is handy, from 5 to 25 volts), through resistor R3 and potentiometer VR1, biasing *on* diode CR1 to a low impedance state. The impedance of CR1 is dependent on the current flowing through it, which is adjustable by potentiometer VR1. During a retrace interval, transistor Q1 is biased *on* heavily by the pulse, which must be positive. This diverts the current through resistor R3 to ground, removing the bias from diode CR1, and, therefore, biasing it to a high impedance state. The TV signal from the cable system or antenna entering input jack J1 passes through R4, C2, and C3 to the output, J2, suffering some attenuation depending on the bias current through CR1. The signal from J2 goes to the TV set. During retrace intervals, CR1 is cut off and the signal passes with little attenuation. During scan periods, signal attenuation occurs since CR1 looks like a low resistance.

Limitations of Simple Descramblers

The previous circuit will *not* decode the audio (if encrypted), only the video portion. It reportedly works on the VideoCipher II System™ used on satellite systems. (The picture has to be reverted.) However, we have had no direct experience with this. The circuit shown in Fig. 3-5, while a good demonstration circuit, is difficult to adjust, is noise prone, and not practical as a descrambler. It is a fine experimental project and is worthwhile trying if a gated synch signal is to be decoded. The cost is very low and the parts are also usable in a more advanced descrambler. However, the circuit requires digging into the TV receiver to get a suitable flyback pulse. By playing with the horizontal hold control (and possibly the AGC adjustment), it can be made to work to some degree. For those readers who would like to try it out and who have some TV service experience, Fig. 3-6 shows a sketch for obtaining a flyback pulse from a TV receiver.

Some methods of obtaining flyback pulses from a TV set are:

1. Wrap 2 turns of *HEAVILY* insulated high-voltage-type hookup wire through the flyback transformer, effectively adding an inductive pickup winding (Fig. 3-6).

OR

2. Wrap 1 to 2 turns of wire (*HEAVILY* insulated) around the lead-to-plate cap of the horizontal amplifier tube (in a tube set) using the power transformer for B+ circuit and filaments.

OR

3. Run one lead from the collector of the horizontal output transistor through a series combination of a 1-meg ohm, 2-watt resistor, and a 0.01-μF, 1600-volt DC capacitor. (This is not recommended if step 1 or step 2 can be implemented.)

Be very careful, with all of these methods, as you are fooling with an RF voltage. Use only *heavily* insulated wire, and operate the TV set from an isolation transformer during these experiments. If you are not

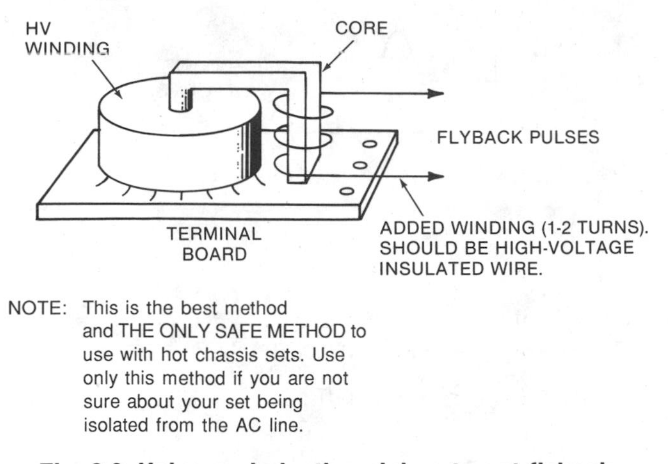

Fig. 3-6. Using an inductive pickup to get flyback pulses from a TV receiver.

sure of your TV set and its internal workings, get a schematic and service manual. *DO NOT* assume anything, it could be extremely dangerous and, if you are lucky, will only cause a $50–$100 repair problem. So, *if you do not know what you are doing, do not try this experiment!*

Variable Gain Amplifiers

The alternate method of video signal modulation is in the use of a variable-gain amplifier, or a mixer circuit. The idea here is to control the gain of the amplifier (which is amplifying the encoded video signal) with another signal (which cancels out the encoding signal originally superimposed on the video signal), leaving only the original video signal that was present before encoding. There are a number of ways to do this, with a couple being shown in Figs. 3-7 and 3-8.

In Fig. 3-7, a standard video RF amplifier circuit is used as a video IF amplifier at 45 MHz (a standard TV practice). The input video IF signal is made up of the standard video signal with an encoding waveform superimposed on it. This is the signal we would get either from the tuner or a previous IF stage. The stage gain is varied by changing the Gate 2 (and Gate 1) bias as shown in the characteristic curve of Fig. 3-7B. This method is normally used for AGC control, but if the bias is a DC voltage with a decoding signal riding on it, we can vary the stage gain and cancel the decoding signal at the same time by a proper choice of the decoding signal level and DC bias point. Ideally, the output will be an amplified version of the input video IF signal, minus the decoding signal. After this point, the signal is decoded and can be handled with standard circuitry.

Video Amplifier Circuits

In Fig. 3-8, a circuit is shown using an LM733-type video amplifier IC. This is a two-stage, differential-input, wideband video amplifier designed for use in a wide variety of video applications. It has a gain bandwidth of over 120 MHz and can be set up for 40 dB or more gain at a 4-MHz bandwidth by a suitable choice of peripheral components. A differential amplifier has the property of producing an output signal that is the product of the gain of the amplifier and the *difference* between the two input signals. By making one input signal zero (grounding that input), the amplifier can act as a conventional amplifier. The output can be either balanced or single-ended with respect to ground, and can be, in many cases, set up to provide two equal output signals that are 180° out of phase. The circuit of Fig. 3-8 does this. The encoded video is applied to pin 14 of the LM733 amplifier. A suitable decoding signal is then applied to pin 1. Both of these signals are capacitively coupled into the LM733 so as to affect the DC biasing of the amplifier. A signal that is proportional to the difference of the two inputs appears at pin 8 of the amplifier. A signal that is inverted appears at pin 7. Therefore, the designer of a circuit has the option of selecting either a positive- or negative-going output signal. This is particularly useful when a signal that is intermittently inverted (such as an SSAVI signal) is to be decoded. A switching circuit that is controlled by a signal from the decoder can automatically switch video polarity, as required.

The two aforementioned circuits, or similar ones, can be used as video or RF modulators. While these circuits are a bit more complex than the previously discussed attenuator circuits, they can provide gain and are somewhat ''cleaner,'' in the fact that they are easier to set up, use higher signal levels, and their characteristics are more linear and easily controlled. This is important in the sine-wave system, where a waveform mismatch between the decoding signal and the encoded video waveform will result in an incomplete cancellation or a residual modulation due to the harmonic components present in the decoder because of nonlinearity, etc. This mismatch has the effect of producing a horizontal shading that would be most obvious on a low-contrast picture (Fig. 3-9).

Synch Pulse Restoration

Synch pulse restorers/regenerators are important in some encryption systems. When the synch signal is

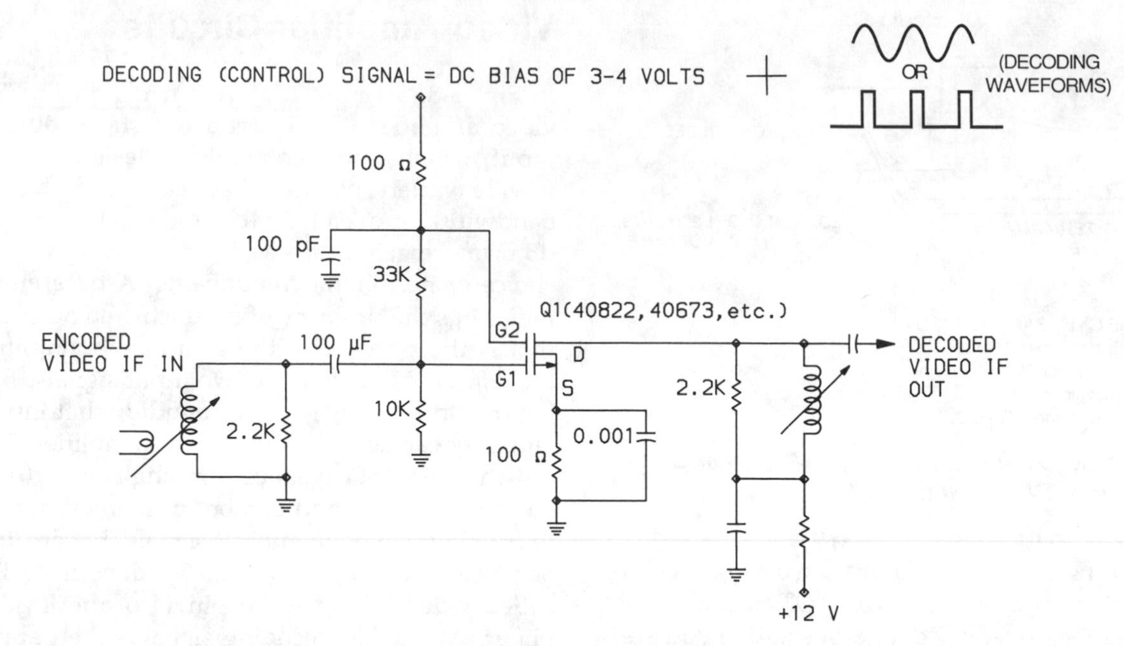

(A) *Circuit schematic.*

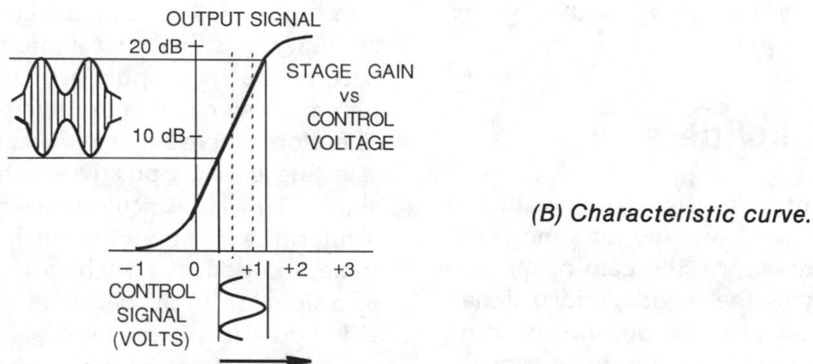

(B) Characteristic curve.

Fig. 3-7. Typical 45-MHz video IF amplifer.

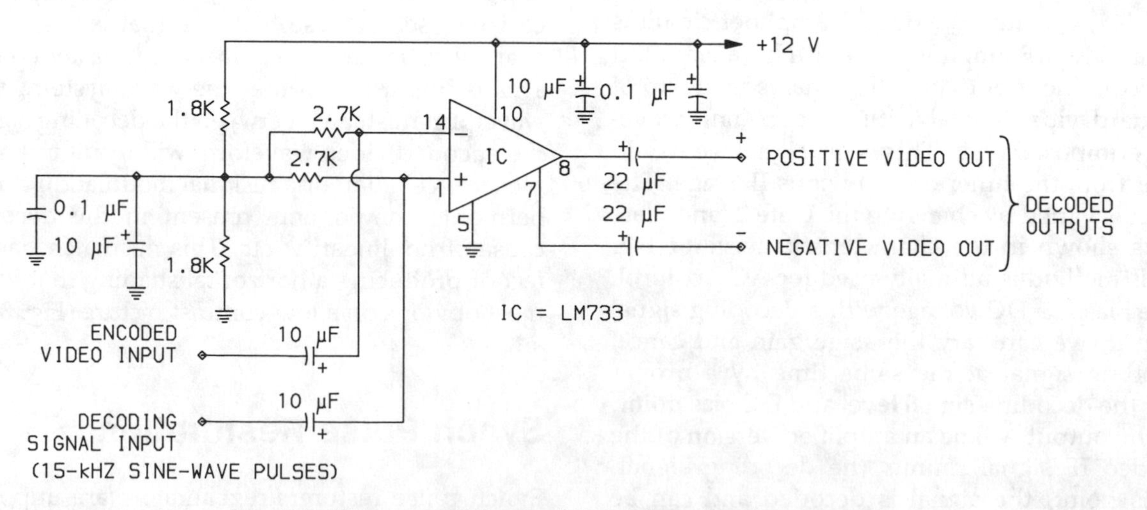

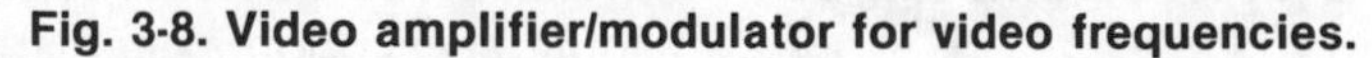

Fig. 3-8. Video amplifier/modulator for video frequencies.

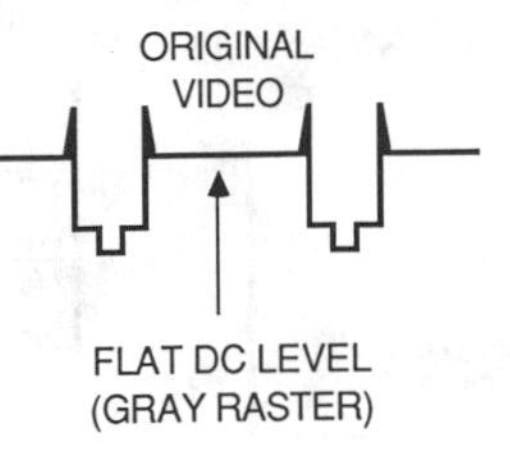

Fig. 3-9. Effect of an incomplete decoding-signal cancellation.

suppressed (or missing entirely), a new synch pulse has to be generated at the receiver. This pulse should be identical to the synch pulses that were originally present in the video signal. Sometimes it is a simple case in which the pulse present in the encoding signal is merely reduced in amplitude. In this case, the original synch pulse is increased in amplitude, with respect to the video signal, in the decoding process. In other cases, it is necessary to construct a new synch pulse using ''hidden'' information transmitted either in the video baseband or, sometimes, as on some cable systems, on a different RF channel. In cases where the synch pulses are sent on a different channel (outband system), a separate receiver that is tuned to this channel can be used to obtain the missing synch pulses. Generally, however, a locally generated waveform, which we call the decoding signal, has to be recovered or constructed in the decoder. Several circuits for doing this will now be discussed.

The generated pulse must have proper amplitude and pulse width, and must be properly positioned in time (phase). The decoding information extracted by the decoder circuitry may be a sine wave or whatever modulation was originally employed. It may not be in proper phase or time relationship, and the waveform may not be that which is required for canceling out the encoding waveform superimposed on the video signal. Fig. 3-10 shows a circuit for regenerating a synch pulse from a sinusoidal waveform obtained in the decoding process.

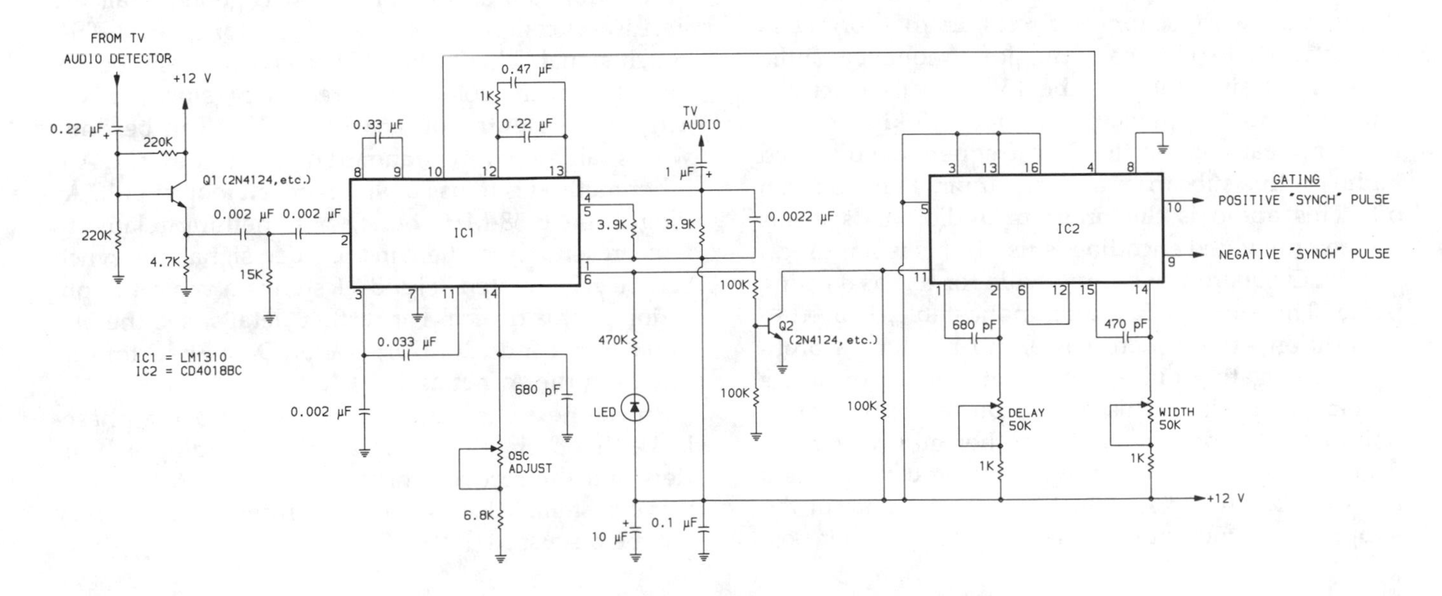

Fig. 3-10. Synch pulse recovery for gated synch system.

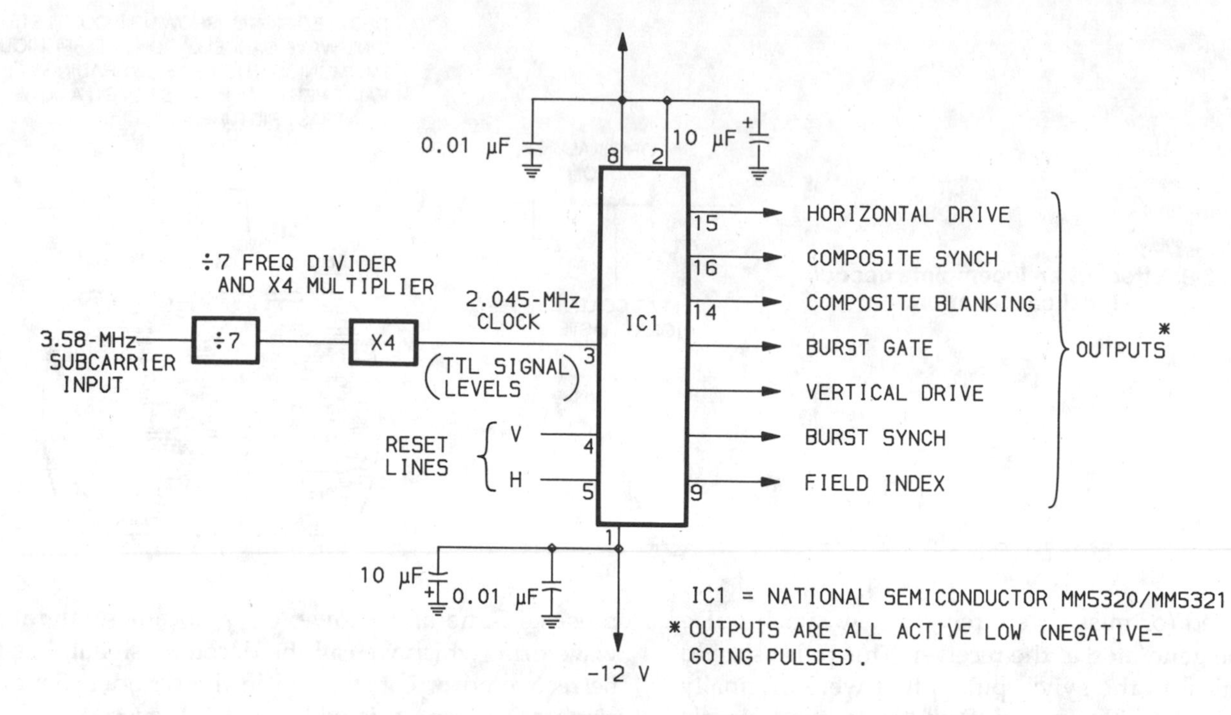

Fig. 3-11. NTSC synch synthesizer.

Circuit Operation

Briefly, the circuit operates as follows. Integrated circuit, IC1, is a chip designed for FM stereo demodulation. While it normally is designed to operate at 19 kHz, minor component changes make it operate at 15.75 kHz (nominal). The actual frequency of operation can be from 15.3 to 16.4 kHz (and sometimes is varied). Audio from the TV audio detector is coupled into emitter-follower Q1. The emitter circuit feeds IC1 through a few capacitors and resistors that form a filter network to eliminate the low-frequency audio component and still pass the 15-kHz subcarrier signal. The VCO frequency (nominally 15 kHz) square wave appears at pin 10. The frequency-modulated audio on this subcarrier appears at pin 4 and is taken off. This audio is the program audio. (In systems where only video encoding is used, this is not used.) The VCO square wave is too wide for use as a gating pulse. Therefore, IC2, a dual monostable, is used to "condition" this square wave and produce a properly timed gating pulse. One multivibrator produces a delay to position this gating pulse at the proper point on the video signal. The other multivibrator is triggered by the first multivibrator, and produces a pulse whose width is adjusted to match that of the suppressed synch pulses. Gating pulses, either positive or negative, are available at pin 10 or pin 9 of IC2, as required.

Missing Synch Pulses

If the synch pulses are missing entirely, they can be regenerated by a circuit using a National Semiconductor MM5320/MM5321 IC chip (Fig. 3-11). This chip only requires a 2.045-MHz clock to generate all the necessary components needed to form an NTSC synch signal. The 2.045-MHz clock can easily be obtained from the color burst frequency, since it is exactly four-sevenths of 3.58 MHz. This can be done with straightforward standard digital IC devices. Another method is to use a phase-locked loop circuit. As long as the 3.58-MHz burst, or something related to it, is present somewhere in the video signal, the synch can be regenerated. Fig. 3-11 shows a typical application of this device. For further details, see the National Semiconductor Corporation Data Sheet for this device in the Appendix.

In the next chapter, we will discuss using phase-locked loop circuits to extract the decoding subcarriers and the recovery of the sine wave in the sine-wave system. Also, some SSAVI decoding circuitry will be discussed.

4

More Scrambling/ Descrambling Techniques

In Chapter 3, circuits used for descrambling were discussed. In this chapter, the phase-lock loop circuitry used in descrambler applications will be examined. Several representative approaches will be covered.

Basic PLL Circuits

The widespread use of FM stereo has resulted in a number of self-contained FM stereo demodulator IC devices. These devices generally contain an input amplifier, a phase detector, a VCO, some form of phase-lock detector for audio muting or stereo lamp switching, a decoder matrix, and a built-in power-supply regulator, enabling the chip to operate from a wide variation of supply voltages. These ICs are usually cased in a 14- or 16-lead DIP package. Due to mass production, they are inexpensive and readily available either from experimenter-oriented electronic supply houses, or as replacement units sold by radio-TV parts suppliers. Many devices are available that do not require coils or other difficult-to-find components. These devices are exactly what we need to use as a descrambler component. They will regenerate the 15- , 31- , 40-, or 62-kHz subcarriers used in the gated synch, sine-wave, and SSAVI systems. In applicable cases, they will also demodulate the hidden audio subcarriers; thus doing two jobs at once.

Theory of PLL Circuits

Basically, a phase-lock loop circuit operates by comparing a signal of a nominally desired frequency—usually that of a voltage-controlled oscillator (VCR) that has a definite tuning range—with a reference signal. Usually the reference signal is the input signal. Both the input signal and the VCO signal are applied to a phase detector (Fig. 4-1). The phase detector has an output signal proportional to the phase difference between the input signals. This phase detector output is amplified and applied to the VCO. The VCO is set up to shift frequency in such a direction as to reduce the phase difference between the input (reference) and itself. Sometimes the VCO is operated at a multiple of the input (reference) frequency. In this case, a frequency divider ($\div$N) is placed in the feedback loop.

Remember the difference between frequency and phase. Frequency is the rate of change of the phase of a voltage, with respect to a reference starting point, per unit time. That is why you often hear of frequency spoken as "radians per second." A radian is 57.3 degrees, and 360 degrees (one cycle) is equal to 2π radians ($6.28 \times 57.3° = 360°$ very closely, where $\pi = 3.14$ approximately). This is also the reason for the 2π factor appearing in many AC circuit equations, such as $X_L = 2\pi f L$.

A phase detector will produce an AC output signal if signals of two different frequencies are applied to the inputs. If the same frequency is applied to the inputs, a DC output voltage proportional to the phase difference results. Therefore, once the loop is locked, only a phase error exists between the two input signals. This is constant and, therefore, the frequency error is zero. Conversely, an automatic frequency control (AFC) system always has a frequency error, since the output of a discriminator is a DC voltage that is *frequency dependent*. You will notice in Fig. 4-1 that the amplifier has an integrator in it. This is necessary for various reasons, such as noise rejection, stability of the system, and lock-and-capture characteristics. It is critical in nature and that is why, when

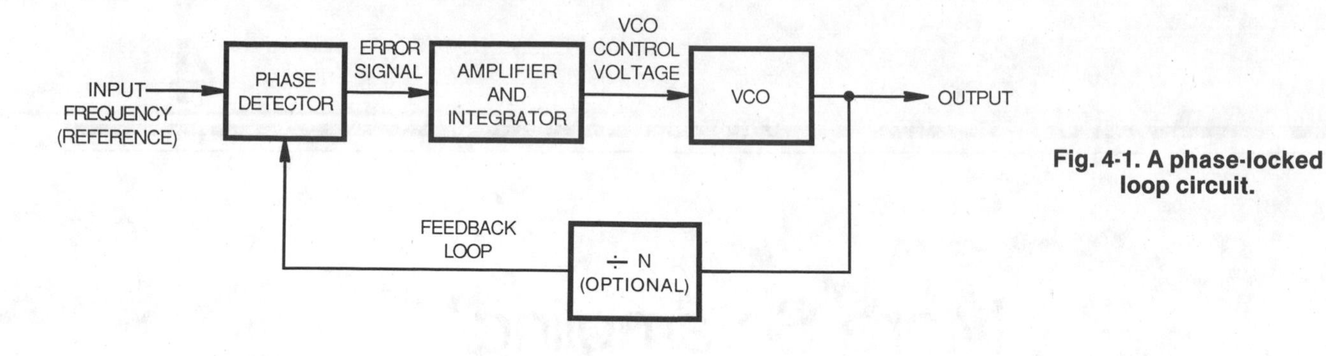

Fig. 4-1. A phase-locked loop circuit.

using any of the phase-lock loop IC devices, the manufacturer's recommendations should be followed.

Subcarrier Regeneration Circuit

Fig. 4-2 shows a circuit for subcarrier regeneration using an LM1800 stereo demodulator IC. It is very similar to the circuit that would be used for FM stereo detection. With subcarrier regeneration circuits, we do not have to worry about stereo separation, since we are only recovering one mono channel. We want to be able to get the pilot signal, since this is sometimes needed to regenerate the synch signal (gated pulse, etc.). FM stereo applications usually only use the pilot signal (19 kHz) as a means of stereo indication, but, in many of these IC devices, the lamp circuit is already included. We do not have to use both the L&R outputs, since we are dealing again with only mono audio. The lamp driver/stereo detector is useful for decoder switching purposes, providing a means to switch in the decoder automatically. The external components used may be of different values due to different frequencies used in the various systems.

This circuit may be used at any suitable frequency, and while designed for 19- to 38-kHz operation, it should work up to over 100 kHz in this application. We do not have any guarantee of this from the manufacturer, but most devices of this type are fabricated using inherently high-frequency transistors that can

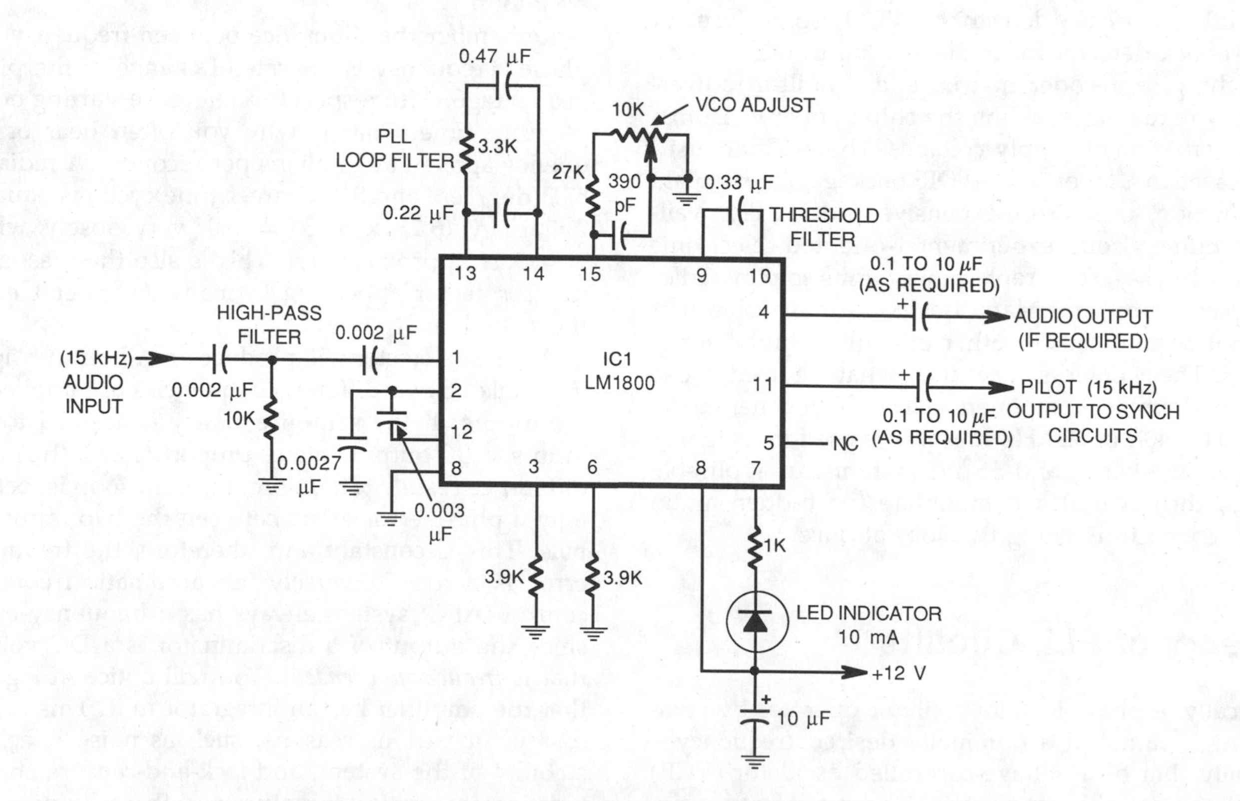

Fig. 4-2. A subcarrier regeneration circuit.

operate to several megahertz. (Of course, some experimentation may be needed to achieve this.)

Gated Pulse Techniques

In the gated pulse system, where the 15-kHz subcarrier has the audio portion (if encrypted) frequency-modulated on the 15-kHz subcarrier, the circuit can perform both audio recovery and subcarrier regeneration. In the sine-wave system, a 62.5-kHz carrier is used for the audio. Here, we would use this circuit simply for audio recovery. In this case, the input high-pass filter should be changed. Use 470-pF capacitors instead of the 0.002 μF capacitors, and also use a 100-pF capacitor in place of the 390-pF capacitor in the VCO circuit (pin 15). Alternately, the resistances in the VCO section (27K resistor and 10K potentiometer) should be changed to about 6.8K and 5K, respectively. Some experimentation with the value of the THRESHOLD and PLL LOOP filters may also be necessary. It would be a worthwhile experiment to set up this circuit and check out its operation, using an audio or function generator. Alternately, if available, the output of a TV sound discriminator can be used. In this case, the deemphasis components must be removed to prevent the attenuation of the desired audio subcarrier signals in the 15- to 60-kHz range.

Sine-Wave Techniques

In the case of the sine-wave system, the recovery of the synchronizing sine wave is somewhat different. In this system, the required 15-kHz sine wave is amplitude modulated on the 4.5-MHz sound subcarrier. In the conventional TV sound limiter, this AM component is lost in the limiter circuit and does not appear at the output of the sound detector.

Fig. 4-3 shows a block diagram of the subcarrier recovery process. In this system, the 4.5-MHz sound signal is taken off at the video detector (or sound/synch detector, if the set has one), or, if possible, from the sound IF amplifier *before* limiting takes place. The 4.5-MHz signal is amplified and envelope-detected. This envelope-detector output contains the low-level 15-kHz modulation (usually 5–15%) and other unwanted components, such as the induced AM audio from the sound channel. Therefore, a high-Q active filter circuit is used to clean up and amplify the sine-wave signal.

At this point, the phase of the 15-kHz carrier can also be adjusted so that the derived waveform is exactly 180° out of phase with the encoding waveform originally placed on the video signal. In addition, any departure from the encoded waveform shape can be corrected, so that distortion due to nonlinearity, unwanted harmonics, and spurious signals can be eliminated. The recovered signal must exactly match the encoding signal since exact cancellation is necessary or, otherwise, unwanted ripples and shading will appear on the decoded TV picture. In fact, the recovered signal could be purposely distorted, if necessary, to compensate for linearity problems in the modulator or elsewhere in the system. However, it is best to avoid this by using properly designed circuitry with good linearity.

Generally, any circuit requiring critical tweaking and tailoring for each individual descrambler unit is an indication of poor design and should be avoided to save headaches and unreliable operation. Most well-engineered circuits work the first time and are not too critical as to adjustment or method. While there are exceptions to this generality, in the case of any descrambler circuits we are presenting, it is true. If something is very critical in its adjustment and is not stable, this is an indication of something wrong.

Sine-Wave Recovery Circuit

Fig. 4-4 shows a representative circuit that one might use for sine-wave recovery. It is very simple—a 4.5-MHz signal is taken from the TV sound system *before*

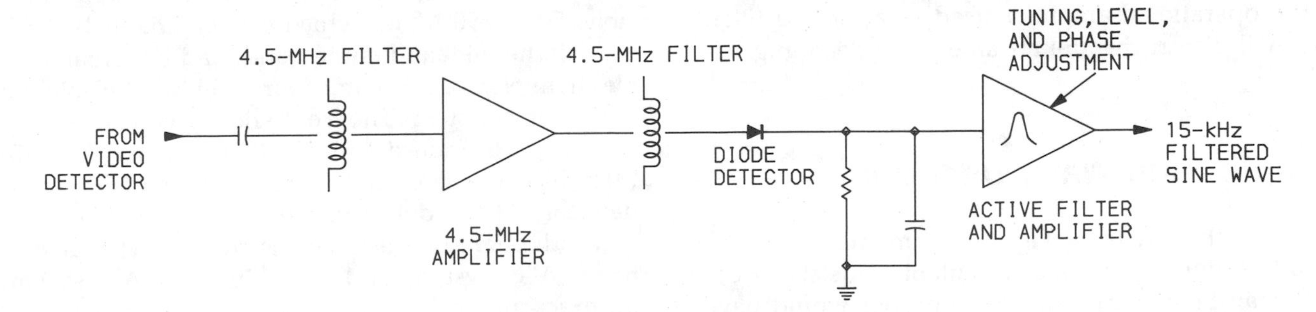

Fig. 4-3. A sine-wave-type 15-kHz carrier recovery system.

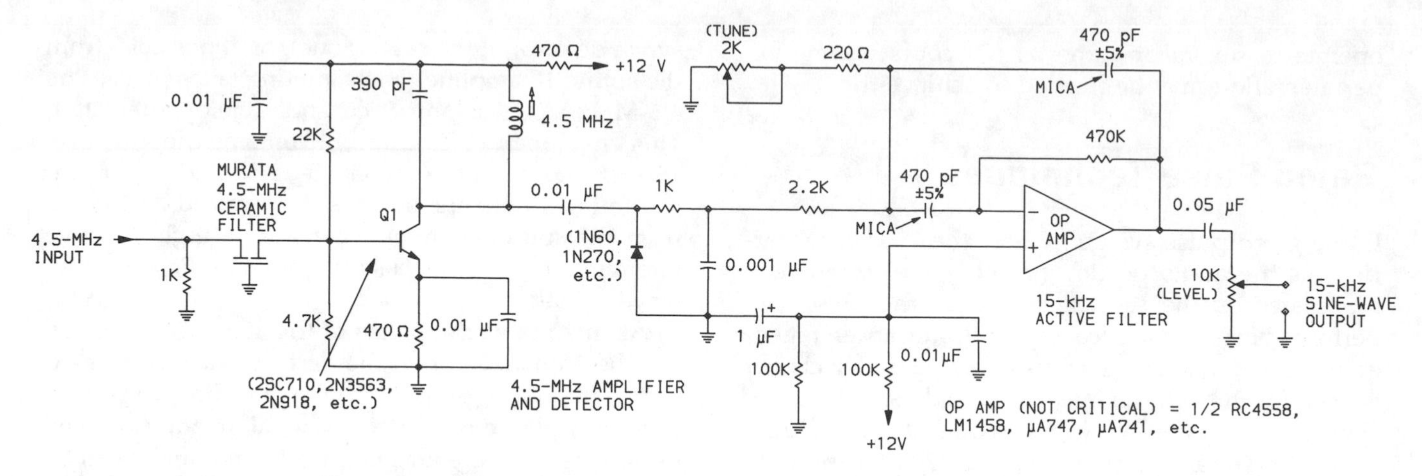

Fig. 4-4. A 15-kHz sine-wave circuit for recovery of decoding signal.

limiting. It is passed through a 4.5-MHz ceramic filter to eliminate "junk," and amplified by an NPN transistor having a gain of 25–30 dB. A 4.5-MHz tuned circuit in the collector of the transistor (2SC710, 2N3563, 2N918, etc.) serves as a load for the 4.5-MHz output signal. This is diode detected (1N60, 1N270, etc.) and fed into an active filter with a nominal gain of about 40 dB (100 × voltage gain). The filter is tuned to 15 kHz and the phase shift can be varied by adjusting the 2K potentiometer. The active filter has a response like a single tuned L-C circuit. The op amp may be almost any type suitable for audio-frequency operation, such as an RC4558, LM1458, µA747, or µA741, and is not critical. The output is taken from the 10K potentiometer, which serves as a level adjustment.

During operation, the circuit should be checked with a scope for linearity. To be conservative, the 15-kHz signal seen at the output of the op amp (top of 10K potentiometer) should *never* exceed about one half the supply voltage, which would be 6 volts p-p for the 12-volt supply voltage shown. The input signal level at 4.5 MHz will be around 30–100 millivolts. This depends on the level of the 15-kHz modulation of the 4.5-MHz sound subcarrier. Excessive input signal will cause limiting in the 4.5-mHz amplifier and/or the operational amplifier used as an active filter. This will distort the 15-kHz sine-wave decoding signal.

Simple Sine-Wave Decoder

By using the circuit of Fig. 4-4, combined with the circuit shown in Fig. 3-8, a suitable sine-wave decoder can be constructed. This decoder would have to be installed inside the TV receiver. This would require that the technician know where to connect to the receiver, and where to obtain correct signal levels and DC supply voltages. While this is no problem for the advanced experimenter or engineer, it is unsatisfactory from the viewpoint of most users of these descramblers. Every make TV is different, and, in some cases (especially in those TV receivers using integrated circuits and large-scale integrated function blocks), the tap-off points may not be accessible, being inside the IC devices. In this case, it would be impossible to connect a decoder.

In order to get around this problem, there is an easy way out. The approach usually used is basically very simple—the descrambler contains some suitable built-in circuits of a TV receiver. Often, this includes a front-end (either UHF or VHF) mixer, an IF amplifier, video and sound systems, and a modulator to place the decoded signals on a fixed output channel, usually Channel 3 or Channel 4. This decoded signal is fed into the subscriber's TV set, which is left tuned to Channel 3 or 4. All tuning is done with the descrambler unit, which is more properly called a converter/descrambler. Fig. 4-5 is a block diagram of such a unit.

The unit contains a front-end covering from, possibly, 50 to 890 MHz, which covers Channels 2–83, plus all the midband, superband, and hyperband cable channels. Signals come in from either the antenna or the cable system. The front-end converts these signals to the standard 45-MHz IF frequency. The 45-MHz IF is conventional, as is the video and sound detector. At the detector, we have scrambled video and audio signals. These signals are controlled in level by the AGC system, and, possibly, by an AFT system for exact tuning.

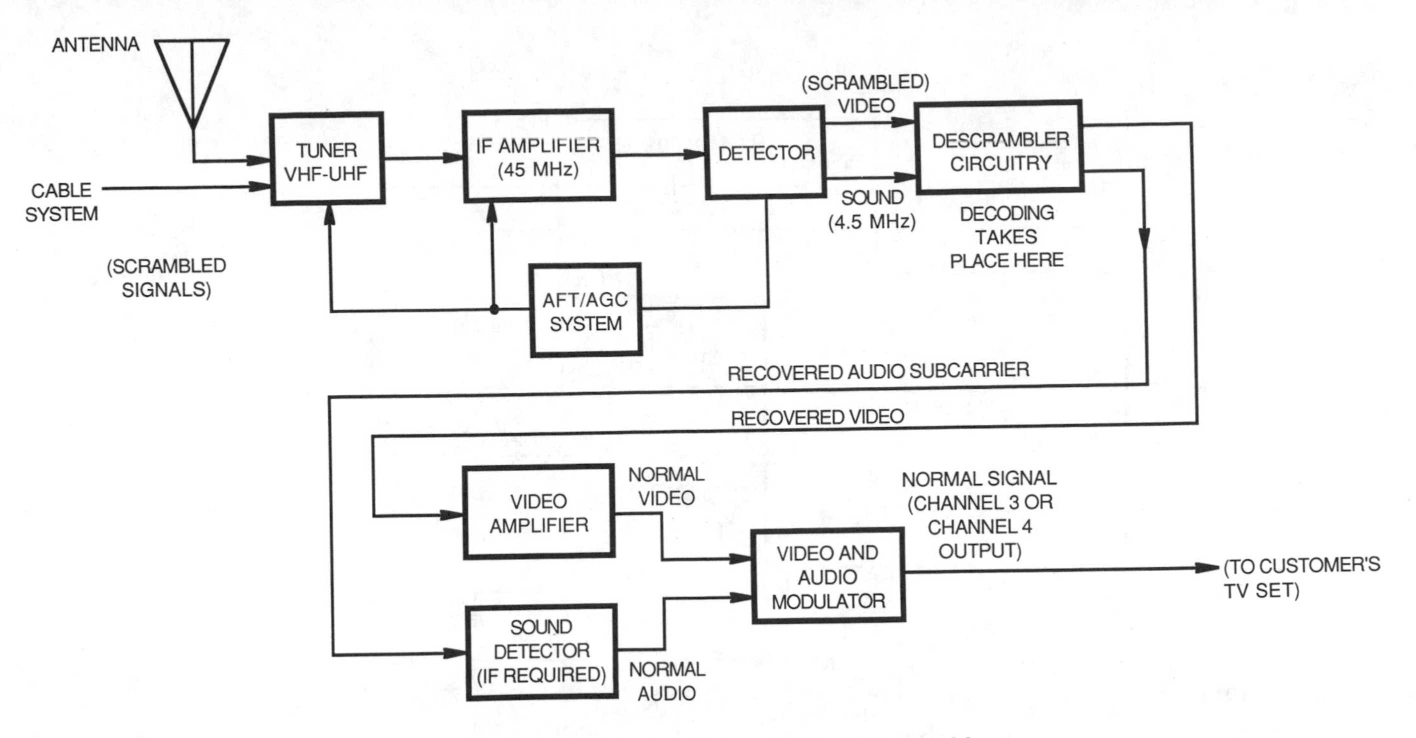

Fig. 4-5. Block diagram of a converter/descrambler.

The signals are then routed to the descrambler circuitry, which is designed to interface with the video IF/detector system. Since all this circuitry is in one unit, everything can be properly interfaced and no external controls are required for normal operation. The only controls are usually the "on/off" switch and channel selector (tuning) control.

Circuitry is also included that will turn off the decoder circuitry when a normal unscrambled channel is received. In this case, the whole system merely acts as a converter, converting all TV channels to either Channel 3 or 4. The output of the decoder circuitry consists of normal video and either the sound signal or the sound subcarrier, which can be detected with another sound decoder (if required). Now we have a normal video and sound signal. These signals could conceivably be connected to a video monitor and an audio system for reproduction.

What is generally done, however, is that these two signals are used to modulate a locally generated RF carrier on Channel 3 or Channel 4, so they can be fed directly into the antenna terminals of any color or B/W TV receiver, with no other connections required. The TV set is tuned to either Channel 3 or 4, as if the signal was from any normal TV station. Bear in mind that in a typical descrambler, the extra circuitry necessary for operation is sometimes more than that needed for just the basic descrambling circuits.

Remodulators

The circuit of a remodulator is shown in Fig. 4-6. It uses a standard IC and a few peripheral components. The IC generates both the picture and sound carriers, and modulates the VHF signal with the applied video and audio signals. It operates from a 12- to 15-volt supply. An alternative method is to use a prepackaged video modulator module. Several electronic firms offer them for about $10.00, or a unit can be salvaged from a discarded TV game or computer.

The SSAVI System

Some of the more complex descramblers make use of digital logic. One example of this is the SSAVI (Synch Suppression and Active Video Inversion) system. This system has four modes of operation:

1. Suppressed synch and inverted video.
2. Suppressed synch and normal video.
3. Normal synch and inverted video.
4. Normal synch and normal video (unscrambled).

The system has the capability of switching between any of these four modes on a frame-to-frame basis, therefore changing the scrambling method 60

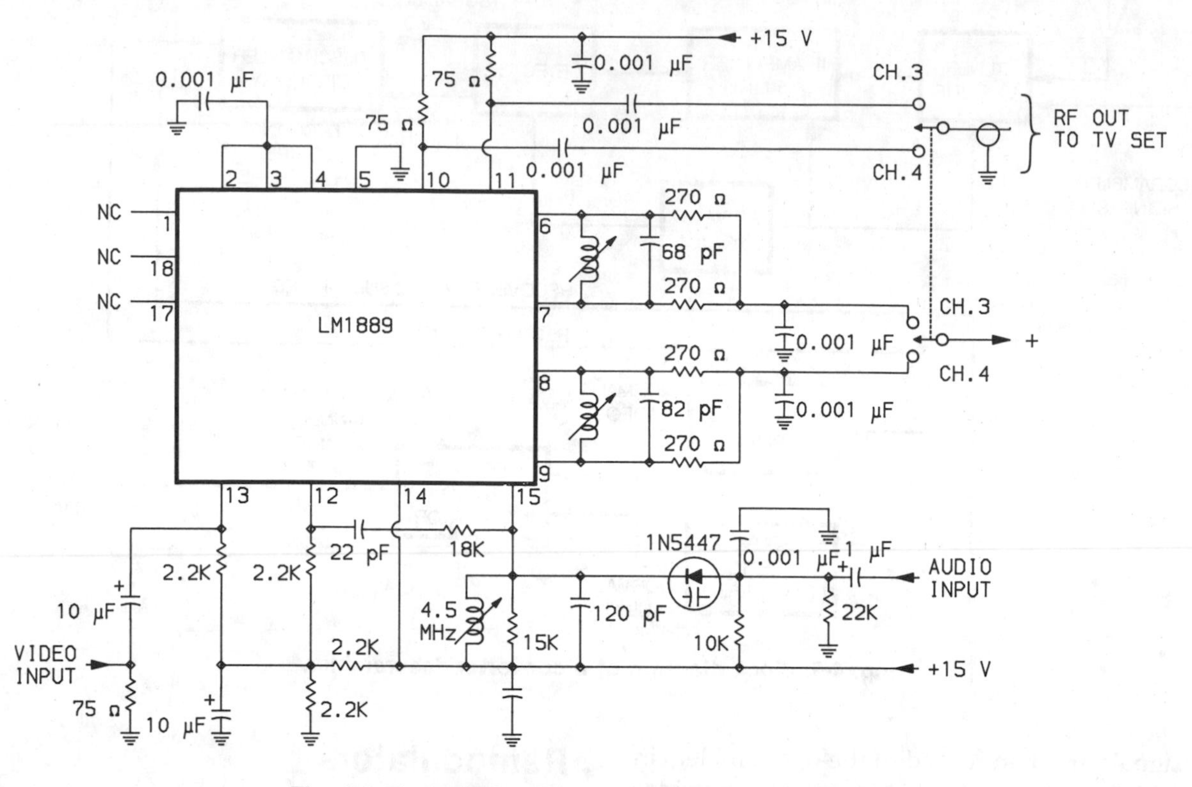

Fig. 4-6. Typical circuit of a sound/video modulator.

times per second, if desired. The audio is generally encoded on a 39.335-kHz carrier (2½ times the horizontal sweep frequency).

No reference signal is sent along with the scrambled picture. The decoder must generate its own synch. Also, the decoder must be able to tell if the video is inverted or normal. Furthermore, the synch pulse is never inverted. Therefore, the decoder must maintain the video portion (scan line) by only re-inverting this video without inverting the synch pulse, which also may be suppressed.

The synch is recovered by using a phase-locked loop. A VCO is used that is running at about 504 kHz—which is 32 times the horizontal sweep frequency. This 504-kHZ clock is also used elsewhere in the system. The composite video, whether scrambled or unscrambled, has a 15-kHz component. A phase-locked loop can lock onto this frequency and use it as a reference to control the VCO frequency, from which synch may be regenerated. Even if the synch is missing for part of the time, it can still be regenerated. As an example, the phase-locked loop in a color-TV receiver only uses 8-cycle bursts of 3.58 MHz (about 2.25 microseconds) every 63.6 microseconds. This is a 3.5% duty cycle—this means that 29 times out of 30, there is *no reference input* to the system. The phase detector output, in this case, consists of pulses at 15.7 kHz, which are integrated, and the resultant DC level is used to control the VCO.

In an SSAVI system, the first 26 lines of the picture are sent without suppressing the synch pulses. The PLL can easily lock onto this information and supply the missing synch information for the rest of the frame, when new synch information is available. The regenerated synch is used to restore normal synch, which is in turn used as a reference (''bootstrap'' operation). The leading edge of the vertical synch pulse is used as a reference from which all operations are timed. During the twentieth line, which is picked out with counter circuits, information is sent as to whether the frame video is normal or inverted.

The SSAVI system requires some rather involved circuitry to do the synch regeneration, pulse counting, and video reference-level restoration. This is necessary to avoid ''switching flicker'' when the video is switched from inverted mode to normal mode. This will be discussed in detail in a later chapter. (The circuitry is actually straightforward but there is a lot of it.) The system also contains capability for addressing individual decoders, several levels of premium service, and anti-theft features. Since these are not really related to the actual decoding process, they will not be discussed. A decoder can be built without them, as they play no part in the decoding process.

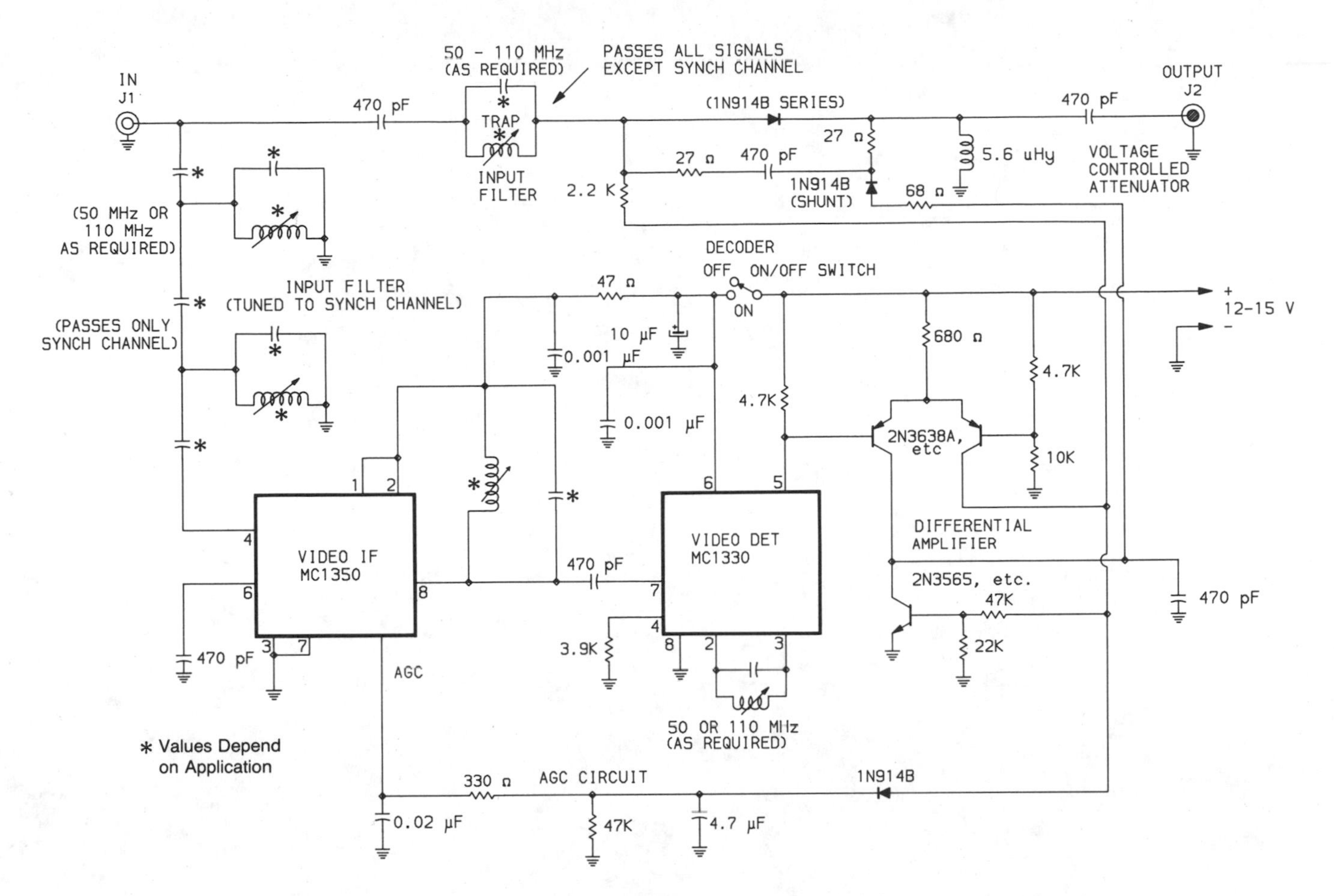

Fig. 4-7. Typical outband decoder circuit.

Outband Decoders

One other system and decoder is worth mentioning. It is called the "outband" system. In this system, used on cable systems, the premium video channel is sent with either suppressed or missing synch pulses. Another channel is used on the cable as the "carrier." This (synch) channel is placed usually somewhere around 50 MHz (below Channel 2), or in the 90- to 110-MHz range (in FM broadcasting). If a TV receiver were to be tuned to this channel, a blank white raster would be received. This signal is just the synch pulses modulated on a carrier. No video is present. In order to decode an outband system, a sound receiver tuned to this carrier channel is required. This can be simply a tuned radio-frequency amplifier, tuned to the 50-MHz or 90- to 120-MHz range, and consisting of one RF amplifier, a video detector, and a synch amplifier. The recovered synch is then used to drive one of the aforementioned video modulator or attenuator circuits. A circuit for such a unit is shown in Fig. 4-7.

In this circuit, the composite cable signal is decoded as follows: The synch signal goes through an input filter (tuned to the carrier frequency) to the MC1350 video IF. This signal is kept from appearing at the output jack (J2) by a trap tuned to this channel. An MC1350 integrated circuit amplifies this signal and feeds it to an MC1330 video detector. A video signal, consisting of only synch pulses, appears at pin 5 of the MC1330 detector. A differential amplifier drives a voltage-controlled attenuator. When pin 5 goes negative, the differential amplifier supplies a bias current so as to reverse bias the series 1N914B diode and forward bias the shunt 1N914B diode, inserting the attenuator in the circuit. During synch pulses, pin 5 goes positive and the attenuator is oppositely biased, thus freely passing the signal with little attenuation. The synch signals (positive polarity) are sampled by an AGC circuit and maintain a constant system gain. No coil data has been given as this depends on the application. A more detailed circuit will be presented in a later chapter.

5

The SSAVI System

A problem with several of the scrambling schemes mentioned earlier is that of security. While the sine-wave and gated pulse systems are widely used, decoders for them can also be easily built or readily obtained. There are several firms in the United States, which we will not name, that sell these decoders. (If you are really curious, look in the classified ad section of any video or TV hobbyist magazine.) As a result, more sophisticated systems, which use a dynamic scrambling algorithm, have been developed. One of these is the synch suppression and active video inversion, or SSAVI, system. This system has four modes of operation, as previously discussed:

1. Suppressed synch and inverted video.
2. Suppressed synch and normal video.
3. Normal synch and inverted video.
4. Unscrambled operation.

The SSAVI Signal

Fig. 5-1 is a diagram of the signal waveform. The main thing to remember is that all operation is referenced to the *falling* edge of the vertical synch pulse. Refer to any TV textbook or the FCC standards for TV waveform details, if necessary.

Subscriber Coding and Security

In the diagram, horizontal scan lines 0 to 9 fit into the normal vertical blanking interval. Lines 10, 11, and 12 are used to transmit the subscriber code number. The SSAVI decoder contains logic for the prevention of unauthorized use of the decoder in another area or on another service. The subscriber code number is matched by the logic in the decoder with a code stored in a programmable read-only memory (PROM). Line 13 is used for checking the incoming market code with the code stored in the PROM. If all codes match and the transmitted program code matches a program code stored in memory, the decoder is enabled (turned ON).

This coding feature prevents a decoder stolen in New York, for instance, from working in another area, such as Miami. Also, this permits any subscriber who has not paid his bill to be shut off. Additionally, various levels of programming can be sold, by cutting the decoder off during unauthorized programs, via the code scheme. No further discussion of this coding feature will be presented as it is not necessary for descrambling the signal. An SSAVI decoder can be operated without this feature by simply bypassing it.

Basic Operation of SSAVI

Refer to the diagram of Fig. 5-1. All the lines from 27 to 260 are scrambled. The 261 and 262 lines as well as lines 0 to 26 are not scrambled.

Horizontal line 20 is a key line. It is used to determine whether or not a frame is inverted (video polarity). This is done by the following process (Fig. 5-2): If the frame is to have normal polarity, the first part of line 20 will be white, while the remainder is black. If the video is to be inverted, line 20 will be entirely white. This line can be sampled and the data stored in a memory (a simple flip-flop), and, then, can be used to select either inverted or noninverted video, as required. In this way, every frame is automatically switched to the proper polarity. Line 20 is also used to establish the proper black-and-white reference levels for either inverted or normal frames.

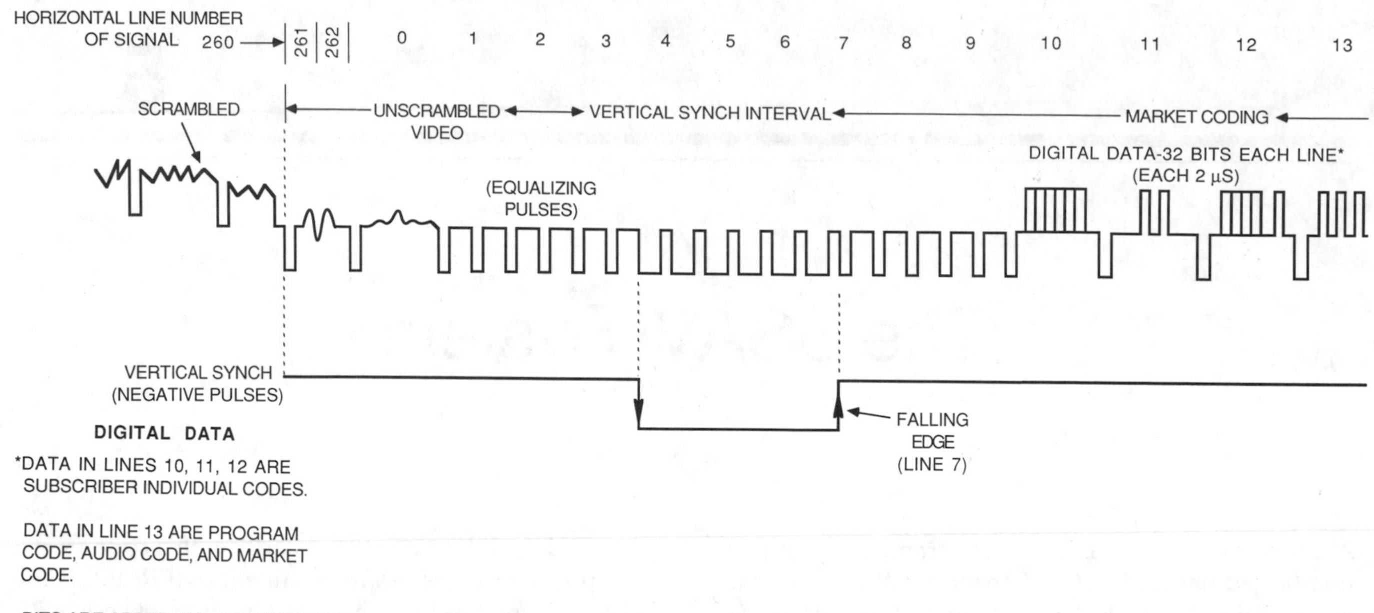

Fig. 5-1. Diagram of the

An SSAVI Decoder

A block diagram of an SSAVI decoder is shown in Fig. 5-3. Scrambled video is amplified by a video amplifier with two inputs. The (+) input is used for the video input. The inverting input (−) is used to add the missing or suppressed blanking pulses. For now, assume these pulses are present. Therefore, the output of the first video amplifier is a composite video signal, corrected insofar that some lines may or may not have "inverted video" present.

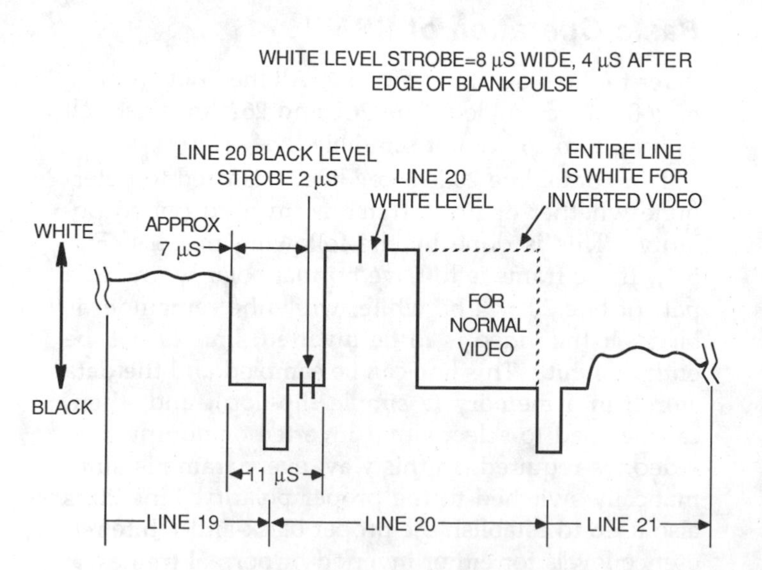

Fig. 5-2. Details of horizontal line 20.

At this point, some of the video is used to drive a synch separator. The synch separator is the same type as used in a standard TV receiver. Part of this synch signal is used to recover the vertical synch pulses with a vertical integrator. (This is simply a low-pass RC network and, also, a gate to "clean up" the signal.) Figs. 5-4 and 5-5 show typical circuitry for this purpose. In addition, the synch pulses are used as a reference for a phase-locked loop. This phase-locked loop is set up to provide a 504-kHz (nominal) signal that is phase-locked to the horizontal frequency, which is one thirty-second of 504 kHz. This 504-kHz signal is used to derive the synthesized horizontal blanking pulses, sampling pulses for black-and-white reference levels, and for detection of normal/inverted video states.

Horizontal Pulse Regeneration

Horizontal blanking pulses may be regenerated with a ÷32 counter from the 504-kHz signal. A 504-kHz signal has a period of 1.98 microseconds. By using five consecutive count states of a ÷32 counter, a 9.92-microsecond pulse can be produced. This is very close to the 10.7-microsecond (nominal) blanking pulse.

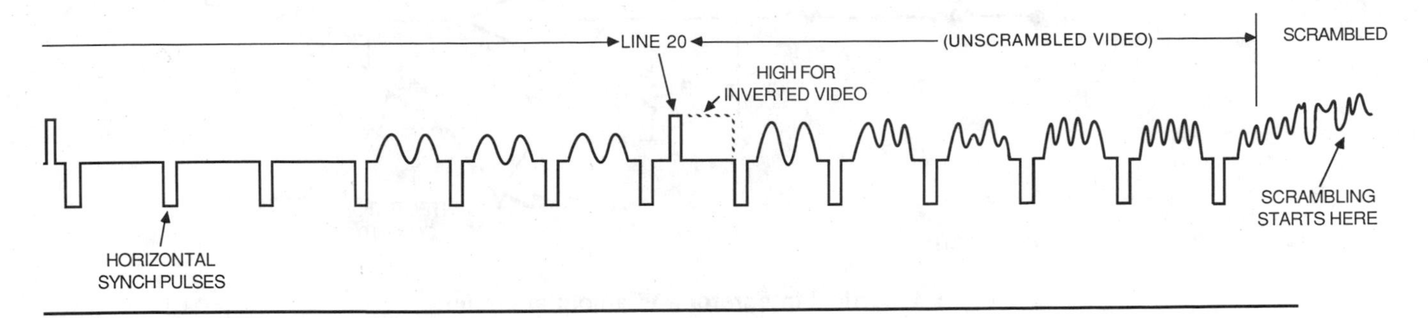

SSAVI signal waveform.

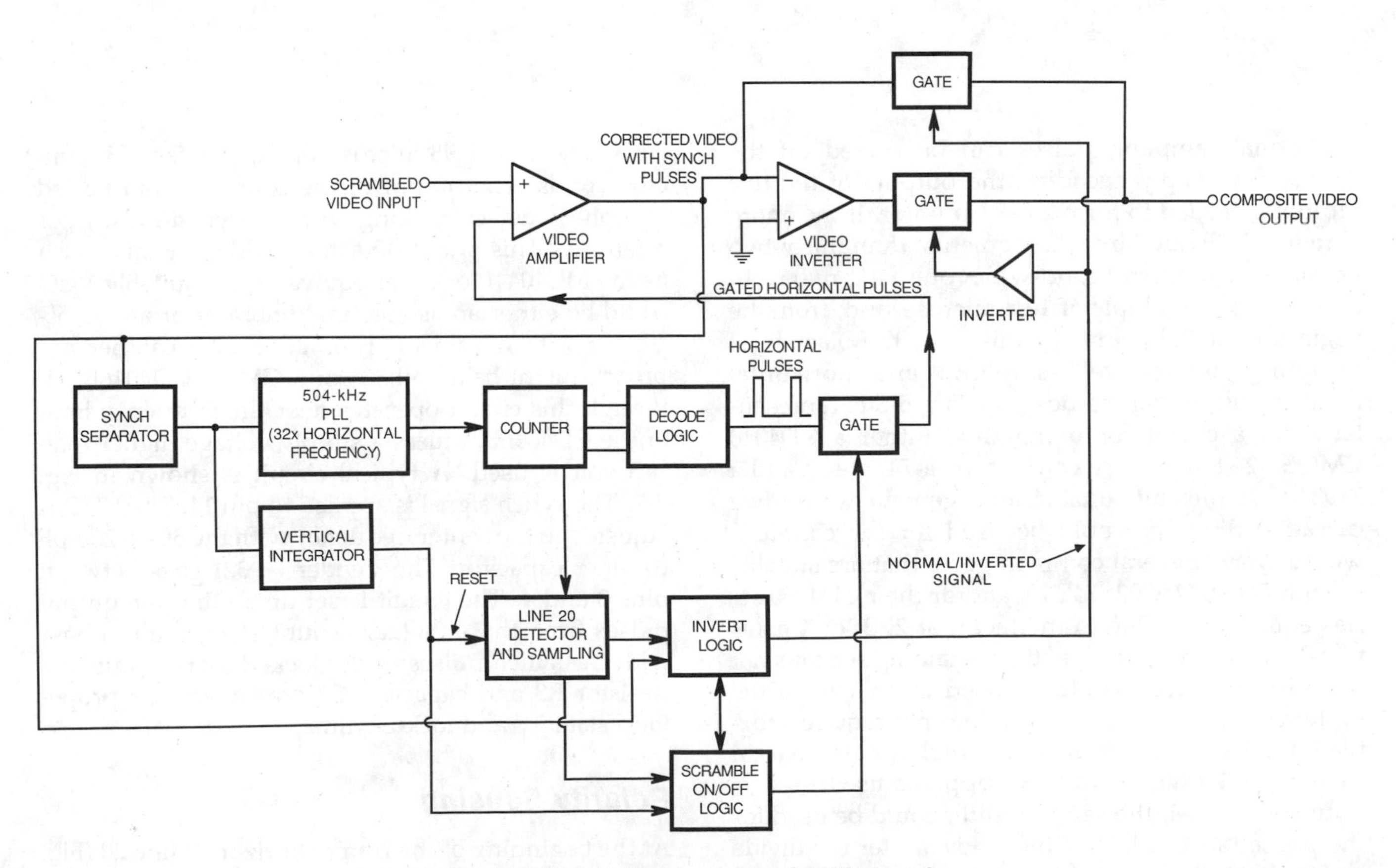

Fig. 5-3. Block diagram of an SSAVI decoder.

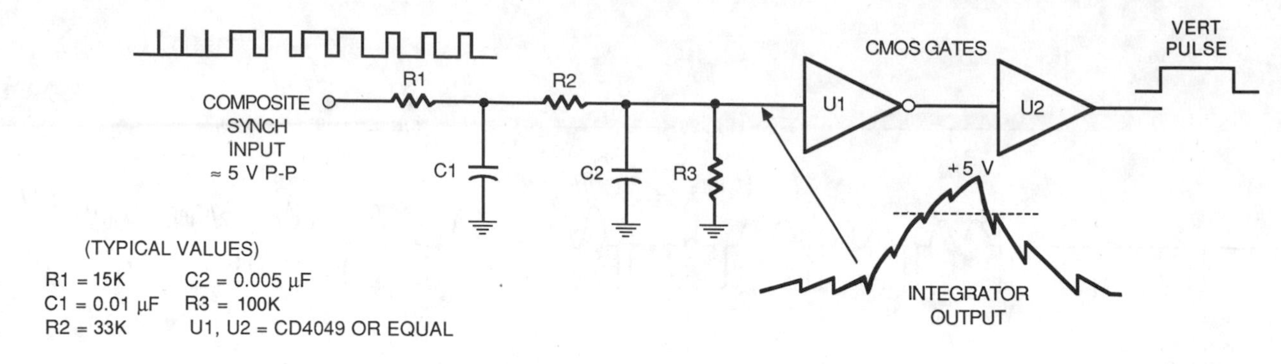

Fig. 5-4. Vertical integrator and amplifier circuit.

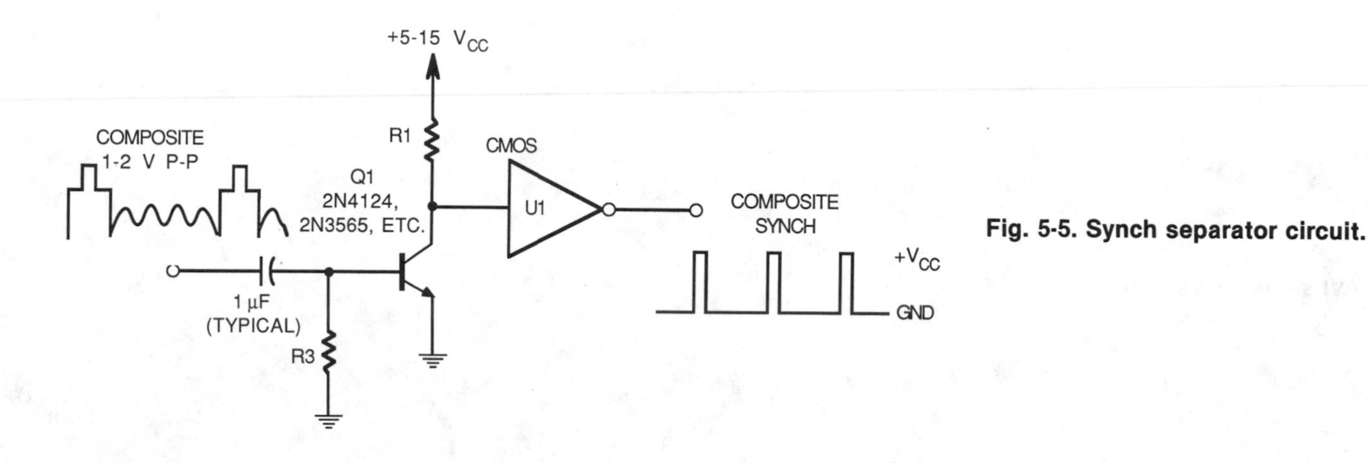

Fig. 5-5. Synch separator circuit.

Additional sampling pulses can be picked off the counter by simply decoding the outputs of the five flip-flops needed to form a ÷32 counter. If the entire scan line is divided into 32 segments, then thirty-two possible 1.98-microsecond-wide pulses can be obtained, at any multiple of 1.98 microsecond, from the beginning of the count. In this way, the black-level sampling, white-level sampling, and horizontal blanking pulses can be derived. Fig. 5-6 is representative of a circuit for doing this. Either a CD4040 CMOS 12-stage binary counter, or a 7473 ÷ 2 and a 7493 ÷ 16 (or any suitable arrangement where five cascaded flip-flops could be used for the counter), would work. Several common gates that are suitable are the 74C00, 74C02, 74C10, and/or their CD4000 series equivalents. Due to the fact that 2, 3, or 4 gates usually come in a package, the remaining sections not used in this circuit could be used elsewhere in the system. CMOS is fast enough, does not require a regulated +5-volt supply, and has a higher noise margin than TTL (if a higher voltage supply is used).

In some cases, the same counter could be used for the phased-locked loop. Since the counter is a divide by 32, the zero state could be detected and would occur every 32 × 1.98 microseconds, or every 63.5 microseconds. The output of the counter could be fed to a phase detector, along with the synch pulses, for reference. This phase detector could be a chip, such as the MC4044P or other equivalent. A suitable VCO could be either an astable multivibrator or an LC oscillator with a varactor tuning circuit. Another approach might be an NE565 or a CMOS CD4046B, although this circuit operates near the frequency limit of the CD4046B, unless a supply voltage higher than +5 volt is used. A typical circuit is shown in Fig. 5-7. The synch signal is applied to pin 14. The VCO is adjusted to the center frequency with the 50- to 200-pF trimmer capacitor. The divider (÷32) goes between pins 3 and 4. The circuit is set up so that the output pulses from the ÷32 (zero count state) are in phase with the synch pulses under locked-loop conditions. Resistor R3 and capacitor C2 are chosen for proper loop stability and lock-up time.

Polarity Sensing

At the beginning of the frame, horizontal line 20 (Fig. 5-1) must be sampled. This can be done by counting

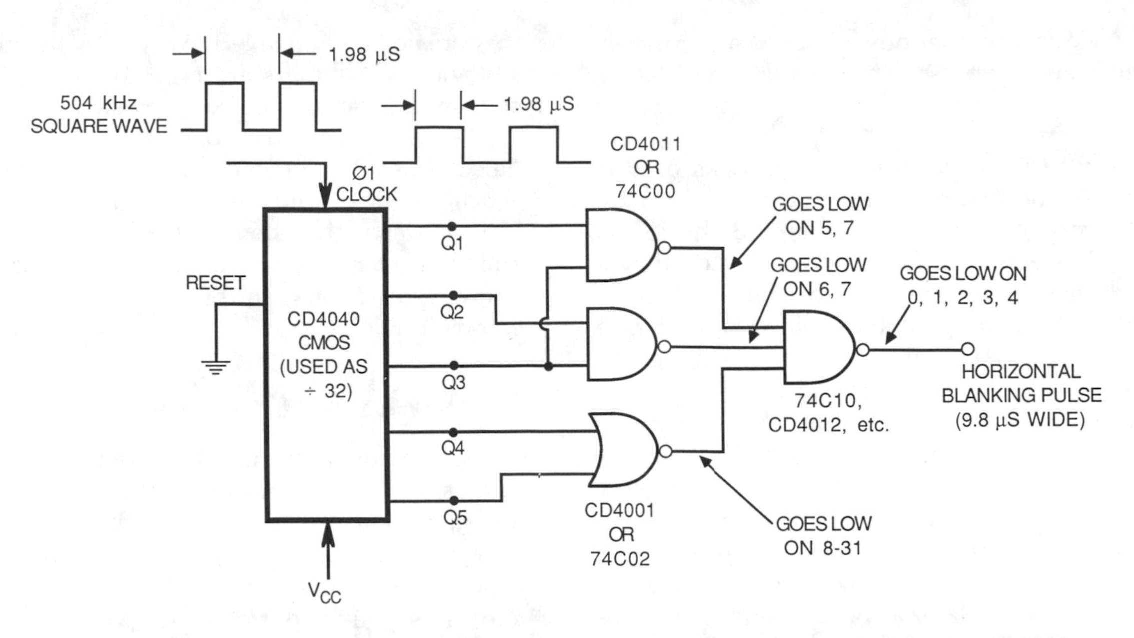

Fig. 5-6. Circuit for deriving horizontal blanking pulse from a ÷32 counter driven at 504 kHz.

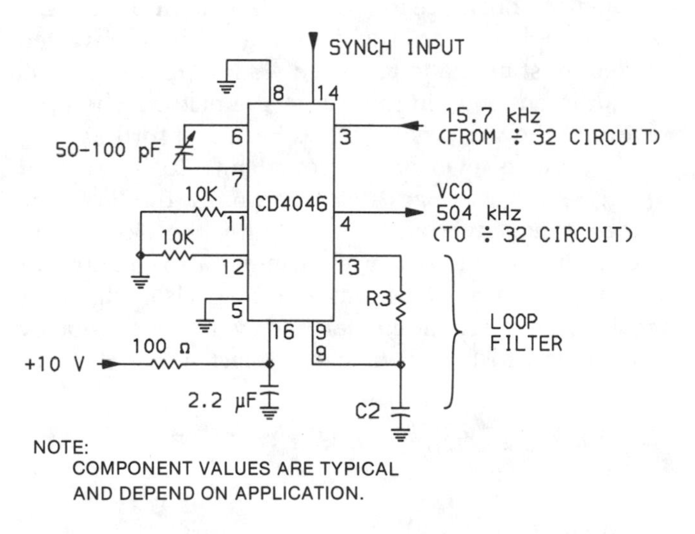

Fig. 5-7. Typical phase detector/VCO circuit (504 kHz).

a suitable number of lines from either the rising edge or falling edge of the vertical synch pulse. This would be 16 or 13 lines, respectively. A binary counter with similar state-detection logic could be used, as previously discussed. Actually, on line 20 (Fig. 5-2), a 2-microsecond pulse that starts about 7 microseconds after the blanking-pulse leading edge is used to sample the back porch of the blanking pulse. This sample is used for establishing correct black levels in the decoder. About 4 microseconds after the blanking pulse, a sample (about 8 microseconds) is taken of the video, which represents the "white" reference level of the system. After the sample, the video portion of horizontal line 20 is used to determine whether that frame of the signal (lines 0–262) will be normal, as signified by a black level, or inverted, as signified by a white level for the rest of the line. This video level is sampled and used to control the inverting logic. The inverting logic is usually a flip-flop, which is used to drive an inverter stage and two gates. These gates select either inverted or normal video.

"Missing" Synch

Starting at line 0 (zero), the blanking pulses are present and used to synchronize the 504-kHz PLL. However, what happens when the blanking pulses are suppressed? How does the circuit "know" what to synch on? In a previous chapter, we mentioned that a PLL can lock on a reference signal that may be periodically missing. The 3.58-MHz color-burst signal is an example of this situation. It is present only 3.5% of the time. The rest of the time, a 3.58-MHz signal, correct in phase, must be supplied to the color demodulators, in the absence of a reference signal. There is an analogous situation here.

Suppose line 26 has just been scanned. While the first 26 lines are being scanned, everything is present. When line 27 comes along, the synch is suppressed or missing. The 504-kHz VCO/divider does not know this and produces a pulse anyway. At this instant, a gate controlled by the scrambler on/off logic is enabled. This pulse is added, in the video amplifier inverting input, to the incoming video signal. The output of the video amplifier now contains an artificially generated blanking pulse. This "fools" the 504-kHz PLL since the synch separator treats this signal as any other pulse signal.

Now, the VCO, in a rough sense, locks onto its previously established reference. As long as the VCO stability is good enough over the next 233 horizontal lines (until line 260 comes around), it will continue to generate pulses at the proper rate. These will be inserted into the video signal in the video amplifier, resulting in a synch-restored video signal at the amplifier output.

As an example, suppose we want the picture to be in synch within 0.1 microseconds after 233 lines. This is about a 0.16% frequency drift. A single frame is 1/60 second (more than 233 lines). This amount of time permits a frequency drift of about 10% for the 504-kHz oscillator, or about 50.4 kHz per second. It is difficult to make an oscillator as poor as this. Therefore, this 15-millisecond lack of reference to the system should be no problem at all. However, one cannot be too careless. The most important thing in a PLL is the VCO itself. The better the VCO, the less work that the rest of the PLL has to do. It is generally of utmost importance to have a clean stable VCO, since the loop cannot simultaneously "clean up" everything wrong with the VCO. This is especially important in PLLs used in frequency synthesis applications in communications equipment.

At line 260, the signal once again has normal synch (note that synch is *never* inverted) and the scrambler on-off logic disables the gate that supplies the regenerated blanking pulses to the video amplifier, since they are no longer needed. At line 7 of the next frame, all logic and counters are reset to the original (zero) state by the vertical blanking pulse.

The scrambler on-off logic is another counter that detects lines 27 and 260. It turns on the regenerated synch on line 27 and turns it off again on line 260. It is a counter with 9 stages (maximum modulus 512), and a decoder circuitry for states 27 and 260, respectively. A block diagram of a suitable logic system is shown in Fig. 5-8.

Polarity Correction

The signal now present at the video amplifier output has restored synch, but the video may or may not be inverted. Synch is never inverted. The color burst is also not inverted since it rides on the suppressed synch pulses. However, the chroma signal is inverted and, thus, is 180° out of phase with the burst reference phase. Therefore, we must invert the video, but not invert the synch pulses, in order to restore the proper color balance to the signal. In addition, the DC reference levels on the inverted and noninverted video must be made to match. Failure to do this will result in an annoying flicker in the picture when polarity of the video is switched back and forth.

The video level of horizontal line 20 is sampled (Fig. 5-2). If it is black (later portion of the line), the video is not inverted. If it is white, the video is inverted in the video inverter. There is also circuitry for setting the correct black-and-white levels so that normal or reinverted video match to avoid the previously mentioned flicker. Two gates select either the cor-

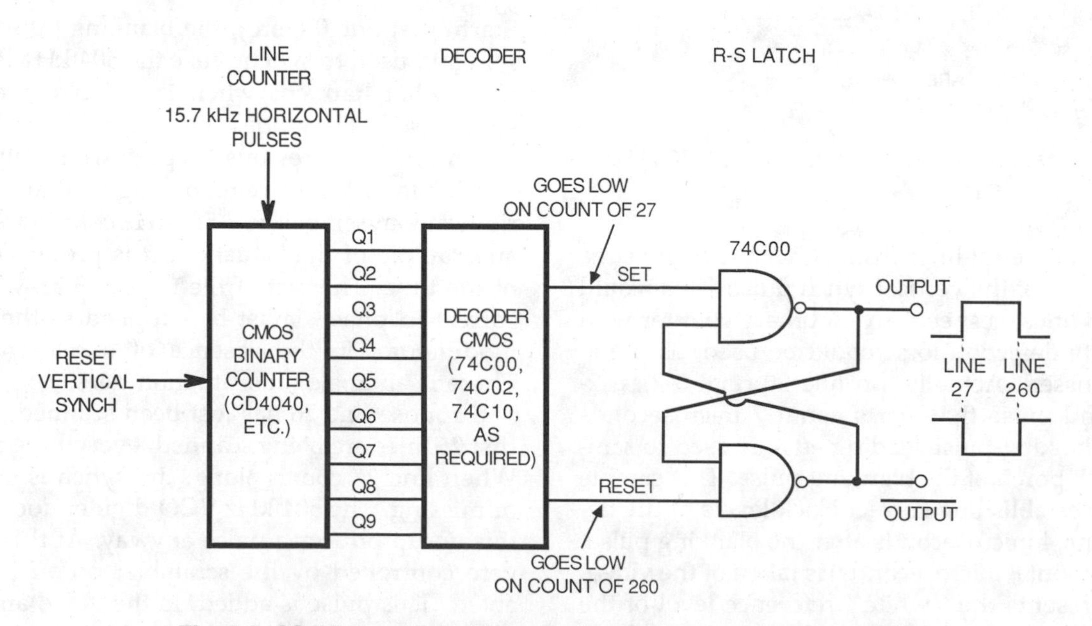

Fig. 5-8. SSAVI scrambler/descrambler counter and logic circuit.

rected video (noninverted) or the output from the video inverter as required. They are summed at a common point; this is the decoded video output.

The invert logic (sampling gate and flip-flop) controls which gate is used, either invert or normal. Since after line 260, nothing is inverted, the invert logic is reset.

Audio Encryption

So far, nothing has been said about the audio. The audio may or may not be encrypted. This is done by choice of the system operator and is optional. If the audio is encrypted, it is done by placing it on a 39.335-kHz subcarrier on the audio carrier (4.5 MHz). Methods of doing this were discussed in a previous chapter.

Conclusion

The SSAVI system is a fairly secure one, due to its addressability and its multiple modes of scrambling. In fact, the mode can be changed every frame if desired. This makes it difficult for pirate decoders to function properly. However, the SSAVI system has only a few modes and suffers from the same problem of having a basically fixed algorithm (four modes). It is complex and, therefore, can be more costly to implement. In the near future, the SSAVI system may be replaced by more sophisticated digital scrambling systems using dynamic methods. Digital methods, due to lower costs and future technology, may become cost-competitive, or even cheaper, and, at the same time, much more secure.

VideoCipher II™

Another system that is currently being used is called the VideoCipher II™ (VCII). This is the system currently used by HBO and other satellite programmers. It is a simpler system than its predecessor, VideoCipher I™*. VideoCipher I™ used line dicing techniques. Five hundred and twelve (512) analog samples of each scan line were sampled, stored in an analog shift register (bucket brigade), and clocked out of the shift register in random order. This was very effective, but it had the disadvantage of high cost and reportedly did not prove to be economical for consumer use. However, future VLSI technology may change that. VideoCipher II™ uses a simpler, different approach. The picture is inverted, synch pulses are removed, color burst is moved to a different location, and the synch pulses are replaced by an 88-bit digital word, representing a portion of the program audio, system data, and error correction. (VideoCipher I™ also encrypted the audio.)

Audio Encryption Using DES

The audio encryption algorithm used is the National Bureau of Standards Data Encryption Standard. It will be discussed in a later chapter. However, we wonder if use of a proprietory universal data encryption scheme is a good idea. If the VCII system becomes widely used and if the public has these decoders in large numbers, someone is bound to figure out the algorithm sooner or later. Will manufacturers be allowed to manufacture these decoders offshore, or even be permitted to ship the decoders outside the United States? Who knows? However, this system appears to be the one that satellite broadcasters will use in the future.

VCII Video Decoding

By taking the video, and using previously discussed synch regenerator techniques, the synch pulses can be derived. Since the color burst has been moved to a different location (DC level shifter), it still can be extracted with a gating pulse. From this color-burst pulse, a subcarrier can be regenerated and, therefore, the synch as well, with a method similar to the SSAVI system. Therefore, VCII video may be very simple to decode. The problem will be the audio channel.

If the audio is missing, what good is the picture? Some things, such as rock concerts, would be useless as an entertainment value without audio. Imagine MTV without audio. On the other hand, X-rated programming may be acceptable without audio. It depends on the program content and the viewer. For this reason, both video and audio scrambling will probably always be necessary.

* VideoCipher I™ is used by network feed systems.

6

Advanced Scrambling Methods— Satellite Techniques

Several systems used for encrypting NTSC video have been described. We have mentioned sine wave, suppressed synch, SSAVI, and some advanced methods, with the advanced systems using such digital techniques as line dicing, where a line is "cut" into several segments and transmitted in random order according to a prearranged sequence. Gated (suppressed) synch and sine-wave scrambling methods are easy to decode and offer only a minimum level of security. Synch suppression and active video inversion (SSAVI) is more complex and therefore more difficult to decode, plus a lot of circuitry is required for decoding. Digital methods currently are somewhat expensive and too complex for widespread use. (This will surely change in the future.)

A system in wide use at present, which may become the (standard) satellite system, is called VideoCipher II™, or VCII. In this system, the video is randomly inverted, normal synch is totally absent, and the audio is encrypted digitally. Video security is moderate, but the audio signal security is very high (hard scrambled).

Representative System

The following description of a system very similar in principle to the VideoCipher II™ and the Oak Orion™ system has been taken from U.S. Patent # 4,336,553. The description will be mainly of the decoding process, although the encoding process will also be briefly described.

The Signal

The scrambled signal for the system is shown in Fig. 6-1. The synch is absent; it is replaced by three 8-bit digital words (one byte each). The first two bytes contain audio information. The third byte is used for synch regeneration purposes and several other possible functions (as may be needed for a particular system). The bit rate here is 4.0909 MHz, with the data stream consisting of 24 bits and, therefore, being about 6 microseconds in duration.

Receiving the Signal

Reception of an encoded signal using this unique system (refer to the block diagram in Fig. 6-2) can be accomplished as follows: The scrambled video from the detector is applied to three circuits. The first is a switched inverter. It is a video amplifier which gives two identical outputs differing only in phase. Suitable clamper circuitry must, of course, be provided so as to set the proper DC reference levels; otherwise, flicker will occur upon polarity inversion. A video inversion detector circuit, which samples a portion of the video signal to determine whether normal or inverted video is to follow, controls the switched inverter. The polarity is normally flagged on a given horizontal line. See Fig. 6-3 for illustration of this. In addition, a data filter circuit is used to strip off the digital data that is used for synch and audio functions. VCII does not use this random-inversion option. The video in VCII is always inverted.

In Fig. 6-3A, a given horizontal line "N" has certain portions either black (high) or white (low). Sampling is done at a predetermined time. The video-level sampling determines the polarity. In Fig. 6-3B, a given horizontal line "N" has a few pulses inserted. During this time, a flip-flop or counter is driven by these pulses. The resulting counter "state" is then used for polarity switching. This method can be used for more functions than can the method in Fig. 6-3A, depend-

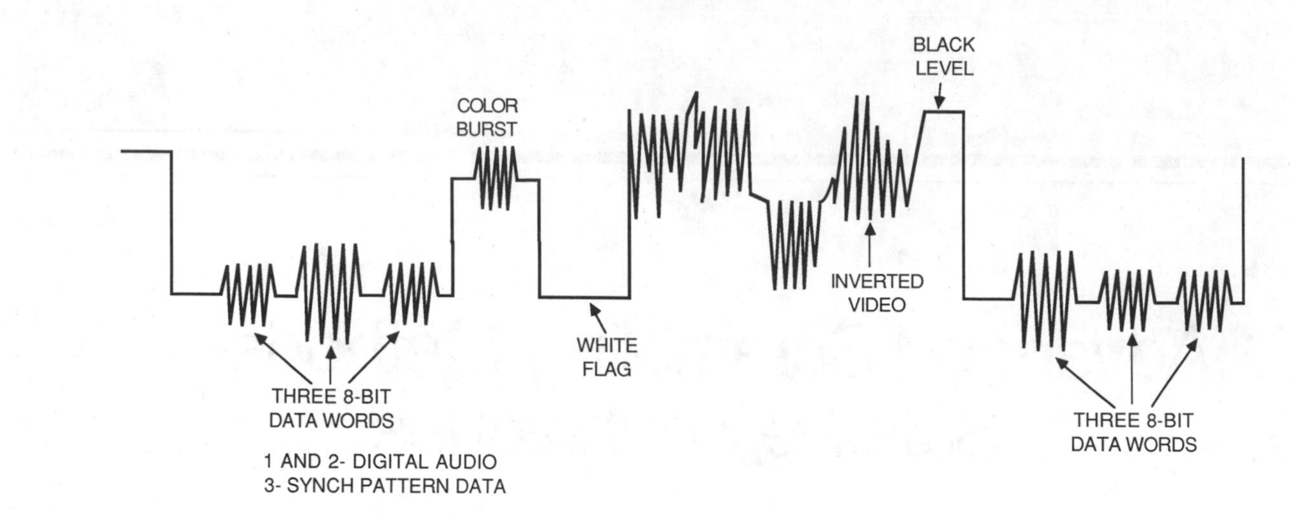

Fig. 6-1. Scrambled signal of representative scrambler system.

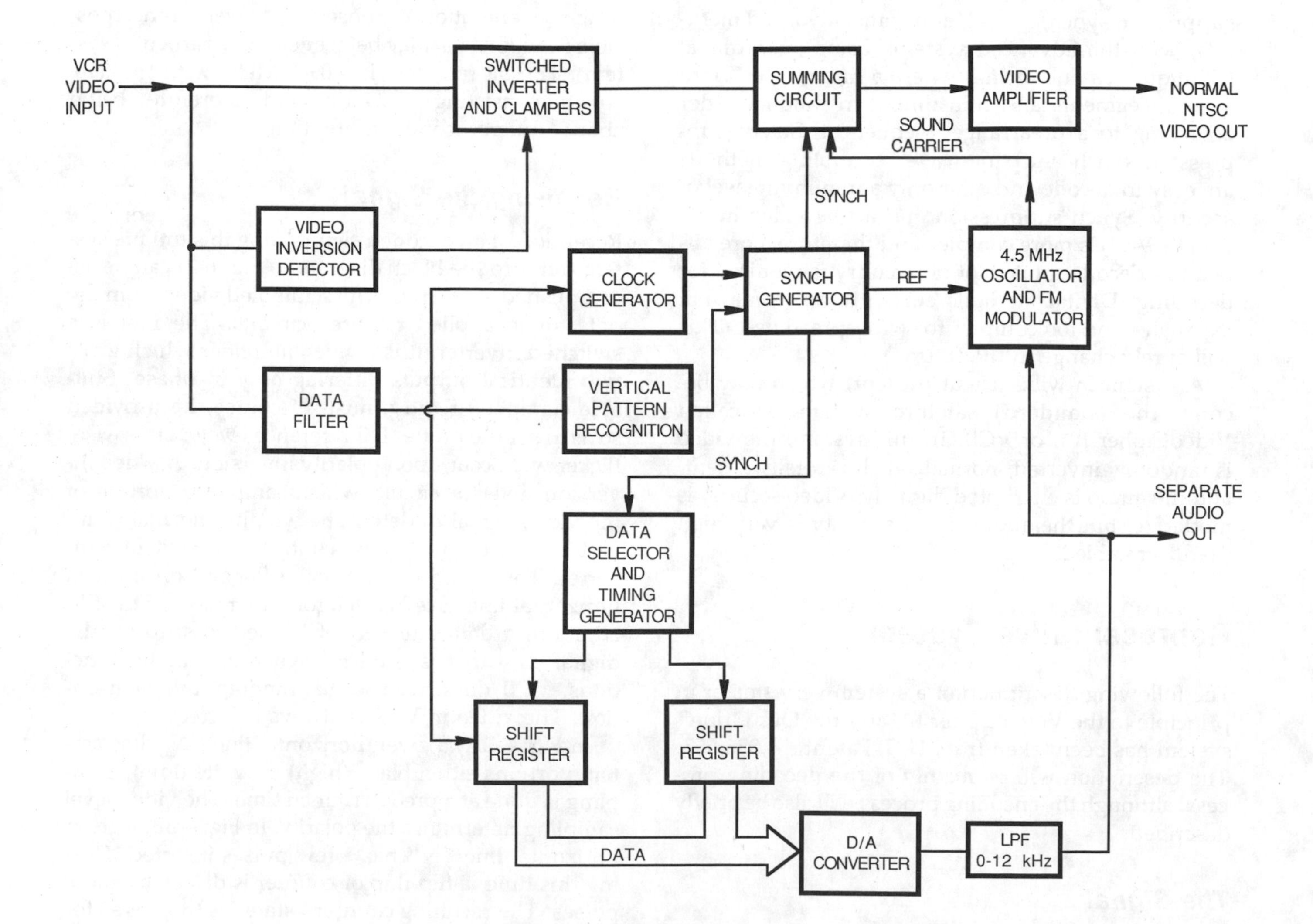

Fig. 6-2. Block diagram of circuit.

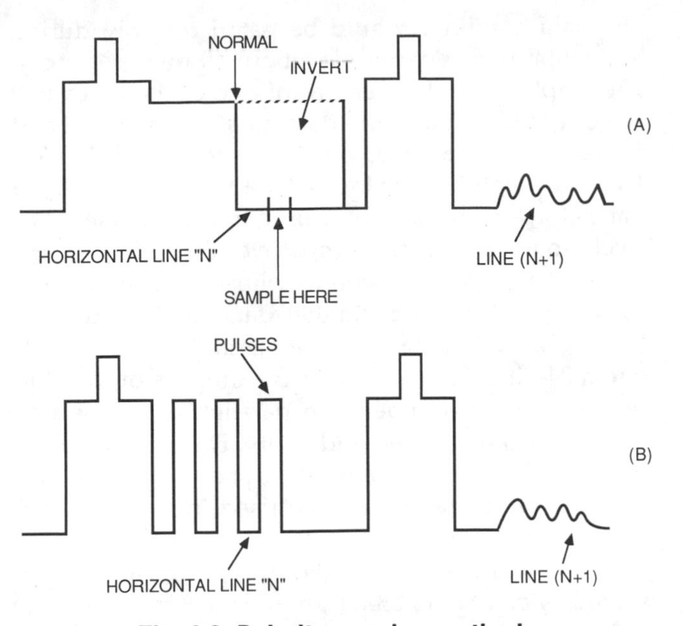

Fig. 6-3. Polarity sensing methods.

ing on the number of possible states that the counter can assume.

The output of the switched inverter (Fig. 6-2) contains corrected and clamped NTSC video signals. The synch information is missing. This video (less synch) is fed into a summing circuit, which combines all the necessary components of the decoded NTSC signal.

Audio Encryption and System Security

The different approach to synch and audio used in this system requires the use of digital techniques. The data filter is an amplifier that is gated on during the horizontal retrace interval. The data may appear as shown in Fig. 6-4. It consists of three bytes (three 8-bit words) with a bit rate of 4.0909 megabits/second. The first two bytes are audio (digitized) samples. The third byte is used for synch-regeneration purposes.

The amplitude of the data bytes is modulated in order to make it difficult for an unauthorized receiver or decoder to synch on them. In addition, a 15-Hz sine wave is mixed with the data bytes so that their bias level varies from one cycle (horizontal line) to another. This discourages clampers and other such circuitry from using these data pulses as a reference. In addition, the sound carrier is modulated with a nominal 15.75-kHz (± 30 to ± 60 Hz) signal. This prevents the chrominance signal from providing a reference for unauthorized decoding. All this is ignored at the receiver. The data-pulse recovery process does not need the 15-Hz signal and ignores it. The data-byte amplitude is similarly ignored by the audio and synch recovery circuits. The third data byte, used for synch in this example, has a given bit pattern for horizontal synch and another pattern for vertical synch. A pattern-recognition circuit is used in the decoder to distinguish them and to provide framing information. Other formats using different bit assignments, of course, are possible. Therefore, the video signal (Fig. 6-1) consists of:

1. Video which may or may not be inverted.
2. A sound subcarrier with 15-kHz (nominal) modulation.
3. A missing synch that is replaced by three 8-bit data words.
4. A 15-Hz sine wave that modulates the level of the data bytes.

Due to the 15-Hz sine wave, random AM is placed on the data bytes, and the interfering sound subcarrier on an unauthorized receiver cannot synch on the video information. The video may be inverted (VCII does away with the sound carrier) and no sound will

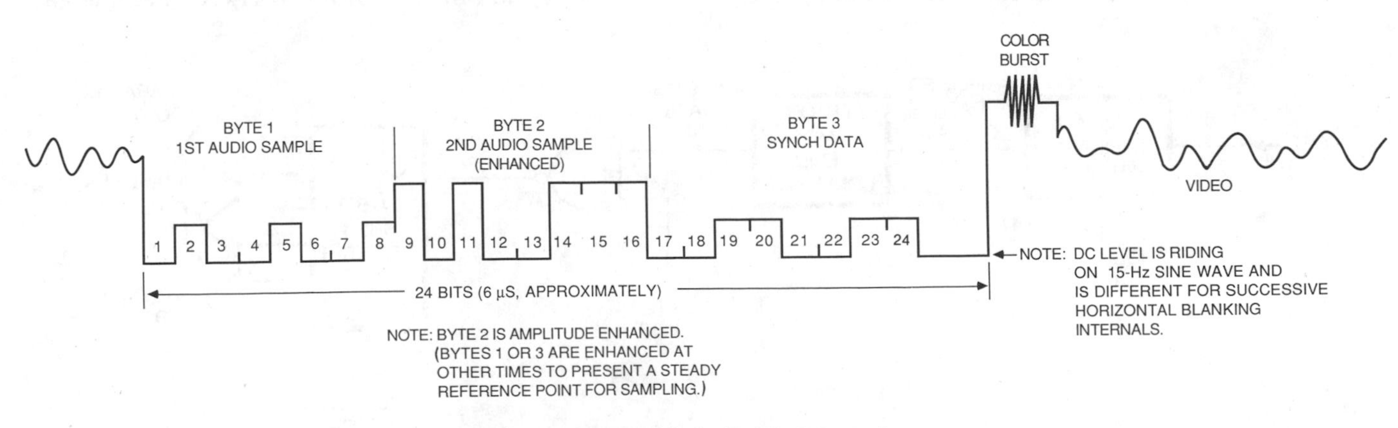

Fig. 6-4. Scrambled signal data pulses.

be heard, since there is no audio on the sound carrier. Scrambling is rather secure. Although video security is not very high, the sound security is extremely high, because a totally different method of audio recovery must be used in the TV receiver. This will be discussed later.

Sampling Theory and Audio Digitization

A well-known theorem in communications is the *Shannon sampling theorem.* This states that a signal having a maximum frequency component of f_{max} can be completely specified by a number of samples taken at the minimum rate of $2f_{max}$. In other words, a 12,000-Hz audio bandpass requires a sampling rate of at least 24,000 Hz. We can sample the audio at a 31.5-kHz rate (twice horizontal frequency). Since we have only 15,750 horizontal pulses per second, we can insert *two* audio samples in this interval. The audio is digitized into 256 levels, which can be completely specified by 8 bits of digital data. Therefore, the first 16 bits represent the audio samples. *Note:* In order to avoid aliasing distortion, the sampling rate should be as high as system constraints permit.

The audio can be digitized by an analog-to-digital converter in the scrambler. The audio can be recovered with a digital-to-analog converter, as will be discussed. During the horizontal synch interval, the digital data appear in serial form. This data can be taken off with a high-pass filter-amplifier. Since the horizontal pulse is about 10 microseconds wide and contains 4-MHz data, an amplifier that will pass a 10-microsecond pulse without undue distortion, and which can handle 4 MHz, will do nicely. An amplifier having a low-frequency cutoff of 10 kHz will reject the 15-Hz modulation on the data stream DC level (which we don't need), and strip off the data pulses. The data amplifier would be gated on only during horizontal retrace intervals (about 10 microseconds). The amplitude enhancement of one of the bytes will be gotten rid of, since the data must be amplified and made logic compatible. Simple limiting will do this. The data amplifier is followed by a TTL or CMOS level translator; again, simply a limiter with suitable DC levels so as to be logic compatible (Fig. 6-5).

The data has now been amplified and made logic compatible. We need parallel data for D/A conversion. Next, this data, being in serial form, is loaded into a 24-bit shift register. At the outputs of the shift register, the data appears in parallel form. The first sixteen outputs are the audio bits, D_0 to D_{15}, and the last eight bits (D_{16}–D_{23}) are the synch and clock data bits. Actually, the bits can be used in any way that the system designer chooses.

We now have two audio samples, each represented by eight data bits. (Ignore byte 3 for now.) The data are now converted to analog form. First, a 31.5-kHz sampling clock selects byte 1. These data are fed to a D/A (digital-to-analog) converter. Next, about 31 microseconds later, byte 2 is fed to the D/A converter. This completes the audio sampling. The shift register will next be loaded with data from the following horizontal blanking interval, and then the conversion cycle is repeated.

Audio Recovery Implementation

Fig. 6-6 is a block diagram of how audio recovery can be accomplished. Fig. 6-7 is an elementary diagram of a method for converting digital signals to analog signals. This method works in practice and could easily be used at audio frequency. Some suitable values might be R = 10 kilohms and N = 0.5 to 2. "N" might be a gain control. A suitable op amp would be an RC4558, an LM1458, or a μA741, etc. Sample gat-

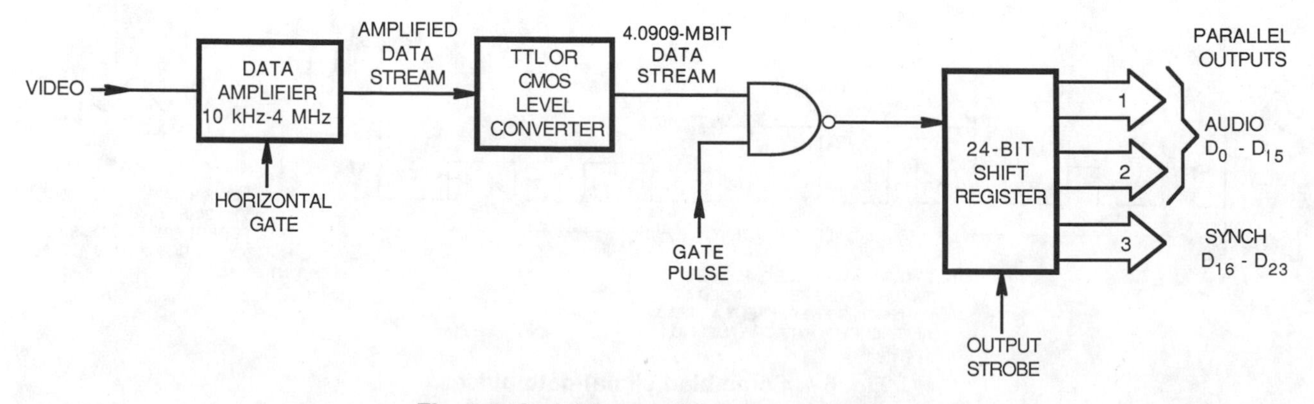

Fig. 6-5. Serial-to-parallel data conversion.

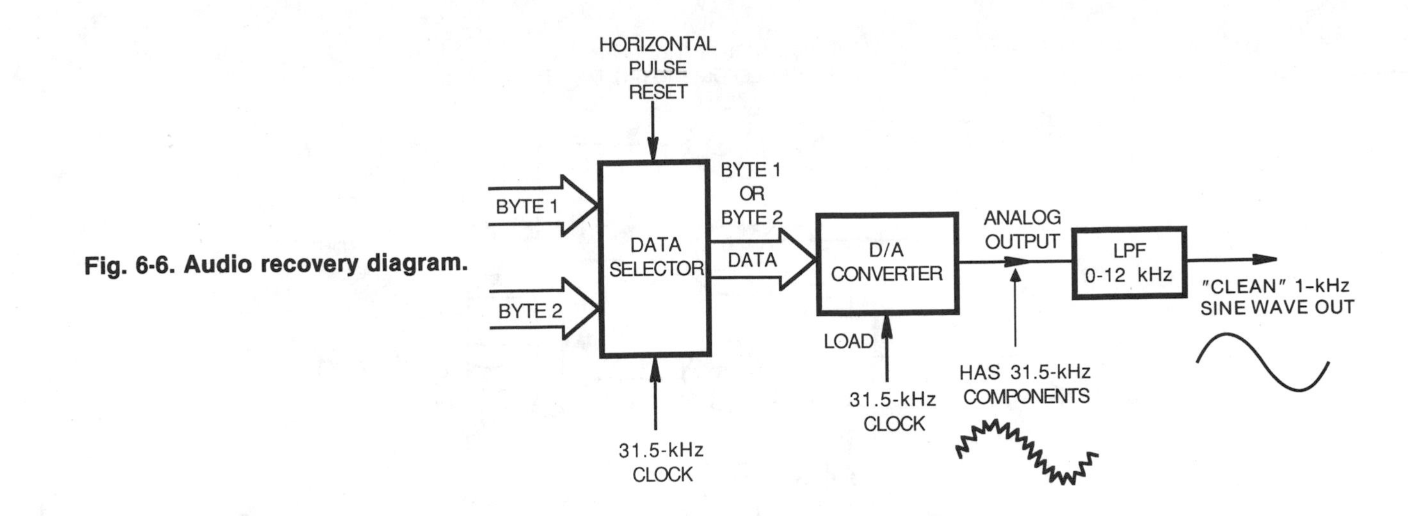

Fig. 6-6. Audio recovery diagram.

ing, while shown as applied to the op amp, would be better done by simply gating the lines D_0-D_7, or by using a suitable latch, such as a 74LS373 8-bit latch, or another suitable logic device. Op-amp gain for each bit is twice that of next lower bit.

$$\text{GAIN} = \frac{E_0}{E_{IN}}$$

$$= \left[\frac{N \times R}{R}(D_7) + \frac{N \times R}{2R}(D_6) + \ldots\ldots \frac{N \times R}{128R}(D_0) \right.$$

$$= N\left(D_7 + \frac{D_6}{2} + \frac{D_5}{4} + \frac{D_4}{8} + \frac{D_3}{16} + \frac{D_2}{32} + \frac{D_1}{64} + \frac{D_0}{128}\right)$$

Thus, a high at D_7 produces 128 times the output that a high at D_0 produces. Therefore, a binary number

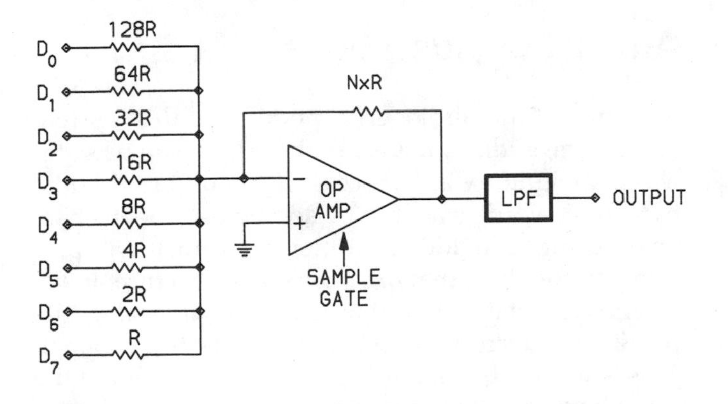

Fig. 6-7. Diagram of an elementary D/A converter.

will produce an analog output depending on the level of each bit.

After conversion to analog, which is done at 31.5 kHz, ripple at that frequency will appear on the output signal. This ripple should be eliminated by using a low-pass filter or amplifier, since the 31.5-kHz ripple may interfere with subsequent audio amplifiers, etc.

The Third Byte

The third byte is used for several purposes. The data stream has a 4.0909-MHz frequency component that can be used to phase-lock a local clock oscillator to 4.0909 MHz. This is twice the required clock frequency for the popular National Semiconductor MN5321 synch generator IC. This chip can be used to generate synch that is phase-locked to the scrambled TV signal. A simple divide-by-2 circuit will produce the required 2.045-MHz clock signal.

Since the third byte is available in parallel form from the 24-bit shift register, this byte can be compared bit by bit, with a hard-wired reference bit pattern. By the way, the third byte could be used for audio, using 12 bits per audio sample. This bit pattern can be used as a "flag" to indicate vertical or horizontal synch intervals or it can also be used for polarity sensing. An Exclusive-OR circuit can be used to make this comparison as shown in Fig. 6-8. An Exclusive-OR gate will only give a true output if both inputs are logically equal. This is exactly what we want. An 8-bit TTL comparator is available on one chip (74ALS518 or 519) to do this. Individual SSI (small-scale integration) chips can be used if desired, but LSI (large-scale integration) helps in keeping things simple and cheap. The 8-bit word is compared

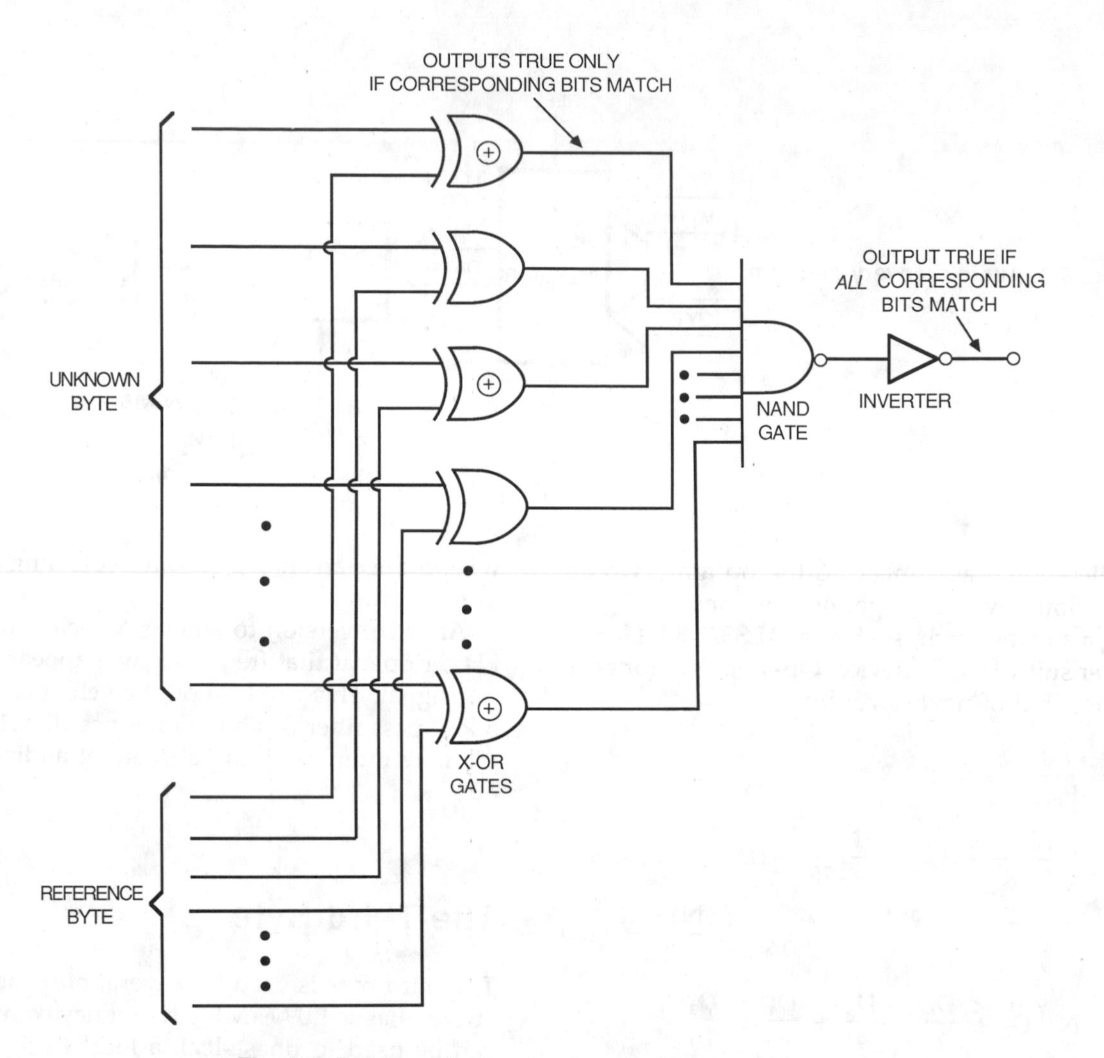

Fig. 6-8. Byte comparison using Exclusive-OR gates.

to a code that is hard-wired in the descrambler. If there is a match, a logic level is generated—high or low, depending on the chip (Fig. 6-8). This logic level can be used to control the synch regenerator, for framing of the picture or for other purposes, such as decoder control (on-off, program billing, etc.). Fig. 6-2 shows this pattern-recognition scheme controlling the synch generator.

Referring back to Fig. 6-2, we now have recovered synch (actually, it's locally generated) and we have also recovered the audio. The recovered audio frequency modulates a 4.5-MHz VCO, which may be phase-locked to the synch generator (not absolutely necessary, but desirable) to ensure accurate sound-carrier frequency. The resultant FM audio subcarrier (sound) goes to the summing circuit, along with the synch pulses. The output of this summing circuit may be amplified and fed to a monitor, or to a TV set tuned to Channel 3 or Channel 4, if a modulator circuit is used, as was previously used with other decoders.

Audio Considerations

A word about audio encryption. As this system stands, the audio is merely digitized. As can be seen, it is a straightforward process. If the digitized audio byte (8-bit word) is altered by adding extraneous bits, or by using it to address a memory which will produce another byte that has a different structure, it will be encoded. If it is then converted back to analog, the resultant waveform will not resemble the original. This is what is being done to the audio content of the scrambled satellite programming. As far as we know, Oak Orion™ does not always encrypt the audio, but VideoCipher II™ does. There are many possible ways to do this. As mentioned before, the NBS Data Encryption Standards (DES) specifies one way to do this. If several audio bytes are encrypted at once, there are many more combinations possible than the 256 combinations that are possible with one byte at a time. The DES algorithm uses 64 bits, with 8 bits reserved for parity checking and other purposes. This leaves 2^{56} possible combinations (about 72 quadrillion). To decipher this many bits would require several decoders, each with its particular 8-bit code, or a large ROM (read-only memory). The audio would have to be decoded with eight different algorithms in sequence, since only every eighth audio byte would have the same coding. Basically, what is done in this example is a changing of the encoding scheme on each successive byte, and more or less than eight algorithms could be used. This method yields very high security. This is only an illustrative example and not necessarily how the encoding is done.

Future high-speed, densely populated, VLSI chips will doubtless make it possible to scramble the video in the manner previously described, at reasonable cost. Then the video security will be very high as well.

"Bootleg" Satellite Descrambler Units

A word about descramblers for satellite use is in order here. We have seen several decoders advertised for sale for the VideoCipher II™ and Oak Orion™ systems, and seen some schematics. While they will decode the video, they will *not* decode the audio. In addition, they are rather unsophisticated and need ''tweeking,'' since they do not make use of the digital synch and framing information. While these decoders can be made to work, they provide no sound decoding and are only for the technically adept ''video pirate,'' since they are difficult to set up. We will discuss some of the circuitry used in them. For other than experimentation, they leave very, very much to be desired. As far as we know *at the time of this writing* (June 1986), no one has successfully pirated VideoCipher II™ audio. However, this could change anytime!

7

Political, Legal, and Consumer Aspects of Scrambling

As soon as home-satellite television reception became more than an experimental curiosity, and more than something that only a few experimenters had been able to accomplish, satellite television attracted lots of attention. In 1980, a satellite dish was a rare sight. Today, they are commonplace, and there are very few individuals who have not seen or heard about them. Previous to 1984, the legal status of the dishes was questionable. Dish owners were considered ''video pirates'' and thieves, stealing programming from pay-TV and cable companies.

In 1984, Congress said it was all right for individuals to receive this programming on private satellite systems in the home, for their *own* personal use. Sales of satellite dish antennas and receivers boomed, and prices of previously expensive components, such as LNAs and dishes, plummeted. An LNA (low-noise amplifier) used to cost over $800. Today, you can buy the same quality LNA for $75.00. The home-satellite industry increased in size to over $1 billion annually at the end of 1985. At that time, there were about 1.5 million dishes in use throughout the United States and Canada.

A lot of motels and bars started adding satellite TV as an attraction, and satellite dishes sprouted from hotels and apartment buildings. By now, quite a few people were watching pay-TV for free and some were even reselling pay-TV programs. Small cable systems could be installed in large apartment and housing developments, permitting use of these free satellite signals without any compensation to the pay-TV services. In addition, some cable-TV systems were either not paying their bills to the program suppliers on time, or not at all.

Then, on January 15, 1986, HBO and its sister company, Cinemax, went to full-time scrambling of their uplink systems. The technology used was the M/A-COM LINKABIT VideoCipher II™. HBO had been testing this system for a year and had at first threatened and then announced that it was going to scramble full-time. This development started a furor of newspaper articles, commentaries, and letters from citizens who were suddenly deprived of their ''free ride.''

Bad News for TVRO

A number of other premium satellite programmers have announced intentions to scramble since then. Some have already done so.

HBO, Cinemax	1/86
WOR, TV	3/86
The Movie Channel	5/86
Showtime	5/86
MTV, VH-11	7/86
Disney Channel	7/86
CNN, Nickelodeon	7/86
CBN	9/86
ESPN	10/86
Arts & Entertainment	12/86
BET, VSH, WTBS	12/86

Note: As far as is known, all are planning to use the M/A-COM LINKABIT VCII technology.

This is quite a list, and represents the ''cream of the crop'' of premium services. Dish owners are now wondering if the skies will go dark, and one unnamed industry figure made a ''wise'' remark to the effect that all those dish owners can still use the dishes as birdbaths. However, this industry figure is obviously a birdbrain.

Arthur C. Clark,* in an essay appearing in the October, 1945, issue of *Wireless World*, had envisioned the day when all people would have access to satellite-distributed radio and TV programming. At that time, scrambling was not thought of. CATV did not exist. This prophetic book did not mention the selfish interests of the entertainment industry; it mentioned just the scientific facts, known then, concerning satellite broadcasting. A number of arguments on scrambling have been presented in the *New York Times, Wall Street Journal, Electronic Engineering Times,* and a number of the industry trade magazines. These will be presented here, without taking sides, as an illustration of the complexities of this issue.

Legality

First, there is the issue of legality. Who owns the air waves? The Communications Act of 1934 gives equal opportunity to all to lawfully use the electromagnetic spectrum. The various broadcasting services are required to show that their existence and right-to-license is in the public interest, and that it shall serve the public interest. Are scrambled signals serving the public interest if, for example, certain people are denied access to programming because satellite programmers say, "We are a business and have the right to choose our customers." Just because someone lives beyond the reach of cable companies and cable connections, they can and are being denied equal access to certain programming.

As an example, consider a dweller in a remote area of, say, the Rocky Mountains. He chooses to live there. One of his problems is the lack of interesting TV program signals. He may not be able to receive VHF or UHF programming, and cannot get CATV service. But, no one is forcing him to live there. Logically, the lack of TV is *his* problem, and he is not being denied access simply because no one will build a TV station nearby. However, suppose he purchases and installs a satellite-TV system? He now has access to the same programming that his city cousins have via CATV. But, say that the programming is scrambled. The programmers, using a limited resource (satellite transponder, positioned in the Clarke belt, and the airwaves), only wish to serve the cable-TV networks and may refuse to serve this satellite-dish owner. Since this would-be customer is certainly part of the public, along with all the other rural dwellers and dish owners, is his interest being served? As far as he is concerned, one less transponder on one of the satellites in the Clarke belt is available to him. The Clarke belt has limited parking spaces for satellites. With 2° spacing coming along, only so many satellites (180, theoretically) will fit. At any point on the earth, a bit more than one third of the Clarke belt is usable. This limits any access to 60 satellites. Quite a few, but still a limited resource. The point here should be obvious, especially since any and every nation on earth has a limited space available in the Clarke belt.

Decoder Costs and Availability

The previous argument has been used by the anti-scrambling faction. The programmers that are *for* scrambling counter this argument by saying that the required decoders will be made available, at a reasonable price and at a reasonable monthly fee to those satellite system owners who wish to receive their programming. For example, HBO advertises that their VideoCipher II™ decoder can be obtained for $395 and, after installation and setting up an account, will be activated to receive either HBO or Cinemax for $12.95 per month each, or $19.95 per month for both. (We understand that a toll-free telephone number has been set up so that individual dish owners can obtain decoder authorization and start an account.) However, a $395 decoder fee, and $13.00 to $20.00 per month, is not what every dish owner considers reasonable. There were reports that only 6000 decoders for HBO were available on January 15, 1986—the day HBO instituted scrambling. Similar sources claimed that 25,000 had been made available and that 5000 per week would be produced. Since the wholesale cost of the VideoCipher II™ system was around $360, dealers were complaining about the poor markup. Whatever the truth was or is, the availability of decoders has evidently become an issue and there seems to be a consensus of opinion that they are not easy to obtain at the time of this writing (June 1986).

Financial Aspects

If viewers are going to pay $12.95 a month per channel, or $19.95 for two channels, and all the services charge the same, the fact that "100 channels are available" is of little consequence. For ten channels, a subscriber would have to pay around $100 a month. While some affluent families may not mind, this ex-

* See: *Voices from the Sky,* Harper & Row Edition, October 1965.

pense is beyond the resources or desire-to-pay of many potential viewers. Clearly, to reap the benefits of the wide choice of programming that satellite TV has to offer, this figure will probably have to come down to something like $40.00 per month for ten to fifteen channels. Of course, it is likely that some special pay-per-view programs will be available at additional charge.

With regard to decoders, who is going to sell them, and who gets the profits? The need for several *different* decoders surely does not look like a practical solution. A practical approach might be the availability of most of the aforementioned premium services for $20.00–50.00 a month, after buying *one* decoder. Most satellite-TV owners would probably consider this a fair and reasonable compromise. Not many people seem to object to the idea that the programmers are entitled to a fee for their programs. Most of the correspondence sent to the newspapers and magazines seems to bear out this fact. The question in doubt seems to be just what is this fee, what is reasonable, and what does one get for their monthly payment. Two or three extra channels where local services are available would not seem to justify the cost of a dish. A rural dweller may have the opposite view since, otherwise, he or she has nothing at all. It will be an interesting several months ahead as all these questions are resolved. Possibly, a new industry of program brokers and competing services will spring up, bringing costs way down via intense competition. In all probability, if it turns out that the scrambling issue can be resolved by the purchase of a $395 decoder as a one-time investment, with a fee of $20.00–50.00 a month for almost unlimited access to programming, most people will be happy. This may very well come to pass due to viewer pressure, mass production of decoders, and competition among the program services for viewer audience.

SPACE

About five years ago (1981), an attorney named Richard Brown started an organization called the Society for Private and Commercial Earth Stations, otherwise known as SPACE. The organization was started at a time when satellite television was in its infancy and it had the purpose of aiding both the satellite television industry *and* the dish owner. SPACE has been a successful organization and a friend of the satellite-TV industry and the dish owner alike. Its first big victory was in 1984 when it successfully lobbied in Washington to exempt individual homeowners from existing laws that made the interception and probable use of commercial satellite signals illegal. This "broke the dam" and the sales of dishes boomed. As a result, SPACE membership increased along with its clout.

SPACE has taken on the CATV industry on various issues, such as scrambling. It has been able to win over 60 congressmen, many of who are rural dwellers, toward signing a bill calling for a moratorium on scrambling (HR 1769). SPACE is also behind the bill (HR 1840) which will assure reasonable rates for dish owners. One congressman fighting for fair practices is Timothy Wirth (D–CO). Senator Barry Goldwater has generally been on the side of the satellite-TV industry.

Other Complaints

As this scrambling issue progresses, it is said that an easy way to get rid of your competitor is by making it illegal for him to do business in your area. Or, by keeping the signals in a form such that only the CATV companies can use them, competition can be reduced. Thus, the scrambling issue continues to boil.

The CATV industry has also been accused of forcing the program suppliers to scramble their programs, or having them thrown off the systems. (No scrambling—no need for your service.) In this way, it has been said that the CATV industry is trying to "take over" the program suppliers and control the dish owners via unfair restraint of trade and monopolistic practices. It has also been stated that the CATV industry has been using scare tactics on consumers, such as warning them that their satellite dishes would soon be useless and the signals would be received as a meaningless blur.

There were also letters that charged the satellite-dish industry with confusing consumers about scrambling, withholding the true facts, distorting facts, and providing misinformation. This includes stories that "black boxes" will soon be available to descramble the signals. Also, that reception of scrambled signals might not be very good, or that several companies will never scramble, etc. All these rumors are getting more and more publicity. After all, they concern an issue that is very close to all of us—our TV programming and the variety of entertainment available.

Had satellite-TV scrambling been the rule five years ago, in all probability, this escalating fight between the CATV industry and the satellite-dish industry would never have occurred. Scrambling would have been taken for granted and, possibly, the satellite-TV industry, as we know it today, may have never been.

Possibly more effort would have been directed toward the KU Band DBS services. Who will ever know, now?

Other issues have been cited in the various aforementioned publications. One such issue has been the accusation that the CATV industry has been using monopolistic practices. The CATV industry, after all, has never been in great favor of home satellite dishes. If homeowners can put up a satellite dish and receive 100 channels, who needs CATV? Why pay for something that can be received for free, and with more variety as well? The CATV industry and the local CATV companies have been accused of promoting, for example, zoning ordinances in their service areas, which prohibit or severely restrict satellite dishes in those areas. They said that the satellite-dish industry wanted to sell expensive systems with no regard for the customer. If you cannot compete, you cannot win. In opposition, the STV industry claimed that the cable companies were the villains. They said that the CATV industry was clearly running scared, seeing the future obsolescence of cable TV, as small dishes and DBS are environmentally "cleaner" and require no cables, pulses, and other hardware, and they can serve a much larger area.

The Current Situation

As things stand now (Spring 1986), the predominant scrambling system seems to be the M/A-COM LINKABIT VideoCipher II™. However, it is expensive for the consumer, and, presently, good only for HBO and Cinemax (March 1986). If several other services do as rumored and use this technology, it will tend to set an industry standard. Possibly, it will be eventually phased out to give way to some future system, like the once popular 8-track tape was replaced by the cassette tape.

If VCII does become popular, it may still eventually be defeated. The video scrambling method is easily decoded, and several "companies" claim to have decoders that will decode the M/A-COM system, or the similar Oak Orion™ system used on the Canadian ANIK satellites. However, the audio is also scrambled and it is much more difficult to decode. But, it only stands to reason that if VCII becomes a commonplace system, someone may "crack" it. This could be a bright 16-year-old tinkerer, a Ph.D. student, or, possibly at some future date, a disgruntled employee of M/A-COM who feels he has to "get back" at the company. Who knows? There is food for thought. Whoever figures "it" out might become quite wealthy selling "bootleg" decoders.

The NBS DES algorithm has 2^{56} (about 72 quadrillion) keys. But, considering that VCII decoders sell for $360 wholesale, and, therefore, have to be made for about $90–$120 in parts, have to be addressable, and have to decode the video signal as well, they can't be *too* sophisticated at that price. Semiconductor chips can be "reverse engineered" and, possibly, the security of the VCII system may not be as good as thought or claimed. This is, however, merely speculation, but we know of one organization attempting to crack it. Time will tell.

If the system price is reasonable and the monthly fees are low enough, it may not be worthwhile to be a "bootlegger." We hope that this will be the case, and that scrambling will be fair for both the satellite-dish user and the program supplier, and that scrambling will result in a better quality of programming, both technically and with regard to program content.

Interference

Another and more ominous issue on the scrambling scene is that of deliberate interference. A recent article in the *New York Times* mentioned this problem. Also, one piece of correspondence written by an unknown "someone" mentioned that "if HBO can scramble, maybe we can help them out." The meaning of this is pretty obvious. A dish owner who is unhappy about the HBO situation can get an old radar transmitter or even a high-power klystron amplifier, and set up a station for running several kilowatts in the uplink band (5.9–6.4 GHz), and get into a satellite transponder with an interfering signal. A 12-foot dish has approximately 42-db gain at 5.9 GHz. If a *pulsed* 10 KW (or even 50 KW) signal were used, we would have a peak ERP of about 100 megawatts. A 100-megawatt pulse could do a lot of "damage" to an uplink signal.

Many satellite uplink transmitters are in the 10-watt to 1000-watt class. Microwave tubes in this power class are not cheap. However, someone who knows where to get a surplus one, a new or used one (and they are around), or even steal one from his company, etc., could conceivably wreak havoc with any satellite program. If the interference was kept at a very low duty cycle (for example, one or two seconds every few minutes, and at random times), it would be very difficult to locate the culprit. Unlike HF or VHF, microwaves can be easily focused into a narrow beam only a few degrees wide. With a typical narrow-beam antenna, a search aircraft at 30,000 feet would have to locate an area about ½ mile in diameter. At an airspeed, say, of 250 mph, this is only about six or eight seconds flying time, and the interfering signal would have to be present at that *precise* time. Thus, such a small area would be very difficult to locate. If the satellite programmer increased power, the jammer could just as easily do the same. Paying subscribers would not tolerate this for very long. The thought of this type sabotage is rather ominous. Possibly, the satellite could even be shut off by pirate command codes. We might have satellite counterparts to the notorious "computer hackers" who managed to obtain data from supposedly safe data banks. It is an interesting point to think about. Obviously, some measures will have to be taken to prevent this type of sabotage.

8

Digitizing Audio and Video Signals

The encryption of audio signals, using an analog method, has been discussed earlier. The audio was taken in its unscrambled form, modulated using a balanced modulator, and placed onto a subcarrier in the 15- to 100-kHz (or thereabout) range. By demodulating the subcarrier, which by itself is inaudible, recovery of the audio program material can be accomplished. Circuitry for doing so, using commercially available stereo-FM demodulator chips, was discussed.

Introduction

Analog systems are effective but they are relatively easy to defeat. They are suitable for low to moderate security applications. However, a more sophisticated digital method has been developed and is used on satellite audio systems, such as Oak Orion™ and VideoCipher II™. This method requires digitizing the audio signal. Then, the audio, in digital form, is placed in the horizontal blanking interval, which has had the horizontal and vertical blanking as well as synch information removed.

The video signal would appear as shown in Fig. 8-1B. The aural (sound) subcarrier either has no modulation, or some slight modulation on it that will interfere with the use of the chroma burst as a reference. It is no longer actually required and can be eliminated if desired. In order to produce this "encoded" signal, the audio must be digitized. Typically, this is done by the process diagrammed in Fig. 8-2, which is representative.

Basic Theory

It is a well-known sampling theorem that a signal of length "T" and frequency bandwidth "fm" can be completely specified by "2fmT" samples of the signal. Alternatively, it may be said that if T = 1 second, then 2 × fm (or twice the bandwidth) samples per second are required to specify the signal. This means, that if we have a typical 12-kHz TV audio bandwidth, we must sample it at a 24-kHz rate (or higher) to completely specify the signal. A convenient audio-sampling rate is 31.5 kHz, or twice the horizontal scan frequency. It is advantageous to sample at as high a rate as possible for the purpose of reducing aliasing distortion and easing filtering requirements.

Typically, the audio is digitized by generating discrete binary numbers to represent the analog level. If we have a binary word N bits long, we can specify 2^N discrete levels. It would intuitively be necessary to specify a large number of levels to reproduce minute changes in analog level. One hundred levels would take care of 1% uncertainty (−40 dB), and if 256 levels were used, uncertainty would now be less than about −46 dB, which is adequate for TV audio. A single data byte can do this, since 8 bits would give 256 possible data bits.

We only have horizontal blanking pulses, however, at a 15.75-kHz rate. How can a 31.5-kHz sample rate be produced? Simple! Each interval contains two bytes rather than 1 byte. This way, 31,500 bytes per second are available, which is adequate for the 12-kHz audio baseband required. In addition, a third byte is inserted in the blanking interval. This byte is a coded digital word used to determine where the horizontal and vertical references (starting points) are located. This code can be used to ensure proper vertical and horizontal timing and, additionally, can be used for other purposes.

Audio Digitization

In digitizing an audio signal, the scheme of Fig. 8-2 can be used. The signal is amplified, level is set for

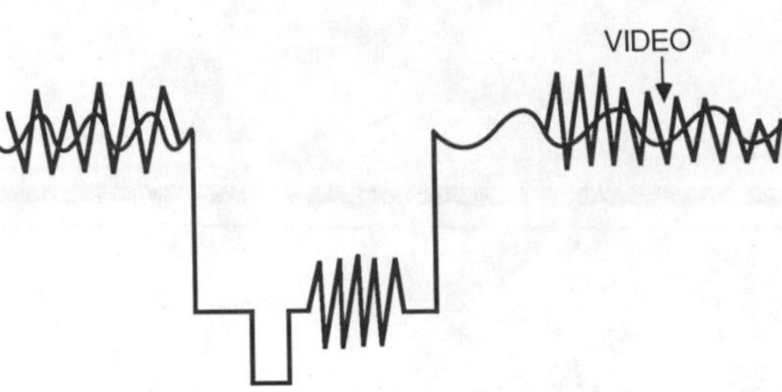

(A) Uncoded (NTSC) video signal.

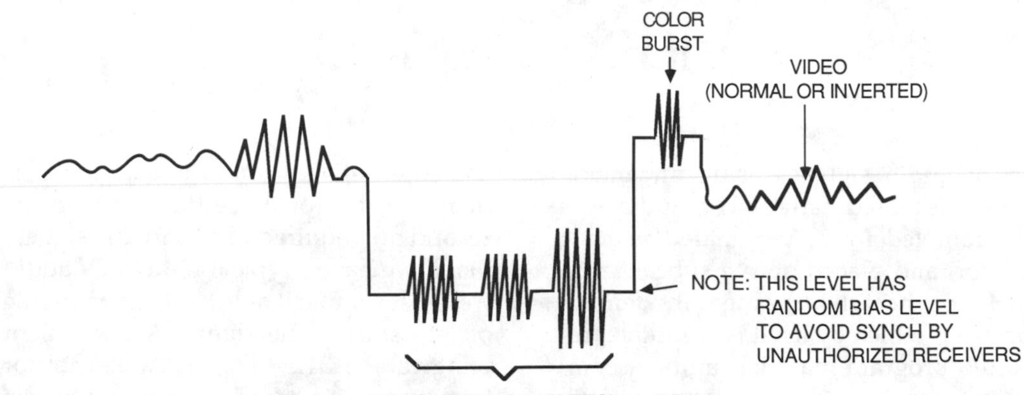

(B) Encoded video signal.

Fig. 8-1. Video signal.

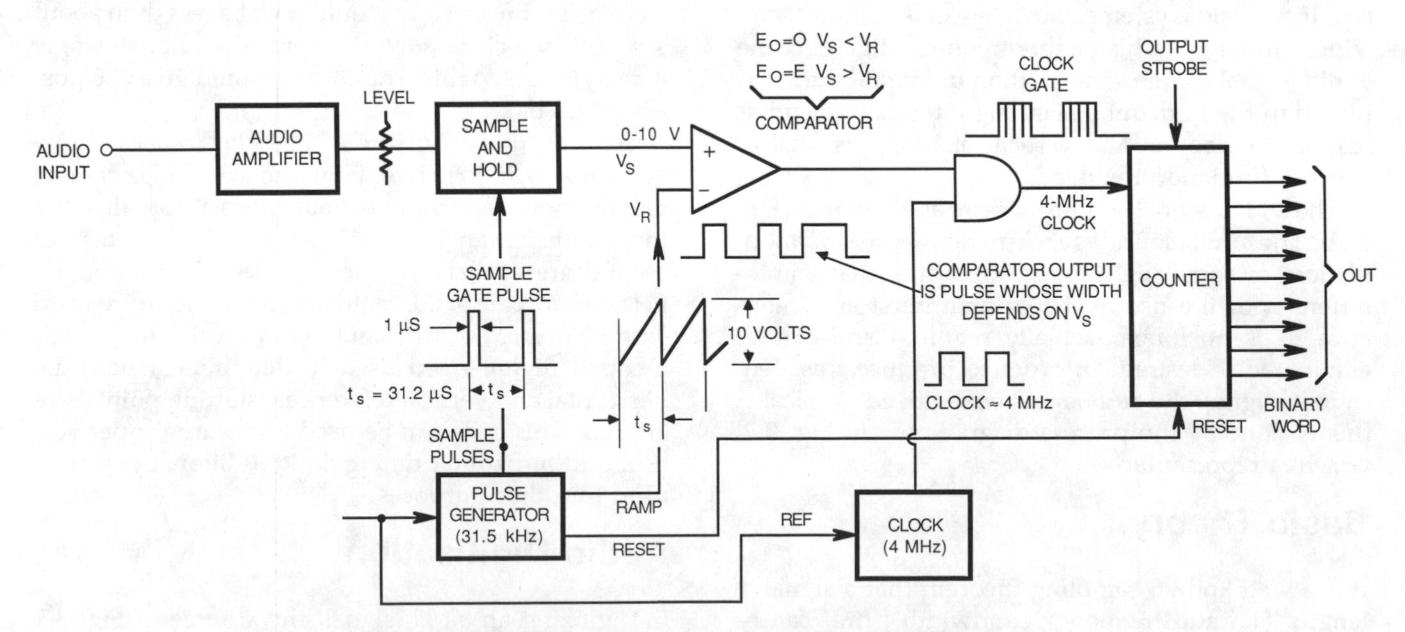

Fig. 8-2. Diagram of circuit used to digitize an audio signal.

proper dynamic range, and sampled every 31.2 microseconds (a 31.5-kHz rate). The analog value is stored in a sample-and-hold circuit until the next sample is taken. Let us assume that it may be any value between 0 (zero) and 10 volts. This analog level is now fed to one input of a comparator. The other comparator input is fed with a linear ramp (sawtooth waveform) whose amplitude varies, starting at zero (slightly after the sampling interval), up to 10 volts (just before the next audio sample is taken). If the audio sample is large (say, 7 to 10 volts) in amplitude, the ramp will have to rise to this value before the comparator output voltage will drop to zero. If the analog sample is small (about 1 volt), the comparator will drop to zero when the ramp exceeds 1 volt (the comparator output is a logical 1 when $V_S > V_R$ and a logical 0 when $V_S < V_R$). Therefore, the comparator output is a train of pulses with a frequency of 31.5 kHz, and a pulse width from nearly 0 to 30 microseconds, depending on the sample amplitude. (Some time must be reserved for sampling and resetting the ramp to zero.)

This variable-length pulse represents the analog value of the audio-sample amplitude. A narrow (< 5 microseconds) pulse represents low values (0 to 2 volts). A wide pulse of 25 microseconds would represent 8 or 9 volts. (Ideally, we should get about a 3-microsecond pulse-width per volt in this instance.) Now, the pulse has to be converted to some binary value. This can be done by using the pulse as a gating pulse for a counter that is clocked by a much higher clock frequency. If we had a 4-MHz clock, 120 clock pulses would be counted in 30 microseconds. By using two separate (alternating) systems and the full line-scan time (63.5 microseconds), it is possible to count up to 240 clock pulses, thus generating a full eight-byte binary word. This is possible because each byte is only needed every 63.5 microseconds, and there are two bytes.

Thus, the counter can be reset to zero, the high-frequency clock signal gated by the variable-length pulse, and the width of the variable-length pulse will determine how many cycles of the high-frequency clock will be inputted to the counter. The counter will count to a state that is proportional to the length of the variable pulse, whose width depends on the analog value of the audio sample. Therefore, a binary number that is proportional to the analog value of the audio sample appears at the output of the counter. This binary number is the digital equivalent of the sample, in parallel format.

Next, the binary number is stored in parallel format in a shift register (Fig. 8-3). During the horizontal blanking interval, it is clocked out in serial format, and it appears as an 8-bit digital word. The lock frequency is 4.0909 MHz. In a 6-microsecond interval, this permits 24 bits (3 bytes). This is the digital information transmitted during synch pulses. Although the illustration and explanation only show how one byte is formed, two audio bytes are generated. (Remember we need 31,500 samples per second, so we must transmit two bytes in every blanking interval. 15,750 samples per second are used. A third byte, which can be coded signals for the determination of horizontal and vertical synch references and for coding purposes, is also placed in the data stream, but we will not be concerned with this here. It is mentioned merely to indicate what the third byte can be used for.

The audio, in digitized form, is inaudible on the sound carrier, since it is not used to modulate it. In fact, the audio carrier is not needed and could be dispensed with. However, it could be used for ''barker'' audio purposes, or, with proper modulation, as an ''aid'' in scrambling to prevent recognition of the 3.58-MHz chrominance subcarrier.

Demodulation

In order to demodulate the scrambled audio (really not scrambled; digitized is more correctly used), a process which is the reverse of the digitizing process can be used. Fig. 8-4 shows a block diagram of a decoder for this digital audio system. A gate after the video detector in the TV receiver extracts the data pulses, removing the video. The data stream, a low-level (0.1 to 1 volt) square-wave signal having bits as short as 250 nanoseconds (that is 24 bits long and occurring every 63.5 microseconds) is fed to a data amplifier. The signal is amplified and unwanted components removed. The scrambled signal has a 15-Hz low-frequency sine wave phased to continuously shift the bias level of the data pulses, and each data byte is enhanced in amplitude (one of three, each horizontal interval). This is done in order to prevent certain TV receivers from synchronizing by using the data pulses as a reference, instead of the missing horizontal and vertical pulses.

The filter and amplifier remove the 15-Hz component. The pulses are then fed into a level converter to make the data signal TTL-compatible (or, whatever is necessary for the logic circuitry used ahead). This data signal is clocked into a 24-bit shift register (which can be made up of three 8-bit shift registers). One 8-bit shift register contains the first audio sample, the second register, the second sample, and then the remaining third byte (containing the synch informa-

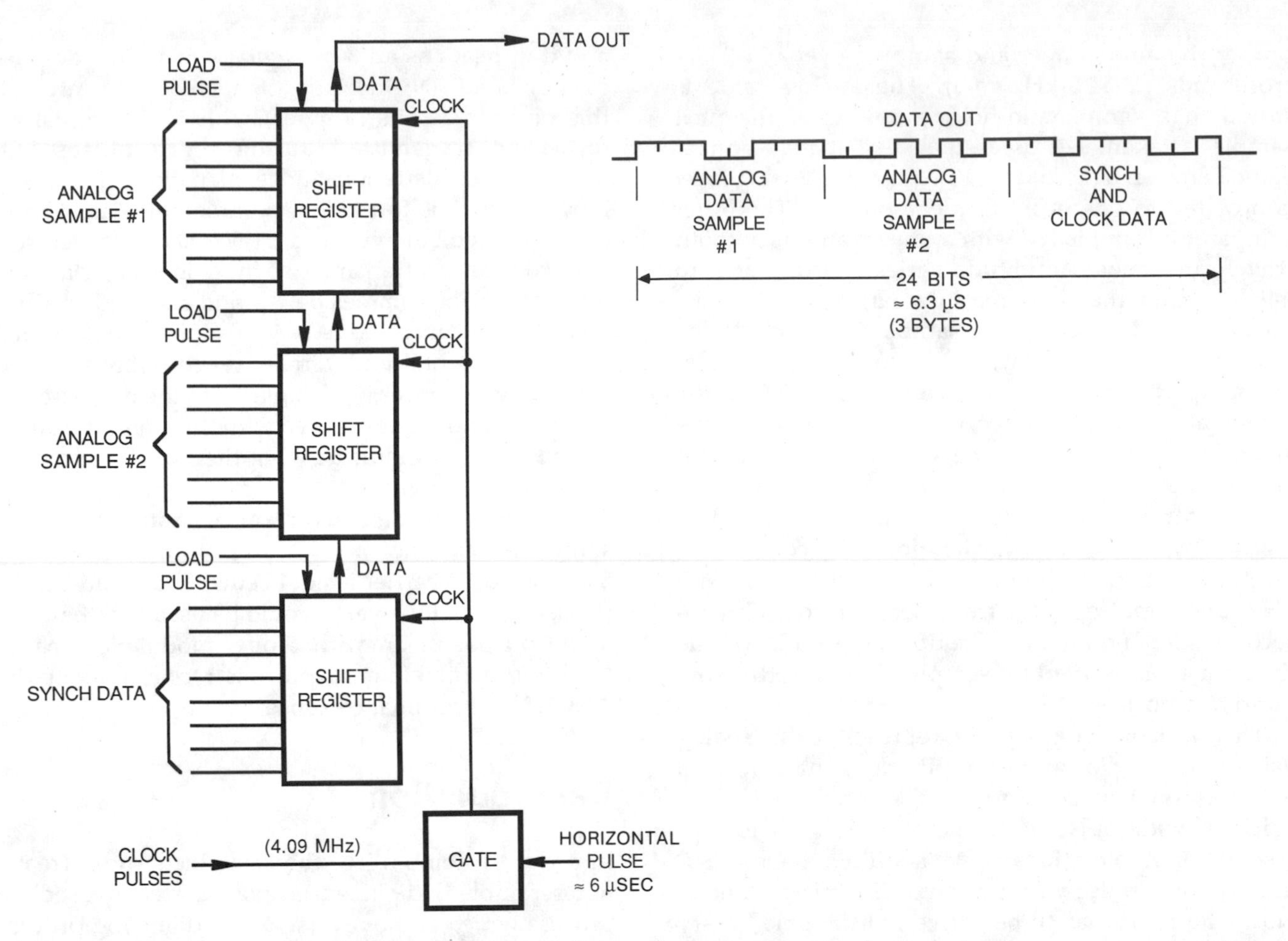

Fig. 8-3. Formation of digital audio data stream and synch data.

tion) is fed to a pattern-recognition circuit. Byte 1 and byte 2 are fed in parallel form to a data selector. This data selector is driven by a 31.5-kHz clock which is derived from the horizontal synch circuitry. Bytes 1 and 2 are alternately selected and fed to a D/A converter. The binary data, in parallel form, are converted back to analog form.

Fig. 8-5 shows an elementary D/A converter. This circuit uses an op amp and a precision resistor network. For any value of R_F/R that is within the limitation of the operational amplifier used, R_F might be 10K and R might be 5K, with a ±15-V supply and TTL logic levels as an example. D_7, the most significant data bit, will have 128 times the effect on the output that D_0, the least significant bit, will have.

Suppose R_F is 10K and R is 5K. Assuming D_7 was high and all other data lines were low, we would get 10 volts out of the D/A converter. D_6 would produce 5 volts, D_5 would produce 2.5 volts, D_4 would produce 1.25 volts, and so on, until D_0, which would produce $5/64$ volt. Each data line produces twice that of its lower neighbor. Any binary number would therefore produce a definite analog voltage, proportional to its value. R_F could be used to determine the constant of proportionality. Consult any basic electronics text for information on op amps if you are not familiar with their operation and circuit analysis.

The output of the D/A converter consists of the original analog signal with a 31.5-kHz component on it, due to the sampling process. The low-pass filter (LPF) removes this, producing a clean analog waveform that closely resembles the original (assuming nothing is malfunctioning in the system). This is illustrated in Fig. 8-6.

Scrambled or Digitized?

Up to now we have been calling the audio "scrambled." It is not really scrambled but merely digitized, and, since we cannot receive it on a conventional TV receiver, we think of it as scrambled. However, it has been digitized in a straightforward manner so that anyone, knowing how the system works, could build

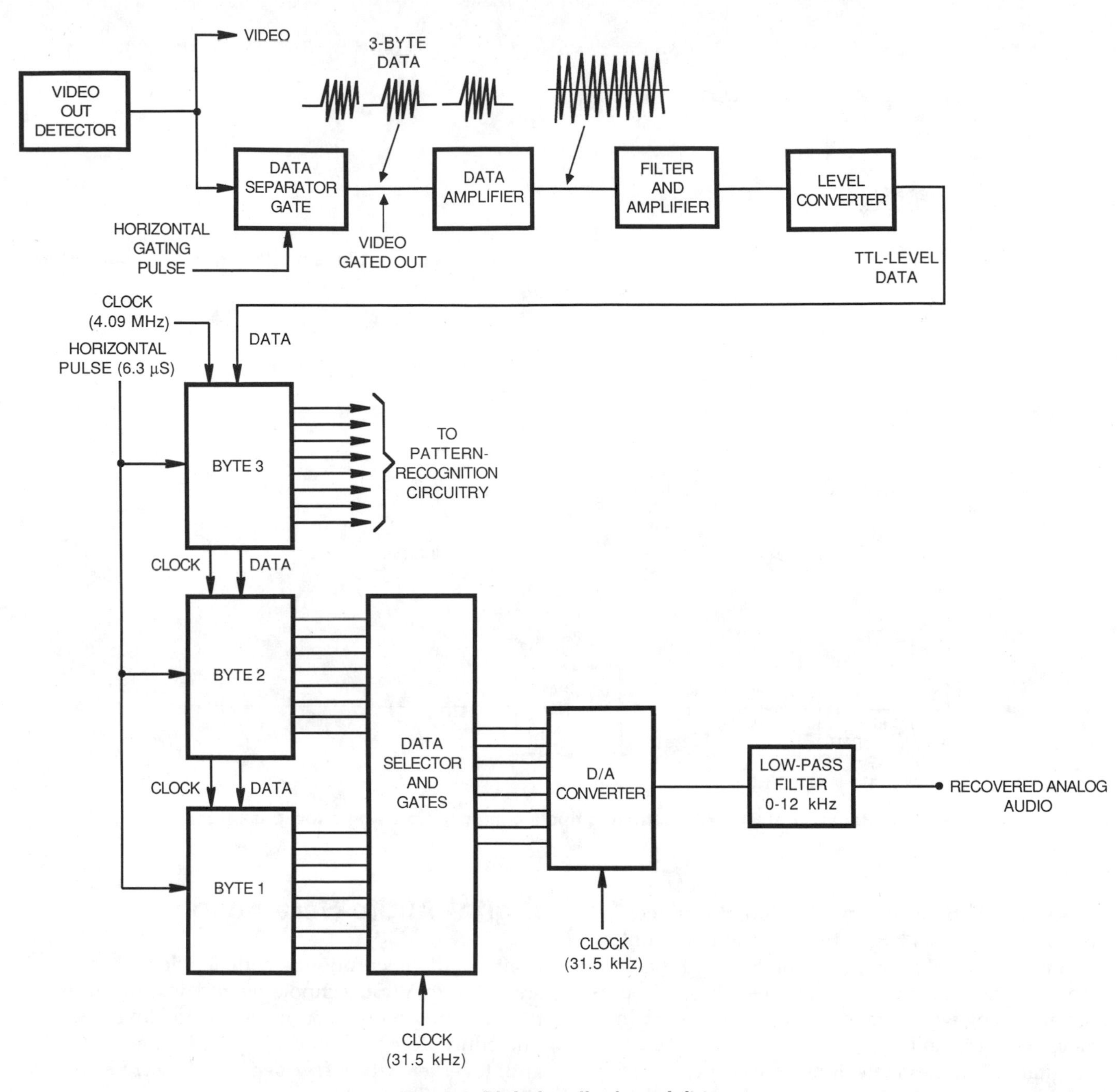

Fig. 8-4. Digital audio demodulator.

a decoder, assuming that he has the constructional ability and had the proper parts available. However, what if, after digitizing the audio, we take the 16 bits in each horizontal blanking interval and "encode" them? This could be done, for instance, by adding a random set of digital numbers to the binary number we have. Or, we could use a read-only memory (or other matrix arrangement) to generate a different binary code corresponding to any binary number. For example, binary 63 is transformed to binary 35, 94 is transformed to binary 181, etc.; for 256 words (16-bit system), there are 256! possible combinations. Or, any other suitable algorithm might be used. This is what the M/A-COM VideoCipher II™ does. The encryption algorithm used in the VCII is the NBS Data Encryption Standard.

In this method, the data is encoded using a 64-bit algorithm (8 of which are used for parity checking). This leaves 2^{56} possible combinations for a de-encryption key. 2^{56} is a rather large number—about 72

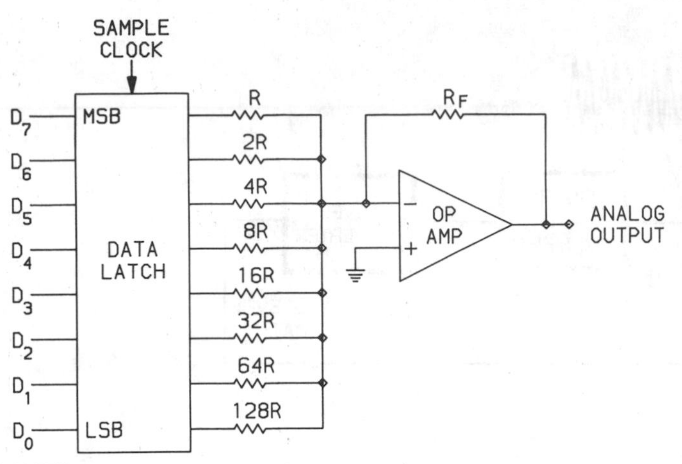

NOTES:

1. D_0–D_7 ARE DATA INPUTS
2. $D_N = D_0–D_7$
3. ANALOG OUTPUT = $D_N\left(\frac{R_F}{R}\right)\left(D_7+\frac{D_6}{2}+\frac{D_5}{4}+\frac{D_4}{8}+\frac{D_3}{16}+\frac{D_2}{32}+\frac{D_1}{64}+\frac{D_0}{128}\right)$
4. D_N IS TYPICALLY 3.5 VOLTS (FOR TTL SYSTEM), OR 5–10 VOLTS (FOR CMOS)

Fig. 8-5. Elementary D/A converter.

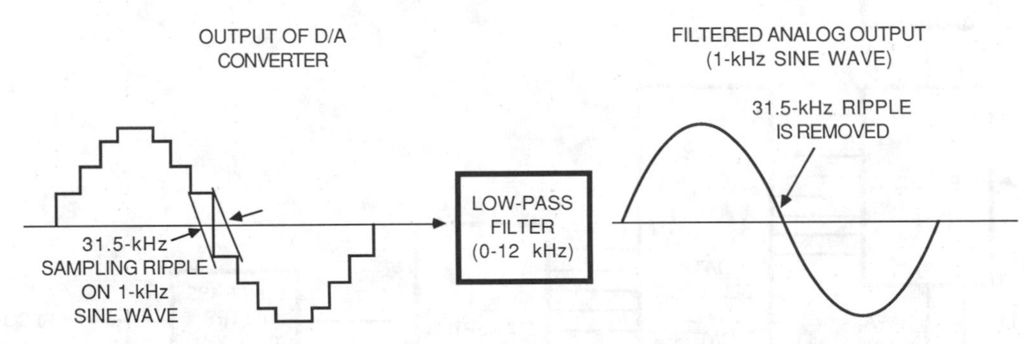

Fig. 8-6. Removal of 31.5-kHz sampling-ripple components using a low-pass filter.

thousand million million (quadrillion); it is stated this way to illustrate this large number. Therefore, unless the correct key is known, it is somewhat difficult to decode the audio. Incorrect data decoding will produce an analog waveform different from the original. This is true scrambling.

Digital techniques make it much easier to scramble the audio. All one has to do to change the method is simply change the algorithm. The algorithm can be implemented using a randomly generated number sequence, and the chance of repeating the number sequence, if the algorithm is changed once a day, is very remote this side of eternity. If a different combination could be tried every second in an attempt to crack the code, it would take about 5.8 *billion* years to try all combinations. This is longer than the earth has been believed to be in existence (4 billion years). This serves to dramatically demonstrate the high security of the encryption system.

Digital Audio Scrambling

Digital audio encryption is both simple and inexpensive, due to VLSI technology, and will probably be the preferred method in future scrambling systems. The entire decoder could be placed on an LSI chip, and, together with a few peripheral discrete components, could probably be mass produced for just a few dollars. At this price, it would be feasible to change the decoding algorithm periodically. With today's LSI electronics technology, throw-away decoder modules, updated every few months, are entirely feasible and may very well be used.

Digital Video Scrambling

For TV applications, if NTSC video is assumed, a bandwidth of 4.2 MHz is generally required in the video-processing circuitry. Most of the commonly used scrambling systems use fixed scrambling algorithms as far as the video is concerned. However, there are several methods requiring memory storage of the video signal—a line, or even an entire frame at one time. In order to use digital techniques, the video must be digitized. Generally, 8 bits or more must be used to describe the signals adequately and to avoid visible deterioration. This would correspond to 256 levels (0 to 255), and each step would be 48 dB below peak video. Since the sample rate must be at least 2 × 4.2 or 8.4 MHz (and more likely 10 MHz would be used), conversion speed time would have to be high. To allow for glitches and pulse settling time, and to reduce aliasing, bandwidths of 20 MHz would probably be used.

A/D Conversion

There are several approaches to A/D conversion. One approach might involve sampling the analog signal at the sampling frequency, using this sample's amplitude in a comparator (fed with a ramp on the other input), and then measuring the time for the sample amplitude to equal the ramp amplitude (Fig. 8-7). The comparator generates a pulse whose width is a function of the sample. This pulse gates a clock into a digital counter, whose maximum count depends on the pulse width. The counter is then strobed, and the binary-value output of the counter is proportional to the analog input.

As an example, if we are expecting a maximum frequency input of 5 MHz and need to digitize the input into 256 bits, our counter requires a clock frequency of 2 × 5 MHz × 256 bits, or 2560 MHz. This is not too practical. For this reason, while this A/D conversion approach is all right for DC or audio application, it is not too practical for the megahertz frequencies found in NTSC video applications.

Successive Approximation Conversion

Another approach is the successive approximation method (Fig. 8-8). In this case, a clock drives a register connected through a D/A converter arranged in a feedback loop. First, the most-significant bit (MSB) is tested. If the comparator decides that the video input-signal amplitude is greater, that bit is registered as a 1 (one) in the register. Next, MSB-1 is tested and if MSB + (MSB-1) is greater than the video, MSB-1 is entered as a 0 (zero). If MSB + (MSB-1) is less than

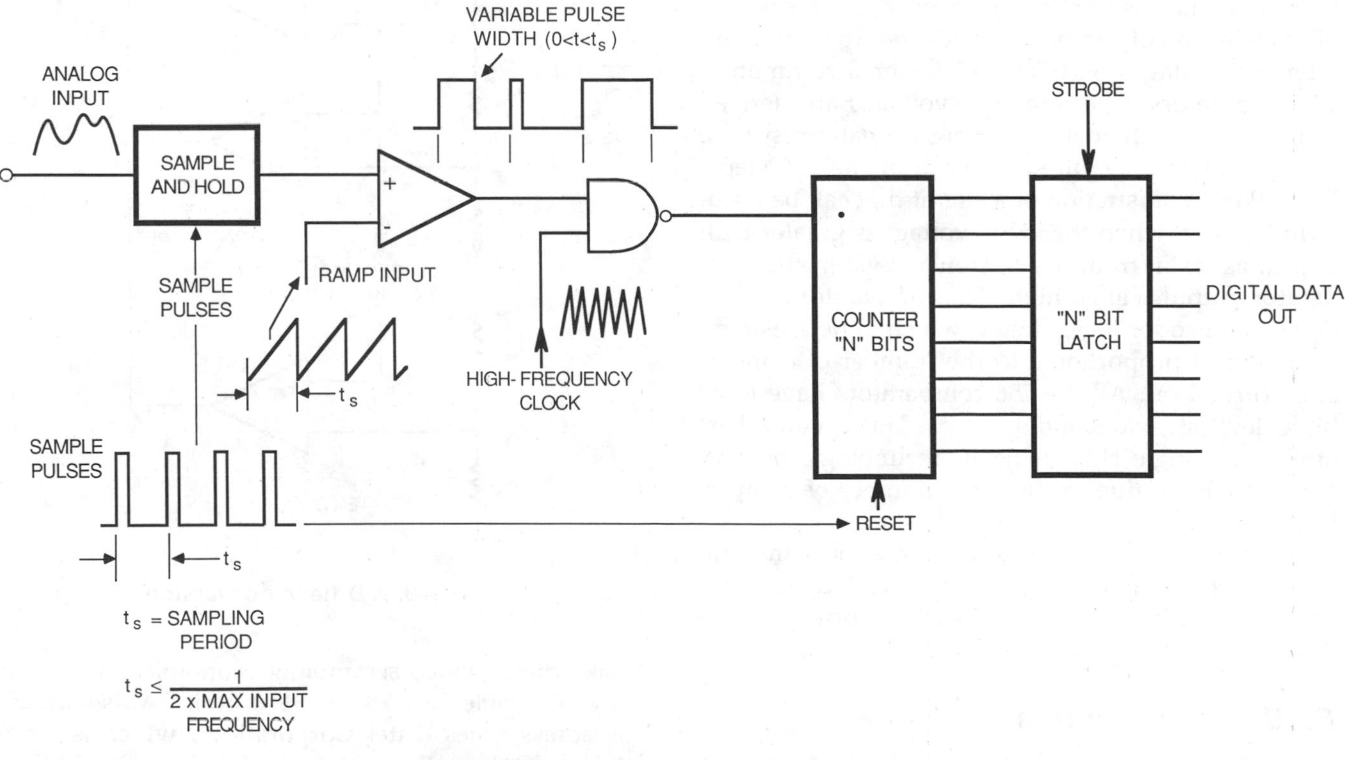

Fig. 8-7. A/D converter using a gated counter.

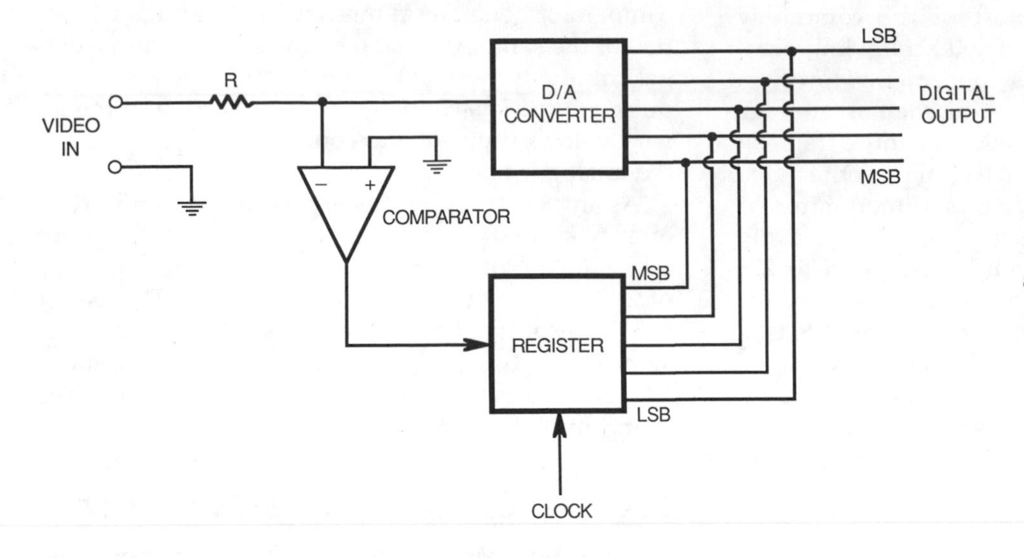

Fig. 8-8. A/D successive approximation converter.

the video, MSB-1 is entered as a logic 1. For an 8-bit register, the whole "successive approximation method" takes eight clock cycles. The clock has to run at least at 80 MHz for a 10-MHz sampling rate. This is a costly approach and requires high-speed logic.

Flash Conversion

The best approach for a cost-effective A/D conversion of video signals is the "flash" converter. This is simply a collection of comparators hooked up to different reference voltages with the video signal common to all comparators. The reference voltages are derived from a voltage divider consisting of many resistors. A comparator is required for each of the 256 steps. Fig. 8-9 is an illustration of a circuit that can be used.

In Fig. 8-9, when the video voltage is greater than V_{N-M}, as taken from the reference divider, the comparator output will go high. The comparator outputs go to an encoder logic circuit, which generates a binary output proportional to the number of comparators turned on. All the 256 comparators have to be high-slew-rate wideband op amps. This circuit is best built on a single chip using IC technology, or with series of chips, due to the large number of comparators needed.

The encoding logic can be built onto the same chip and a latch added on the output if needed. The output would be binary data, 8 bits wide, corresponding to the analog value of the video signal sample.

Forthcoming Approaches

Future improvements in these devices, plus faster, cheaper, random-access memories, will probably make digital video scrambling economical in the future. Available now are 16384 × 1-bit RAMs with 25-ns access times (Intel Corporation), which is more than enough memory to store a few lines of a TV picture. If the data were stored in an 8-bit-wide memory,

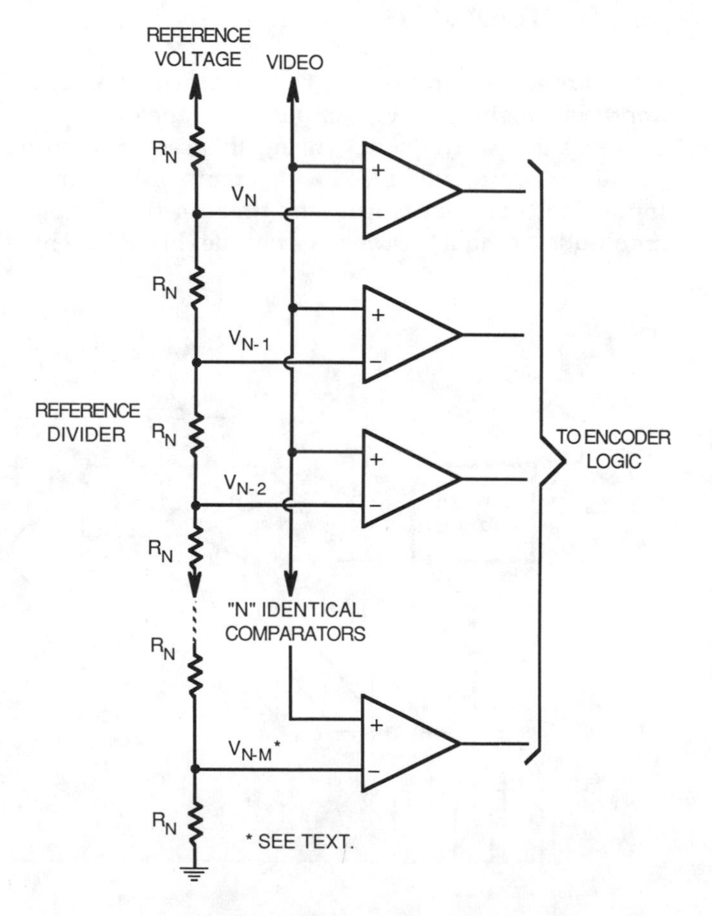

Fig. 8-9. A/D flash conversion.

all 8 bits of each pixel could be retrieved (in parallel form) in 25 ns, which is faster than necessary. This would allow plenty of time for D/A conversion. With 256 pixels per line, for example, two 4096 × 4 RAMs could store 16 lines. Thirty-two such RAMs could store an entire frame in memory. As of this writing, Matsushita Corporation has a 1-megabit RAM which reportedly can store a whole frame. The technology is here, the cost just has to be brought down. Probably in a few years, digital video encryption will be common and may replace the VideoCipher II™ with a still more secure system.

9

Cable and Satellite Decoder Circuitry

The information in this chapter is provided solely for educational purposes. The authors make no warranty as to the operation of the following decoder circuits. It may be illegal for a person, persons, or business to use these decoders to decode pay-TV signals without proper authorization. The authors make no warranty as to the legality of using these or any other decoders, and in no way are responsible for any consequences whatsoever for their use. However, at the time of this writing (June 1986), experimentation for educational purposes is probably permitted. The authors do not necessarily guarantee this. If in doubt, consult an attorney for your particular requirement.

Fig. 9-1 shows three experimental decoders which you can build. Again, these decoders are experimental but will work if properly constructed and aligned. Some experimentation may be necessary for best results. After all, this is the "object of the exercise"—to learn. Although the decoders are somewhat similar, in a few cases, different circuit approaches have been deliberately used to illustrate, as much as possible, various circuits, and, also, to avoid repetition. Feel free to make any necessary changes, modifications, and improvements.

Outband Decoders

An *outband decoder* will be useful in those systems where the CATV system uses a modulated carrier as a means of providing decoding information. The carrier is modulated with synch pulses that are derived from the encoded video. The encoded video generally has no synch information present. This synch carrier is usually located below Channel 2 (50 MHz), or somewhere in or near the FM broadcast band.

The Synch Subcarrier

To find the carrier just mentioned, you need a *block converter.* These converters can be picked up at most radio-TV parts houses or at a store that handles video accessories. A block converter converts all the VHF, midband, and superband cable channels to UHF (Fig. 9-2); it is simply a mixer and a local oscillator operating around 400 to 500 MHz. The sum of the input frequency (50–300 MHz) and the local oscillator (420 MHz, in this example) is between 470 and 720 MHz, or from about Channel 14 to Channel 56. The entire VHF spectrum is thus shifted to the UHF band.

When the block converter is hooked up to a TV receiver tuned to UHF frequencies, each VHF channel will appear on a UHF channel. If you tune below Channel 2, or between Channel 5 and Channel 6, you will pick up the cable channels. If your system uses outband (scrambling), you will receive at one frequency what appears to be a "blank" white *raster.* The TV receiver will lock to this, but there will be no picture visible. This is the "synch" channel. Only synch pulses are transmitted here. No video is present, thus giving a white raster (or blank screen). By knowing the local oscillator frequency used in the block converter, and knowing the channel that the UHF receiver is tuned to, the VHF frequency of the synch channel can be determined. Or, if you know which VHF channel you are watching, the local oscillator (LO) frequency is equal to the UHF channel

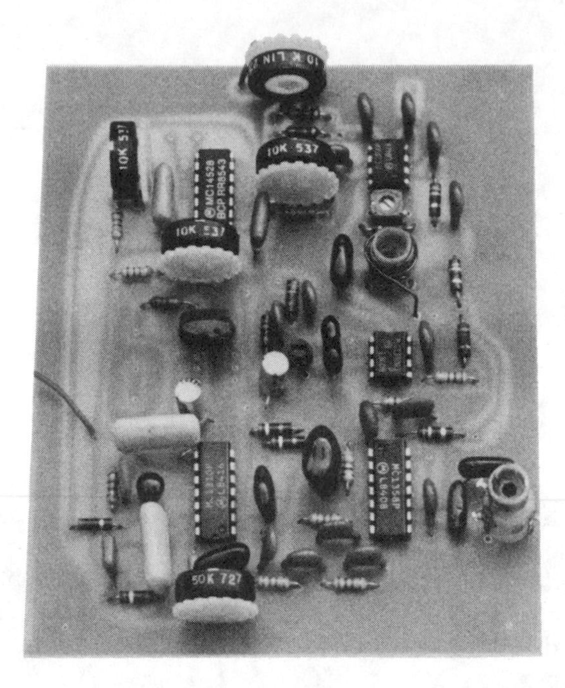

(A) Gated pulse decoder.

(B) Outboard decoder.

(C) Sine-wave decoder.

Fig. 9-1. Three experimental decoders.

frequency minus the VHF channel frequency. For example, Channel 2 is seen on Channel 15:

Channel 2 = 54–60 MHz
Channel 15 = 476–482 MHz
LO frequency = 476 − 54, or 422 MHz.

Using such a method, you can determine the "synch" channel frequency. Fig. 9-3 is a circuit of a typical outband decoder. Refer to this circuit during the following discussion. Fig. 9-4 shows the completed board, with all the components in place. Note that slug-tuned coils are used for ease of adjustment.

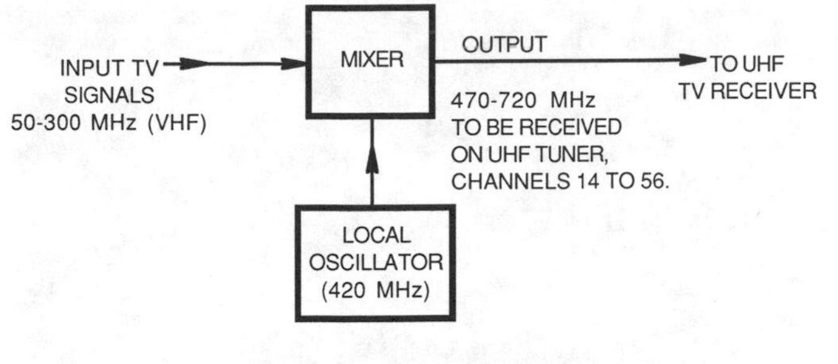

Fig. 9-2. Block diagram of an elementary block converter.

Component Values

When the synch channel frequency is known, suitable values for capacitors C1 through C5 and inductors L1 through L4 can be determined. If the synch channel is not exactly as shown in the chart, you can simply scale the values shown in the chart (as required), using the chart as a rough guide. The component values would not be radically different from those shown in Fig. 9-3, for frequencies not too far removed from 50 or 100 MHz, respectively.

Construction Methods

Coils are wound using a No. 8-32 × 2-inch-long screw as a winding form. The screw is removed and the slug inserted. Table 9-1 gives the coil data. All leads must be kept short.

Insert resistors first when stuffing the PC board, and then install all the capacitors. Next, the ICs and transistors are installed. The coils are inserted last, to avoid damaging them. All solder joints should be checked, and then power applied. Assuming nothing smokes or overheats, proceed with the alignment.

Coil L2 and capacitor C3 are resonated at the synch frequency, while L1 and C1 resonate at the premium

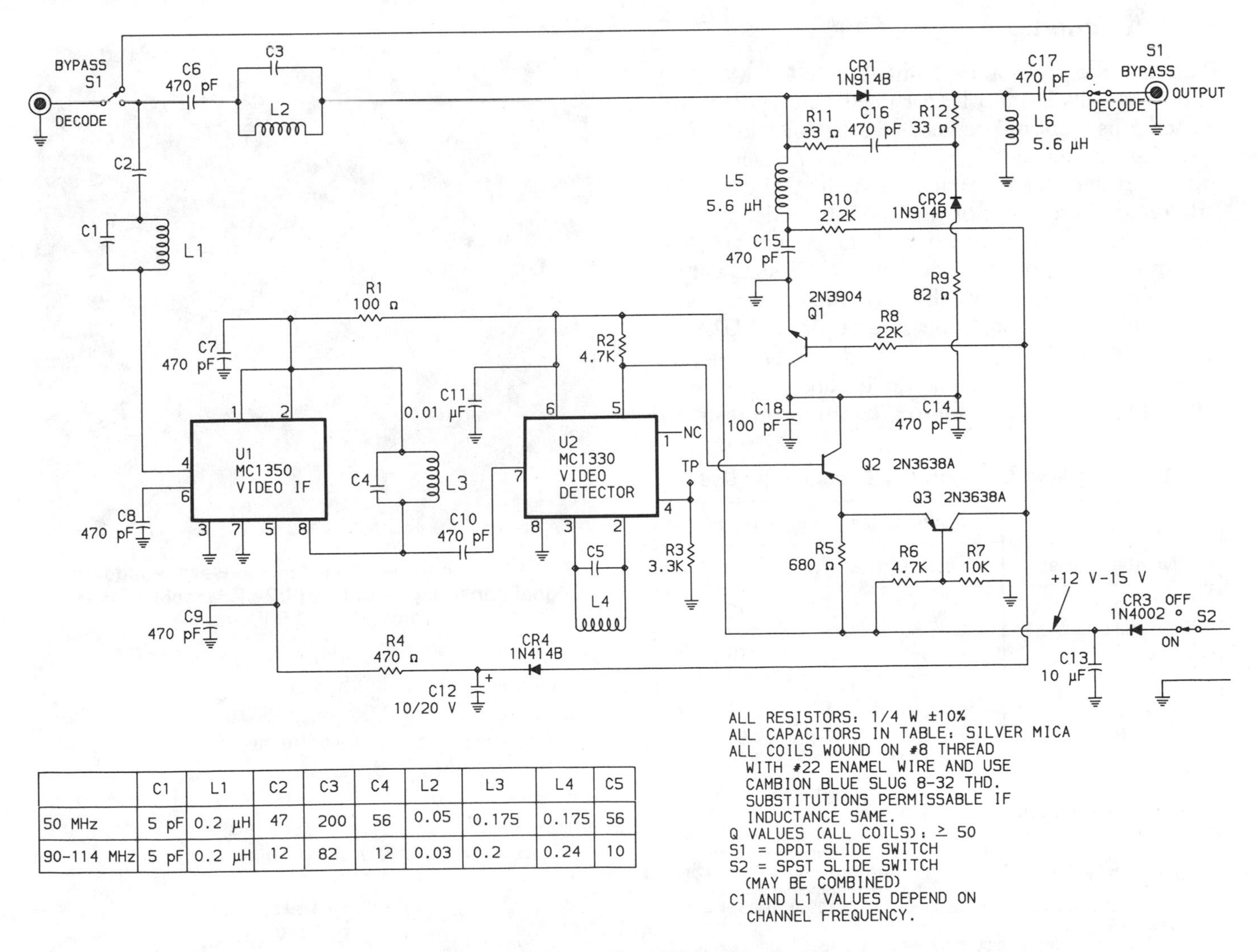

	C1	L1	C2	C3	C4	L2	L3	L4	C5
50 MHz	5 pF	0.2 µH	47	200	56	0.05	0.175	0.175	56
90-114 MHz	5 pF	0.2 µH	12	82	12	0.03	0.2	0.24	10

Fig. 9-3. Cable outband decoder circuit.

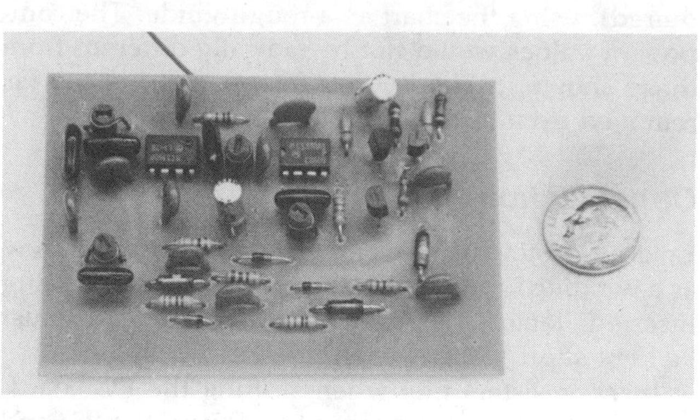

Fig. 9-4. Photo of an assembled outband decoder board.

channel frequency. Components, C2 and L1, and L3 and C4, should series-resonate at approximately the synch frequency.

Circuit Tune-up

Place a scope probe at tie point TP (pin 4 of MC1330), and tune coils L1 and L3 for maximum synch level. Synch pulses should be visible at the collectors of the 2N3638A transistors. The idea is to reinsert synch pulses in the video signal by switching the variable attenuator, consisting of the 1N914B diodes and the 33-ohm resistors, in and out of the circuit. (Fig. 9-5 illustrates the sine-wave encrypted video signal at pin 4, while Fig. 9-6 shows the recovered 15-kHz sine-wave encoding signal appearing at pin 1 of U2A.)

Final adjustments are then made for best picture. The input signal level should be about −6 to +6 dBmv (0.5 to 2 millivolts into a 75-ohm load resistor).

Table 9-1. VHF Coil Data for Descrambler Use

Number Turns	Inductance in Nanohenries	
	Slug Out	Slug In
2.5	30	75
3.5	50	100
4.5	75	170
5.5	85	200
6.5	105	230
7.5	125	300
8.5	155	350
9.5	170	380

Notes:
1. All coils wound with No. 22 enamelled wire.
2. All coils wound on a No. 8-32 NC screw-thread, which is used as a coil form. The screw-thread is then removed and a slug is inserted.
3. Part number of slug is Cambion P/N 515-3225-06-2100 (blue). Other slugs may give more or less maximum inductance and Q values.
4. All inductance (L) values are approximate (±10%); 1000 nH = 1 microhenry.
5. Individual constructors may experience slightly different results due to technique and measurement method.

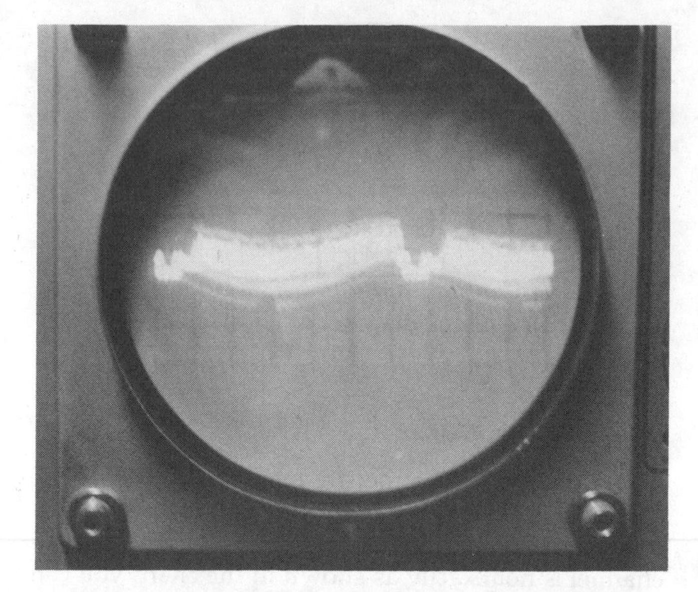

Fig. 9-5. Sine-wave encrypted video signal at pin 4 of MC1330; H = 10 μs/cm, V = 0.5 V/cm.

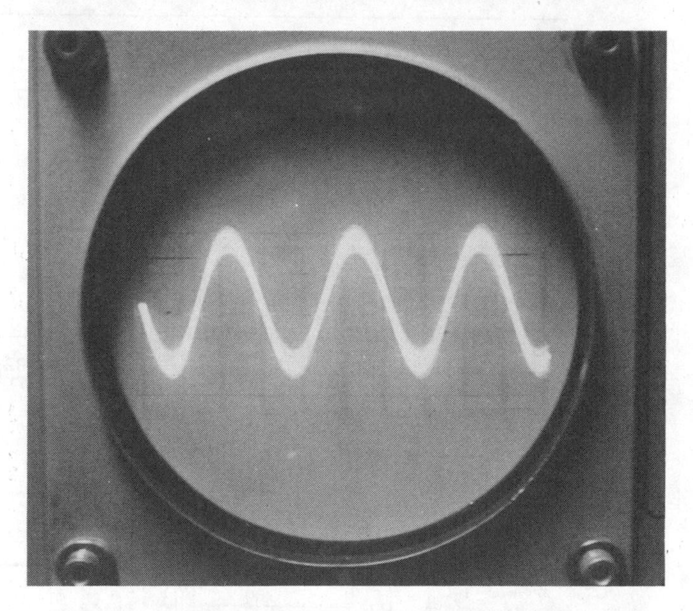

Fig. 9-6. Recovered 15-kHz sine-wave encoding signal appearing at pin 1 of U2A (LM1458); H = 20 μs/cm, V = 0.5 V/cm.

A PC-board layout is shown in Fig. 9-7. The board should be made of 0.062-inch G-10 epoxy fiberglass. Other materials may affect the circuit's operation and require that slightly different capacitances be used in the tuned circuits. The PC board and all the parts that mount on the PC board, plus additional data and tune-up instructions, can be obtained from:

North Country Radio
P.O. Box 53
Wykagyl Station, New York 10804

as Catalog No. OB-2 for $34.95, plus $2.50 for shipping and handling.

Pulse Decoder

The *pulse decoder* circuit shown in Fig. 9-8 is for use in systems where the synch is suppressed (gate synch). Its input signal is the output (Channel 3 or 4) of the cable converter box. This makes the design of the decoder much simpler, since fixed tuning can be used.

Circuit Operation

Incoming signals are amplified by an MC1350 IF amplifier chip. The gain of this amplifier (a function of

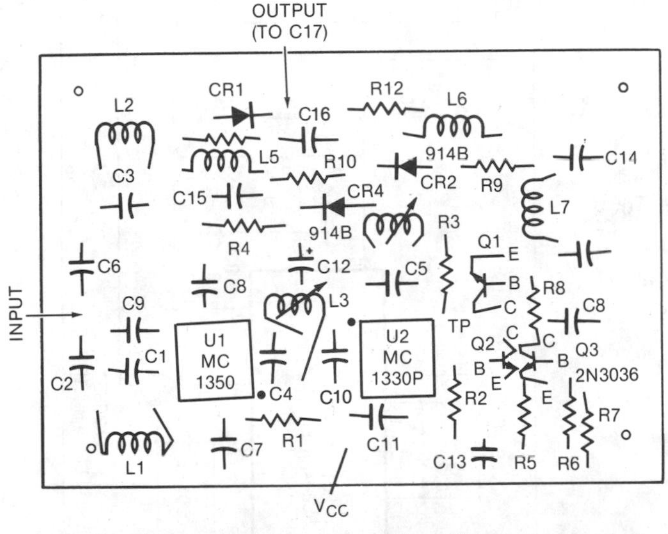

(A) Component location.

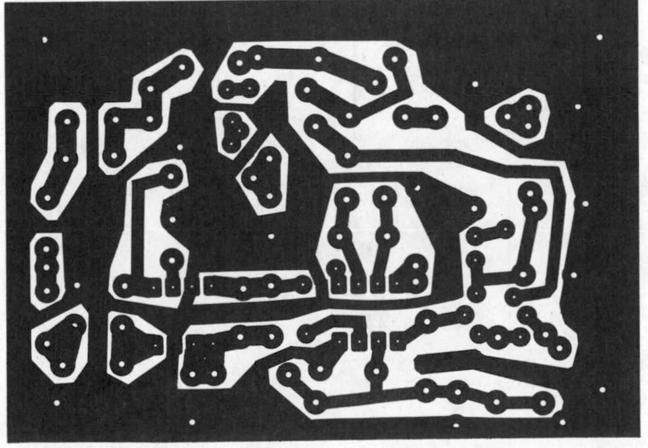

1. FOIL SIDE SHOWN.
2. PC BOARD SIZE: 2¼ × 3¼ INCHES.

(B) Pattern side.

Fig. 9-7. Outband decoder PC-board layout.

the bias on pin 5) depends on the setting of the 10K bias potentiometer. Superimposed on this bias voltage is a pulse that is equal in width to the blanking pulse; this pulse is present only during the horizontal blanking interval (HBI). It is not used during vertical blank intervals (VBI).

A sample of the video signal is taken off at transformer T1 and fed to an MC1330 video detector tuned to Channel 3 or Channel 4, as required. Composite video appears at pin 4 of the MC1330. A 100-pF capacitor couples high-frequency (above 4 MHz) signals to an MC1358 integrated circuit, which acts as a sound IF for the 4.5-MHz subcarrier. Detected audio (in this case, a 31.5-kHz subcarrier) is fed to an MC1310 IC, which functions as a subcarrier demodulator. This is a PLL chip of the kind used as an FM stereo demodulator. The chip uses a 63-kHz oscillator, divided down by two to get 31.5 kHz, and again by two to obtain the required 15.75-kHz reference signal. A high-pass filter consisting of 470-pF capacitors and 33-kilohm resistors is used to pick off the subcarrier from pin 12 of the MC1358 sound IF and detector chip. An 18-μH coil and 68-pF capacitor serve as a quadrature network for the MC1358. The 18-μH variable coil in conjunction with the 68-pF capacitor is adjusted for maximum audio recovery at pin 12 of the MC1358. Fig. 9-9 shows the waveform of the recovered synchronizing pilot carrier, while Fig. 9-10 is a photo of the regenerated synch pulse waveform.

Subcarrier Recovery

The MC1310 IC demodulates the subcarrier. The VCO frequency is adjusted with the 50K potentiometer on pin 14. The 15.75-kHz pulses are taken off pin 10 and then amplified by a 2N3564 transistor used as a pulse amplifier. This pulse amplifier is disabled in the absence of a subcarrier (as during the VBI) when pin 10 of the Mc1310 demodulator goes high. The output of the 2N3565 transistor is fed to a dual monostable multivibrator, CD4528, which generates a pulse that is timed to coincide with the HBI. The 10K delay potentiometer sets the pulse position and the 10K width control adjusts the pulse width. This generated pulse is fed back to pin 5 of the IF amplifier through a 10K potentiometer for amplitude adjustment, to properly vary the gain of the MC1350 IF amplifier. To restore the synch pulse effectively, the gain should rise from 6 to 10 dB during the HBI.

Construction Details

The required coil inductance depends on the intermediate frequency used, which can be 45 MHz (stan-

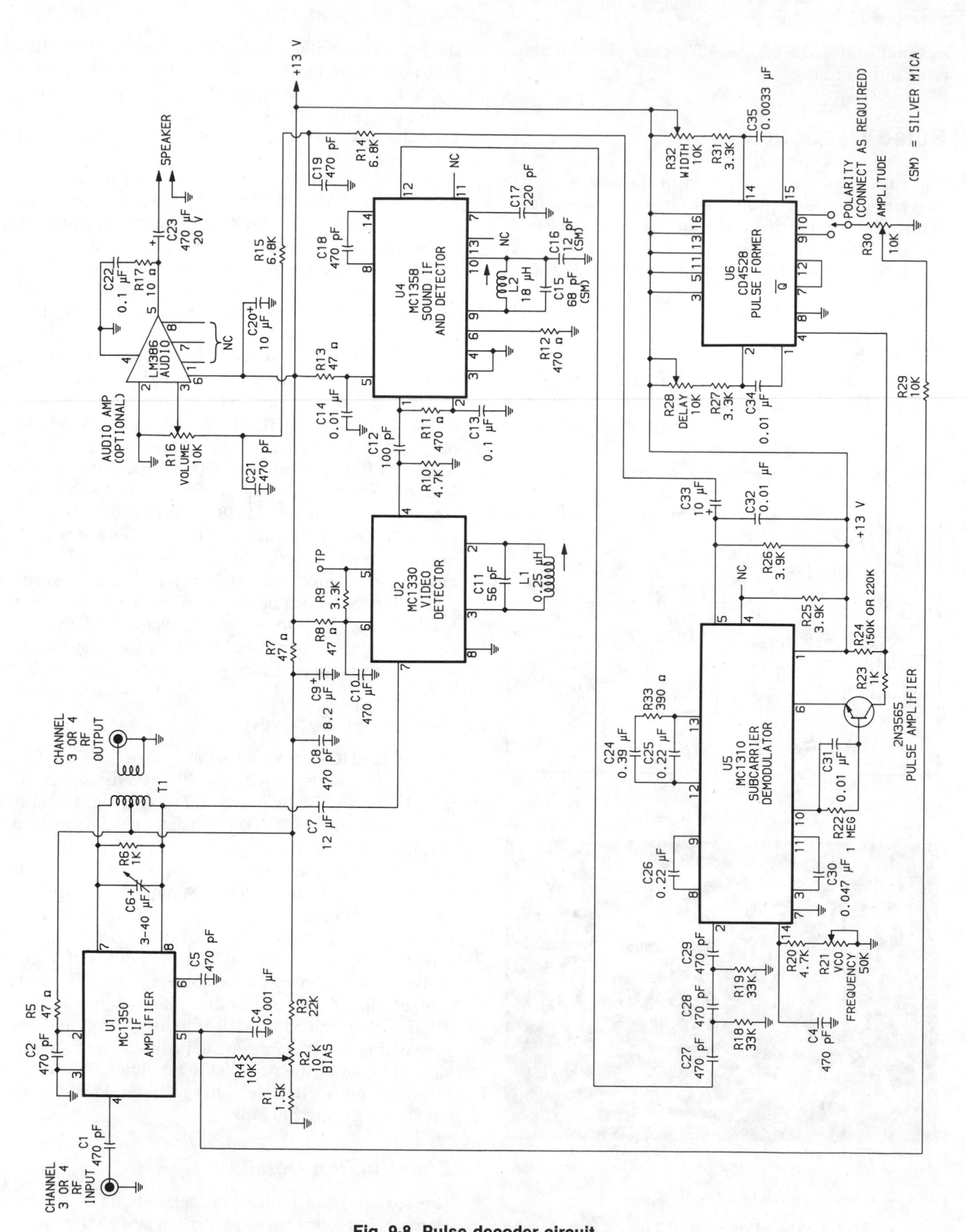

Fig. 9-8. Pulse decoder circuit.

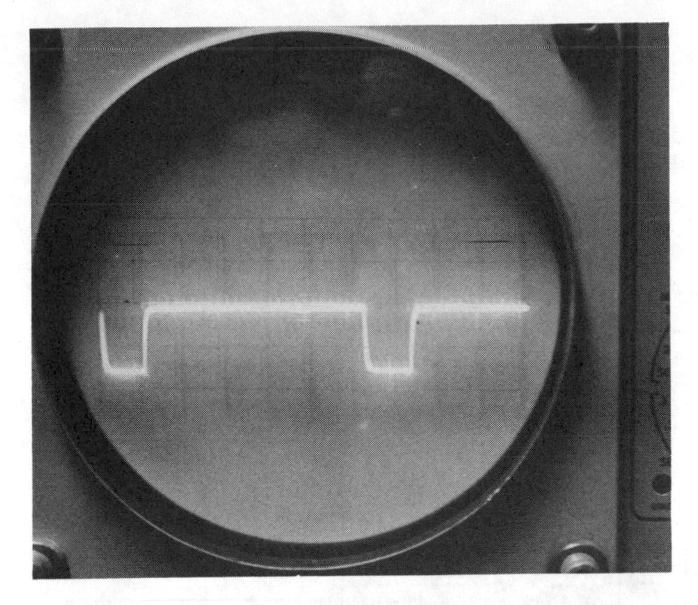

Fig. 9-9. Recovered synchronizing pilot carrier; H = 10 μs/cm, V = 2 V/cm.

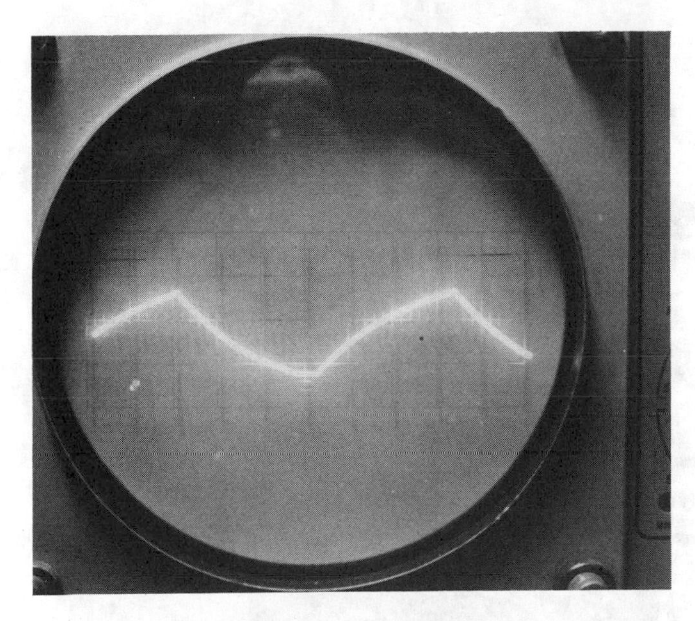

Fig. 9-10. Regenerated synch pulse; H = 10 μs/cm, V = 5 V/cm.

dard TV IF) or equal to Channel 3 or Channel 4 for outboard converter use. Layout of the PC board is shown in Fig. 9-11. A G-10 epoxy fiberglass (0.062 inch) board should be used. Other board materials may require circuit modification due to different dielectric constant and the resulting shifts in distributed circuit capacitances. Component placement is shown in the photo of Fig. 9-12.

Circuit Tune-up

Alignment consists of peaking transformer T1 for maximum signal, and peaking the 0.25-μH coil for maximum signal at pin 5 (TP) of the MC1330 detector. Next, adjust the 18-μH quadrature coil in the sound IF circuit for maximum recovered audio, which, with the adjustment of the 50K audio-gain potentiometer (pin 6), should be around 100 to 500 millivolts RMS, as required.

Next, adjust the 50K potentiometer (VCO FREQ) for a 15.75-kHz square wave at pin 4 of the CD4528 pulse former. Adjust the 10K width control for a pulse about 11 μs wide, at pin 9 or pin 10 of the CD4528 IC. The delay control is set at center.

Connect the RF output to a TV set. Adjust the 10K delay potentiometer for best results on a scrambled picture, and then adjust the 10K width potentiometer. As a start, set the 10K amplitude potentiometer at mid position. Readjust all pots for best results. Some interaction may occur, so readjustments may be necessary. The adjustments depend on input levels, the TV set, and the tolerances of the parts, and experimentation for best results will be necessary.

The PC board and all the parts that mount on the PC board, plus additional data and tune-up instructions, can be obtained from:

North Country Radio
P.O. Box 53
Wykagyl Station, New York 10804

as Catalog No. PD-2 for $49.95, plus $2.50 for shipping and handling.

Sine-Wave Decoder

The *sine-wave decoder* circuit shown in Fig. 9-13 is similar to the gated-pulse decoder, except that, in this case, a sine-wave is derived and then added to the video waveform. Input from the tuner (at 45 MHz) is fed into a 2N3563 video IF amplifier that has a gain of up to 20 dB. The IF signal is then fed to an MC1330 video detector, where a video signal is recovered and appears at pin 4. The 56-pF capacitor and 0.25-μH coil tune the MC1330 detector to 45 MHz. A 15.75-kHz sine-wave signal appears on the video signal. This 15.75-kHz sine-wave signal is then fed to an active filter, consisting of an LM1458 op amp and a feedback network, having maximum response at around 15.75 kHz. The phasing control adjusts the center frequency.

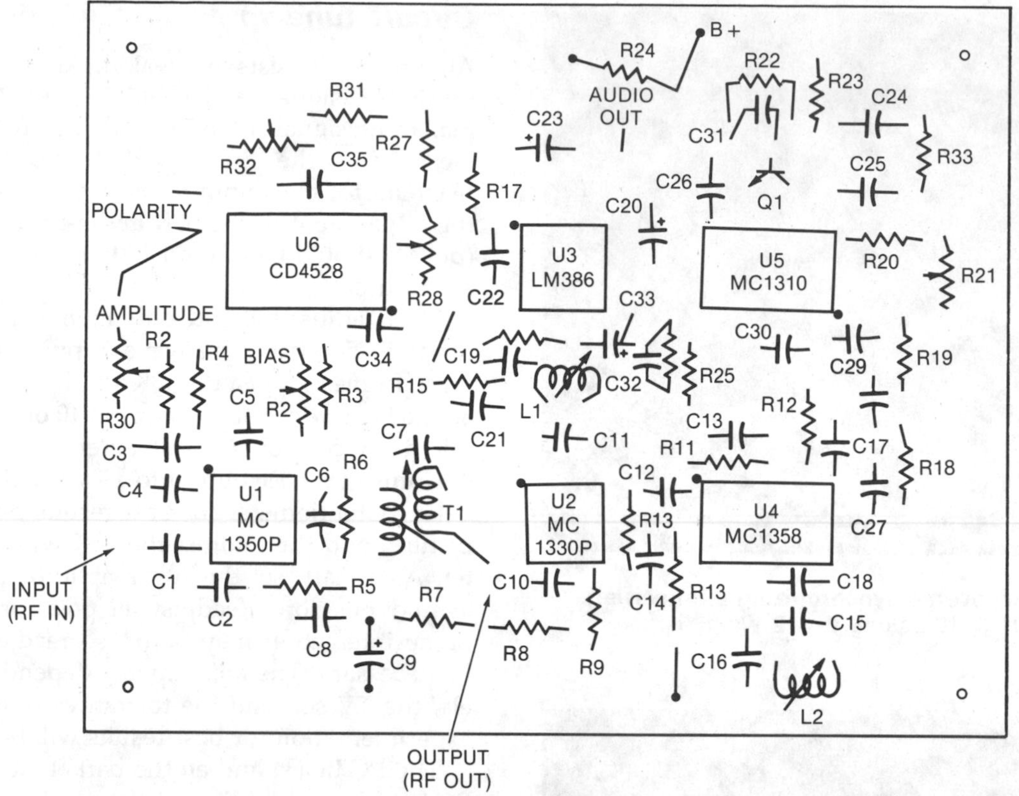

(A) Component location.

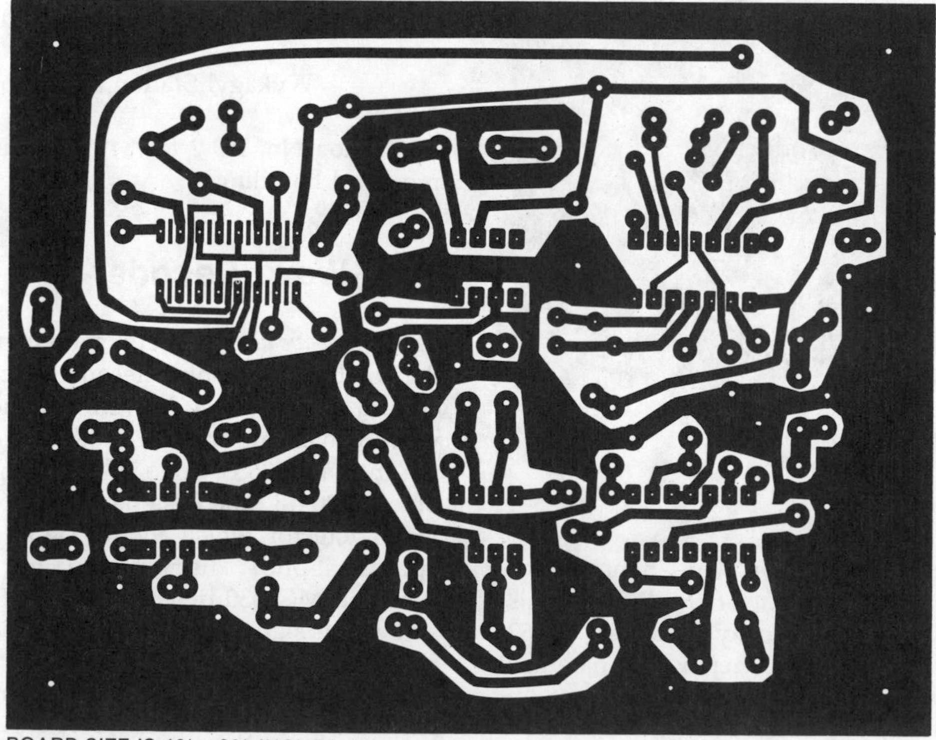

BOARD SIZE IS 4¾ × 3¾ INCHES.

(B) Pattern side.

Fig. 9-11. Pulse decoder PC board layout.

Fig. 9-12. Component side of the gated pulse decoder PC board showing components in place.

The phase shift can be varied using this phasing control so that a sine wave of correct phase can be obtained, as necessary, to cancel the sine-wave modulation on the video signal. Amplitude can be adjusted with the 5K potentiometer. The 15-kHz signal is fed through a 5.6-μH RF choke and a 1K isolation resistor, along with a DC bias, to PIN diode MPN3404. These signals vary the RF impedance of the MPN3404.

The IF signal from the tuner is partially shunted to ground through the MPN3404 diode, depending on the instantaneous current through the diode. By proper adjustment of the phasing and cancellation controls, the IF output signal can be descrambled.

The sound signal is frequency-modulated on a 625-kHz subcarrier. An NE565 IC acts as a PPL FM detector, whose output is fed to an LM386 audio amplifier and speaker. If the TV sound is *not* encoded, the audio recovery circuits (lower half of the schematic) can be omitted.

Fig. 9-14 is a diagram of the construction and component arrangement of an interface box for the experimental sine-wave decoder.

Circuitry Tune-up

Fig. 9-15 is a photo of the sine-wave decoder, ready for alignment and testing. Circuit alignment consists of adjusting the IF gain control initially to maximum, tuning the 0.25-μH coil for maximum video at pin 4 of the MC1330 video detector, and adjusting the 18-

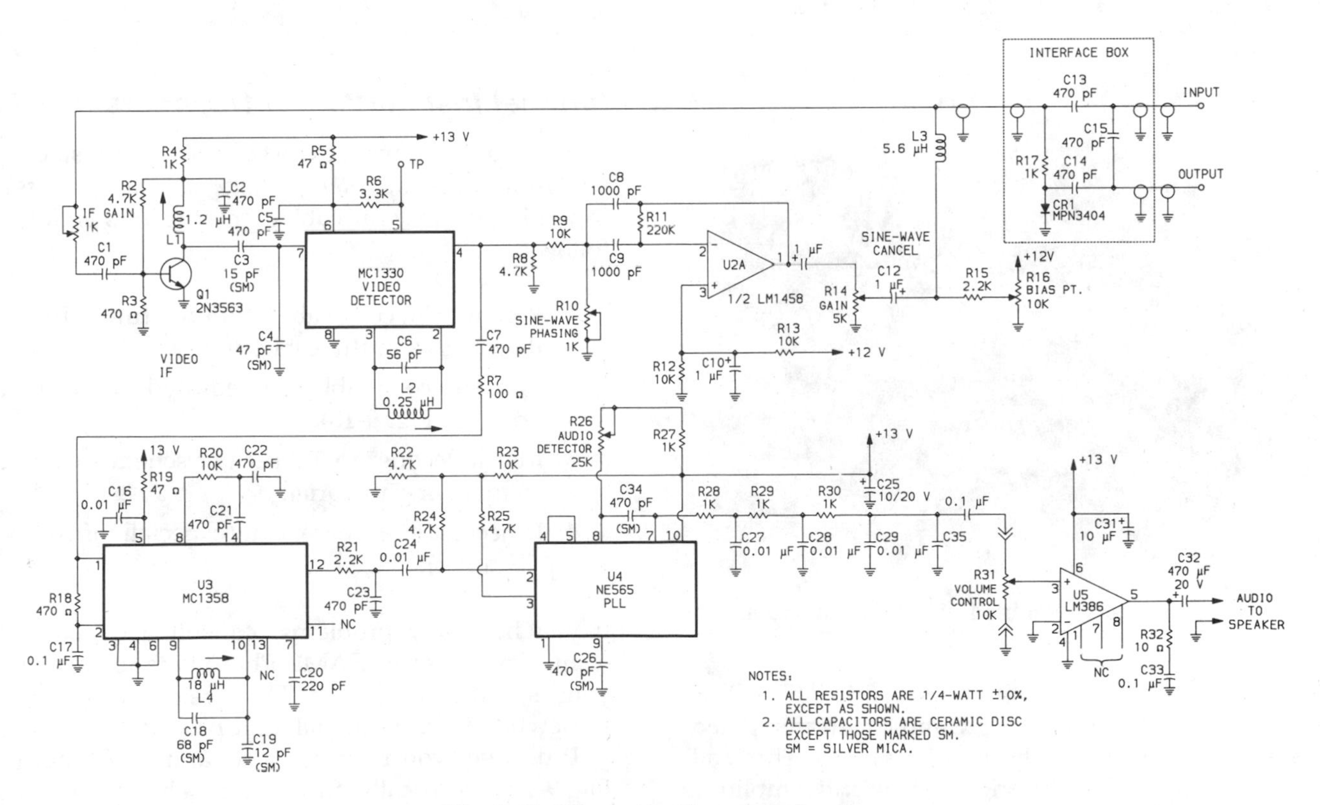

Fig. 9-13. Sine-wave decoder circuit.

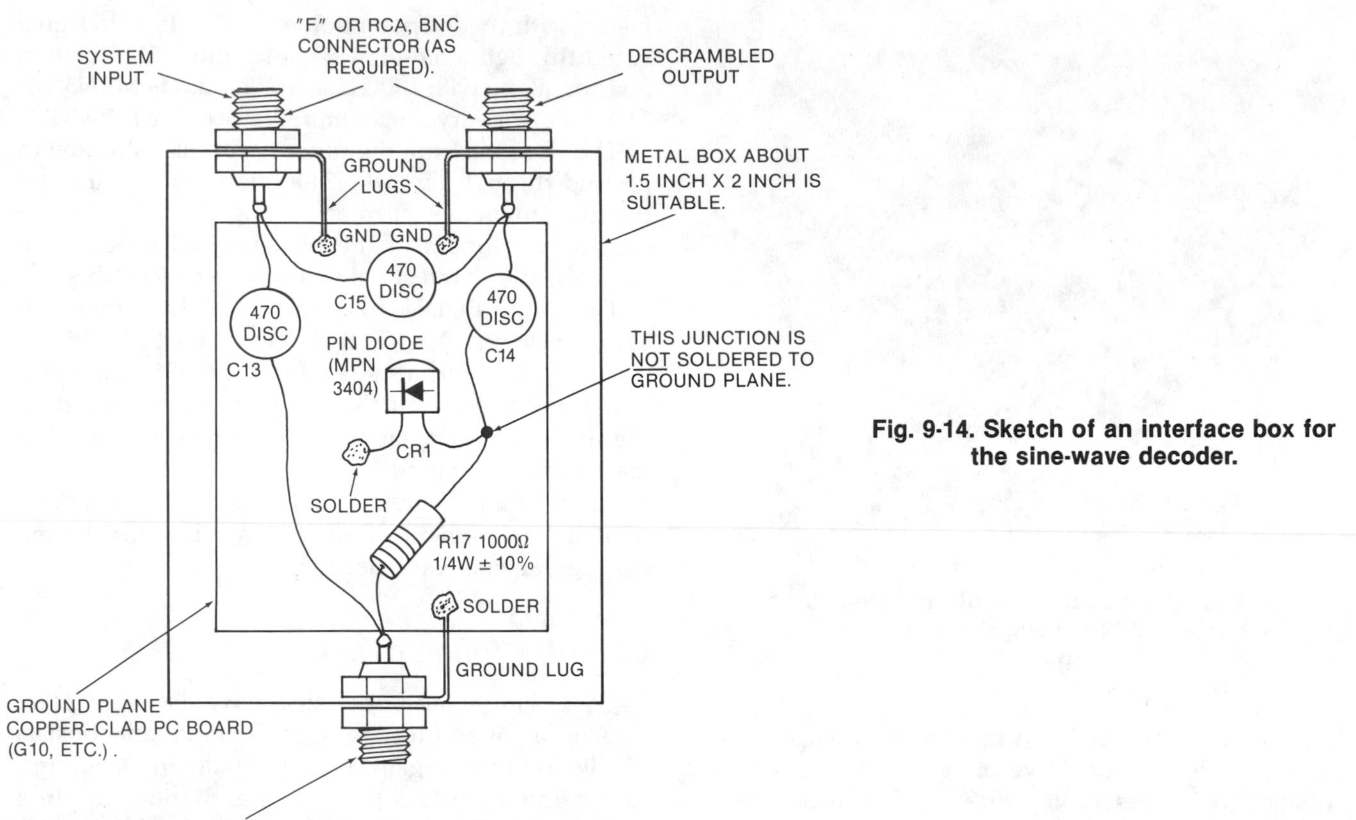

Fig. 9-14. Sketch of an interface box for the sine-wave decoder.

Fig. 9-15. The assembled sine-wave decoder.

μH quadrature coil (pins 9 and 10, MC1358 IC) for best received audio. The 25K audio detector potentiometer is adjusted for best audio as well. Then, adjust the sine-wave phasing control and amplitude control for best reception. Some experimentation will be necessary.

Internal Installation of Decoders in a TV

You may possibly want to install a decoder inside your television receiver. *After you have obtained permission,* in writing, from your cable operator, this is done as follows:

1. Remove the coax cable from the tuner. Remove the plug from the cable, if any.
2. Make up coax cables, as required, for reconnection (see Fig. 9-16).
3. Install decoder in TV set per schematic and accompanying pictorial.
4. If necessary, re-peak coil (especially if coil is in TV tuner).

If you have any problems, consult a qualified TV technician. Refer to SAMS PhotoFacts for details on your specific set. (Each TV receiver may require somewhat different installation procedures.)

If desired, you may use a separate TV tuner (see Fig. 9-17). Electrically, this may be a better choice, especially if your receiver is very compact or does not have the correct interfaces (and/or voltages).

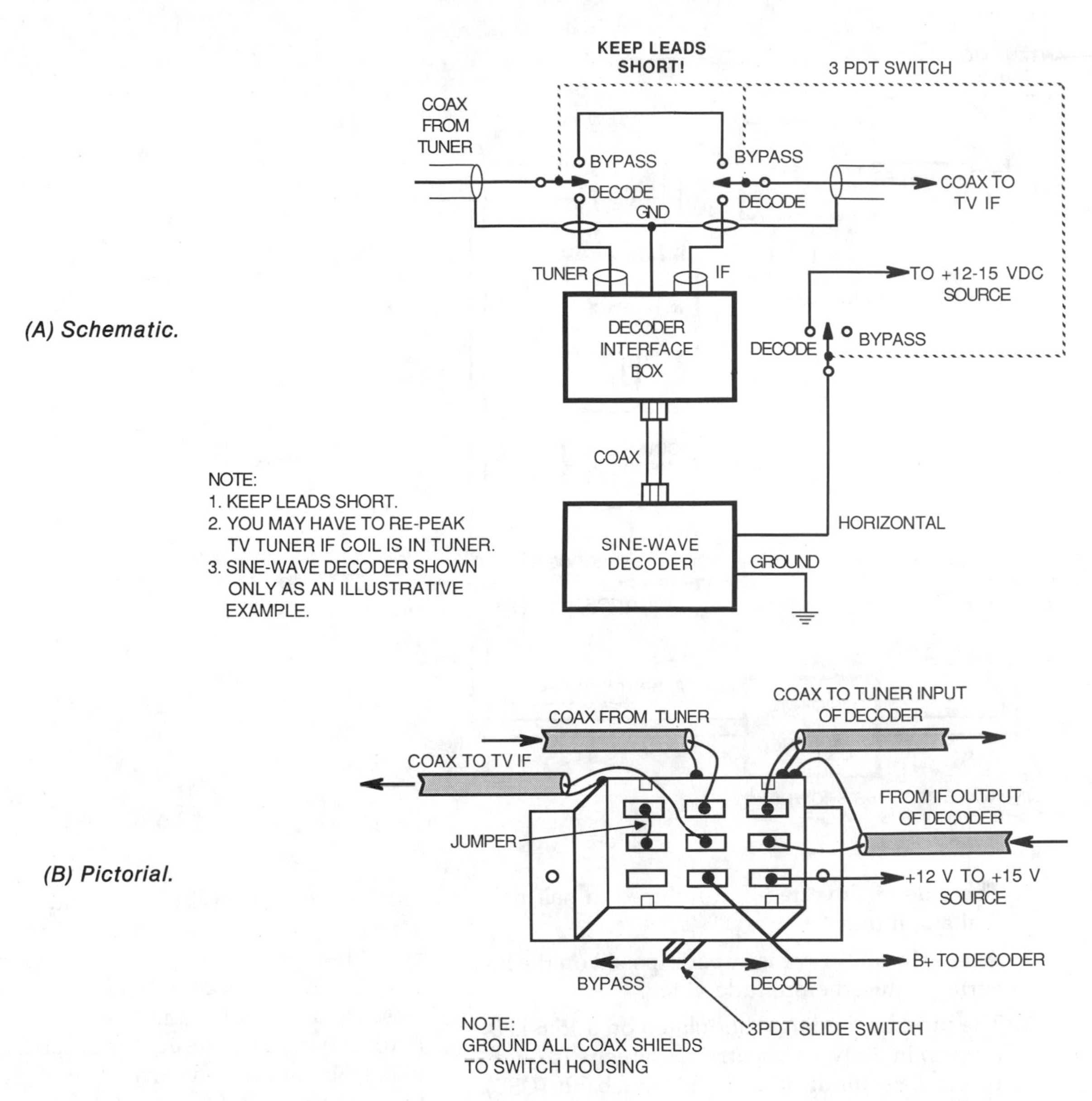

Fig. 9-16. Decoder-TV hookup.

Decoder Kit

A PC-board layout of the sine-wave decoder is shown in Fig. 9-18. Use only a 0.062-inch G-10 epoxy fiberglass board. The PC board and all the parts that mount on the PC board, plus additional data and tune-up instructions, for this decoder can be obtained from:

North Country Radio
P.O. Box 53
Wykagyl Station, New York 10804

as Catalog No. SW-2 for $52.95, plus $2.50 for shipping and handling.

Telease-Maast Scrambling System Decoder

The Telease-Maast system is currently being used by some satellite services as a means of signal encryption. Among those using the system are two sports channels and two X-rated program suppliers. Although this system is not used by the more popular cable-program suppliers, it is worth mentioning and describing its characteristics.

The Telease system is basically an analog video-and-audio scrambling system. It is similar in principle to the sine-wave encoding system mentioned earlier in this book. What happens is as follows:

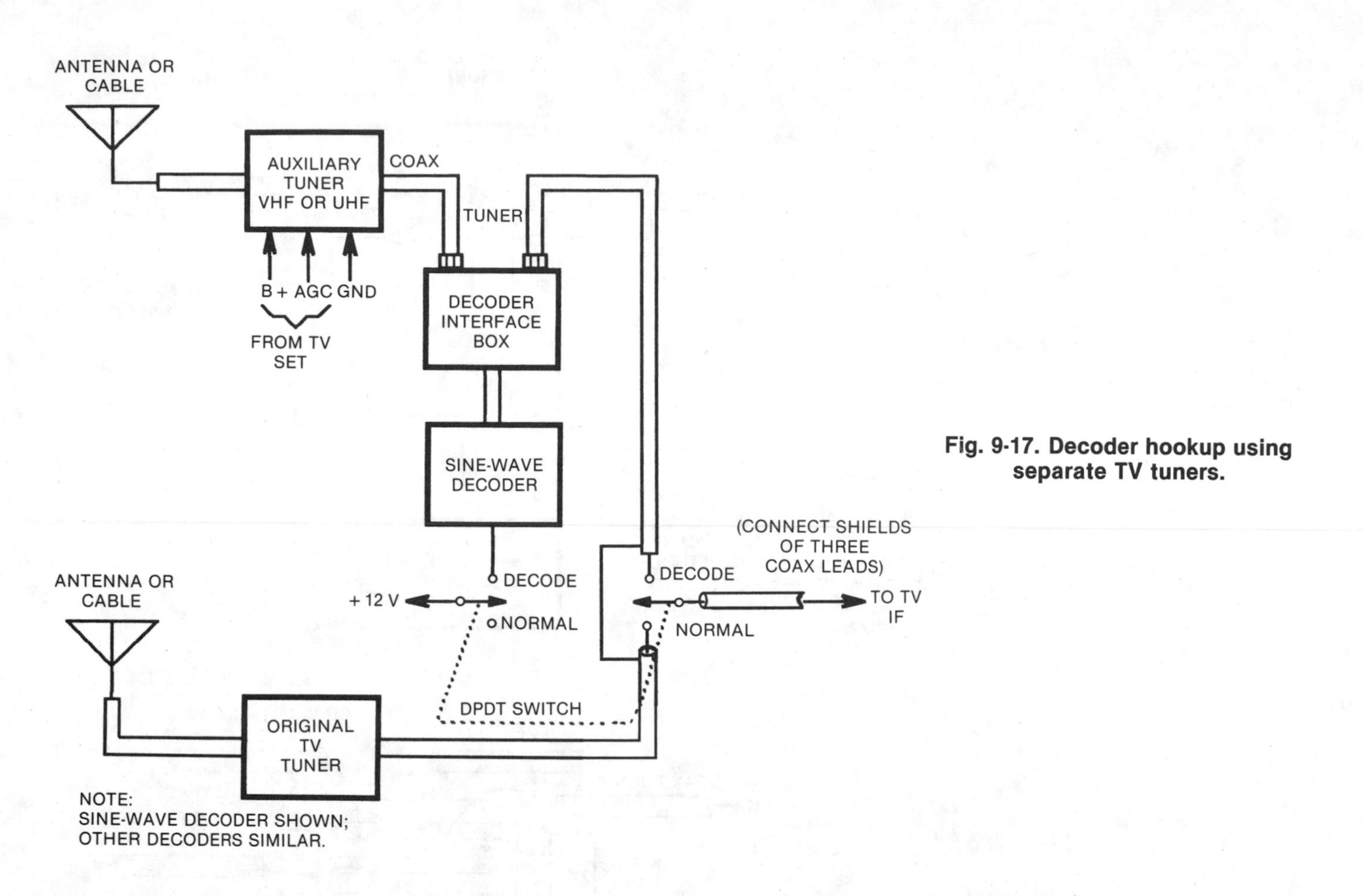

Fig. 9-17. Decoder hookup using separate TV tuners.

1. The video is inverted and reduced to half normal amplitude.
2. A 94-kHz sine-wave is superimposed on the inverted reduced-amplitude video.
3. The program audio is modulated on a 15.6-kHz (approximately) subcarrier, which is suppressed, resulting in a double-sideband (DSB) suppressed-carrier audio signal.

Fig. 9-19 shows a block diagram of a suitable system. It is important to note that the 94-kHz frequency is *not exactly* six times the horizontal-scan frequency. In practice, frequencies used are in the range between 94 kHz and 95 kHz. Table 9-2 gives the frequencies

Table 9-2. Telease Sine-Wave Scrambling Frequencies

Service	Frequency (kHz)
Sports	94.5651
	94.6370
	94.1428
Fantasy Channel	94.0689
Adult Movie Channel	94.1944

used by several services. The audio subcarrier is exactly one-sixth of this frequency. The scrambled video signal has the 94-kHz sine wave superimposed on it.

Fig. 9-20 shows a method for descrambling a Telease signal. A video amplifier and detector are used to directly retrieve the audio subcarrier, which is a 15-kHz DSB suppressed-carrier signal. A 94-kHz (nominal) filter picks off the descrambling sine wave. This signal is used as a reference signal in a phase-locked loop, which has a VCO running at 64 times (nominally 6 MHz) the sine-wave frequency. (The VCO may be a VCXO, which is a voltage-controlled crystal oscillator.) The 6-MHz signal is divided by 64, and the VCO output (divided in frequency by 64) is fed to the other input of the phase detector. The output of the phase detector controls the VCO (or VCXO) such that phase-lock occurs. Therefore, the 94-kHz output signal from the ÷64 unit is exactly equal in frequency to the scrambling sine wave. This sine wave is filtered with an active bandpass filter for both phase and amplitude control, and is combined with the original scrambled video (which has been inverted) so that the sine waves cancel.

This leaves the original video, which is now restored to normal. The 15-kHz DSB audio signal is fed

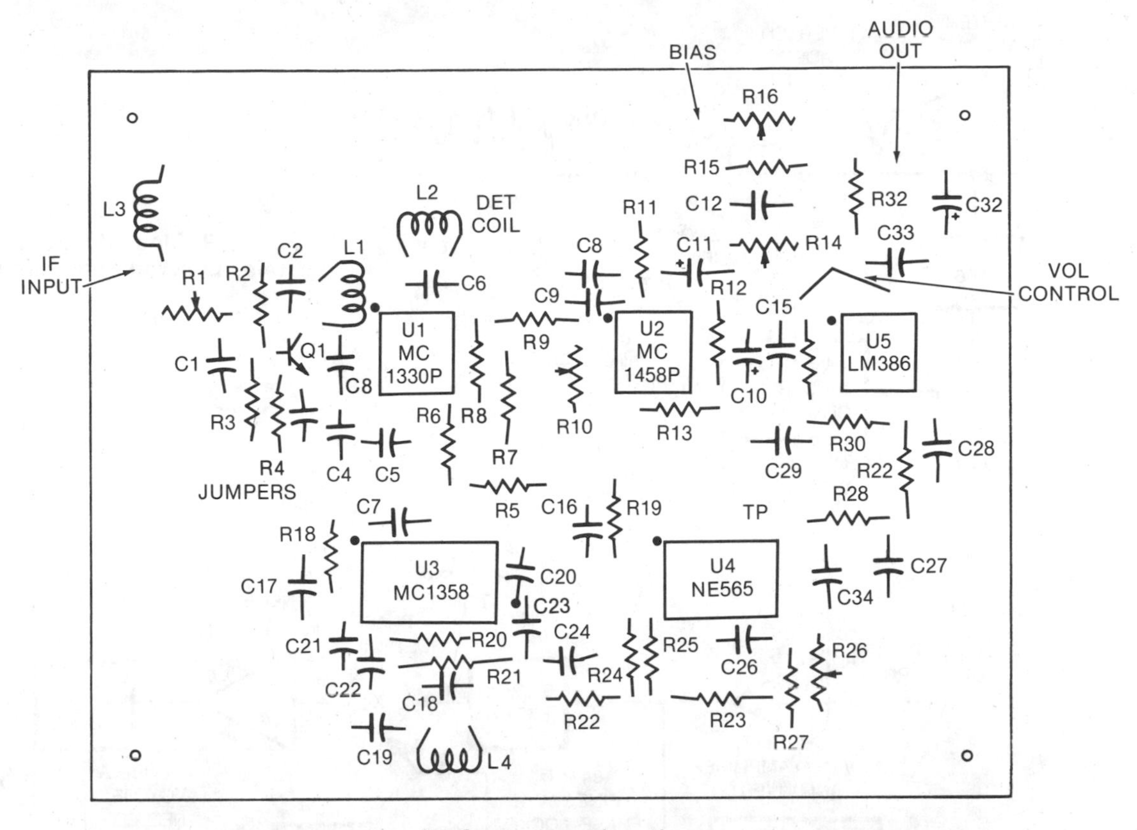

(A) Component location.

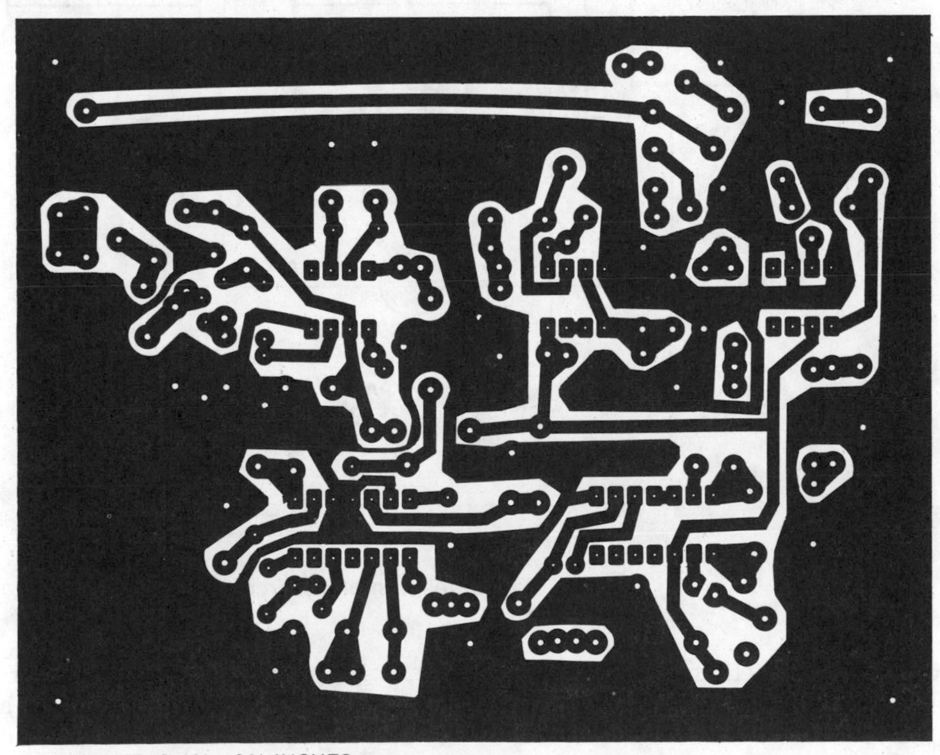
BOARD SIZE IS 4¾ × 3¾ INCHES.

(B) Pattern side.

Fig. 9-18. PC board layout for the sine-wave decoder.

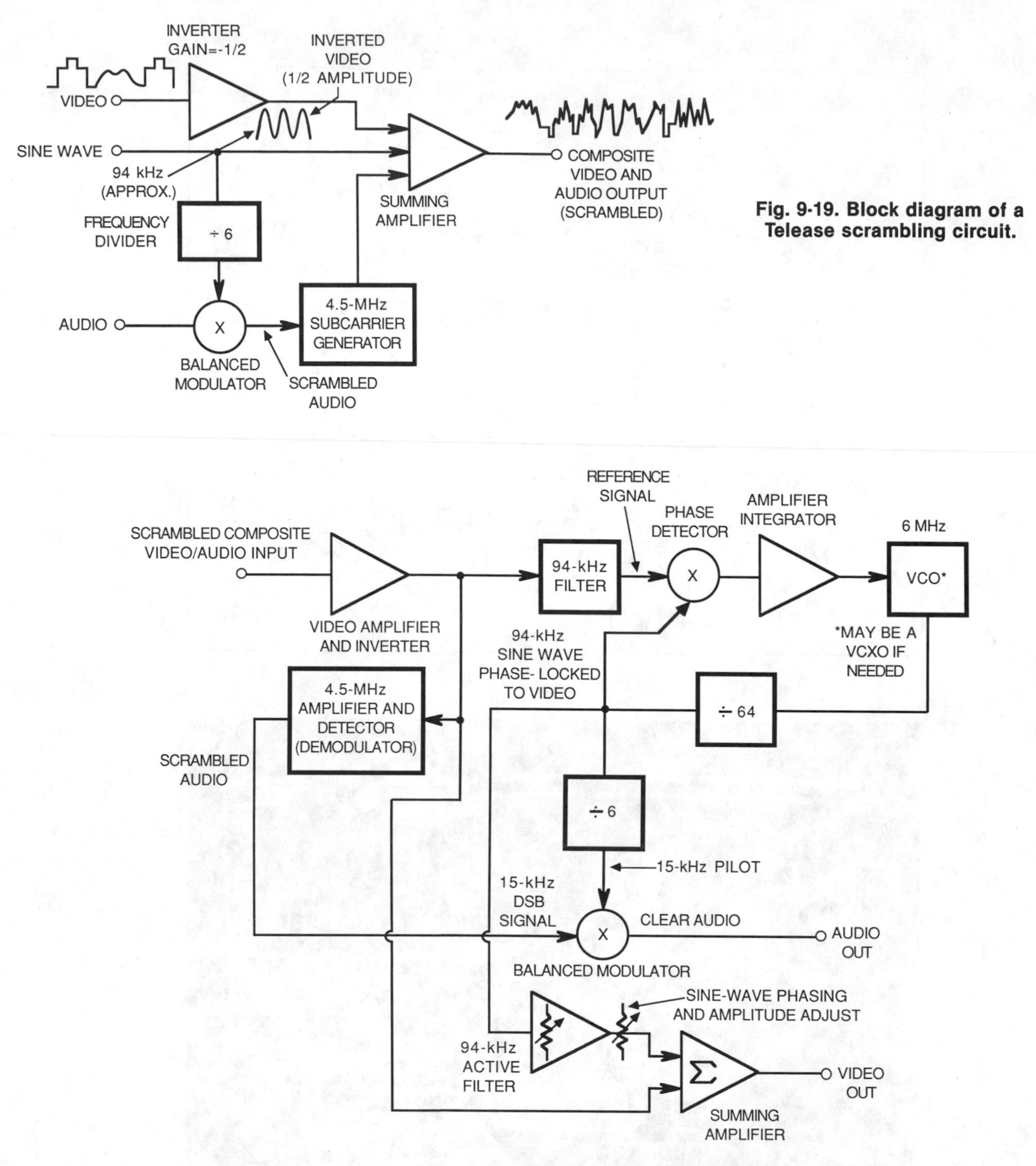

Fig. 9-19. Block diagram of a Telease scrambling circuit.

Fig. 9-20. A Telease descrambling circuit.

to a doubly balanced mixer (sometimes called a product detector). The scrambling signal at 94 kHz is divided by 6 and used as another input to the doubly balanced mixer. The output of the mixer is then clear audio, which is fed either to an audio amplifier or to a TV modulator on Channel 3 or Channel 4 (along with the recovered video signal). In order to get perfect cancellation and no residual bars or ripple in the video output signal, the amplitude and phase of the scrambling sine wave must be matched exactly by that of the recovered sine wave.

For information on the availability of a printed-circuit board and parts, send a self-addressed, stamped envelope to:

North Country Radio
P.O. Box 53
Wykagyl Station, NY 10804

"General" Decoder

This decoder should decode almost any kind of scrambled signal that uses modified (or omitted) synch pulses. Sometimes a signal is transmitted with the video scrambled, by omitting the synch pulses, and, possibly, by using the horizontal blanking-pulse intervals to transmit other information (VideoCipher II™ does this). Possibly, the color-burst information is misplaced. What we want to do in this case is remove the ''junk'' from the scrambled signal. New horizontal and vertical pulses have to be generated. The chroma reference (burst) may have to be restored to its correct place. The video may or may not be inverted on alternate frames or scenes. The entire signal has to be reconstructed from whatever information is present in the scrambled signal.

System Approach

A way to reconstruct the signal would be to remove the distorted (or missing) synch pulses and the misplaced color-burst pulse, and replace them with locally generated signals that are phase-locked to some fixed characteristic of the scrambled signal. What we want to construct is a pulse generator and clamping system to generate pulses and clamping that are all individually adjustable. In this way, the unknown missing components can be locally generated and a ''custom system'' can be set up for various scrambling methods. The idea is that somewhere in the signal there is a reference that can be used.

It is possible to generate a local signal at, say, 3.58 MHz. Then, using a PLL, this signal can be phase-locked to a reference signal on the scrambled video. Remember that 3.58 MHz, or more exactly, 3.579545 MHz was chosen as 445/2, or 227.5, times the horizontal blanking frequency. To phase-lock this signal, a scheme such as the one shown in Fig. 9-21 can be used. Two digital divider chains can be used in a phase detector/VCO setup, as shown.

Reference Recovery

Suppose we know that a reference signal is always present at 4.1164 MHz but there is no 3.58-MHz signal. Dividing the reference by 3.58 (or 3.579545),

$$\frac{4.1164}{3.579545} = 1.15 \text{ (very closely)}$$

Since we cannot divide by 1.15 directly, we could divide 4.1164 by 23. This will be exactly equal to 3.579545 divided by 20. Therefore, $N_1 = 23$ and $N_2 = 20$.

Once we have this signal, the 15.734-kHz and 59.95-Hz (horizontal and vertical) pulse can be derived by using a chip, such as the MM5321 manufactured by National Semiconductor Corp. In this case, it will be necessary to divide the 3.58-MHz signal by 7/4 to obtain the 2.045 MHz needed for the clock input on the chip. Now we can reconstruct the vertical and horizontal frequencies phase-locked to the original signal. However, we still have no way of knowing where the horizontal and vertical synch signals belong on the scrambled signal.

Since the signals are correct in frequency, but still may be out of phase, we can use them as a reference to control a locally generated synch waveform. This can be accomplished by using the recovered horizon-

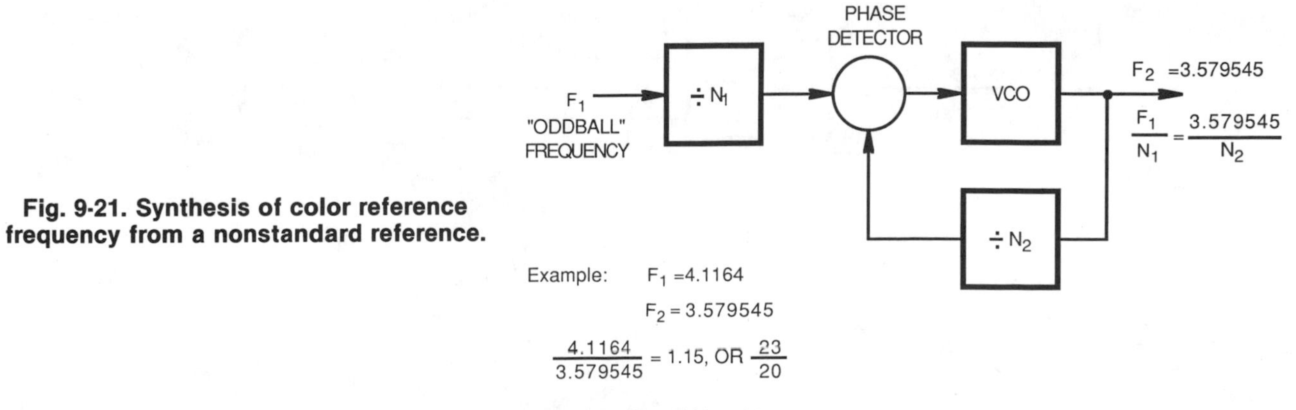

Fig. 9-21. Synthesis of color reference frequency from a nonstandard reference.

tal and vertical frequencies to lock a second synch-generator circuit, whose phase is variable. This can be done as shown in Fig. 9-22. The MM5321 IC has two independent reset inputs. This enables the reset of both horizontal and vertical synch. By using variable delays, the generated synch can be varied in phase. Therefore, we can use these pulses anywhere in the scrambled video, as required, to add or to cancel, as necessary, those components needed to restore the synch pulses. the pulses can also be used to reinsert the color burst in the correct position. We can also generate sampling pulses (phase-locked to the original reference in the scrambled signal) to ''dissect out'' what we want from the scrambled signal, such as the reference signal (we originally assumed we had this), using the ''bootstrap effect.'' A system to decode signals of this nature will now be discussed.

A General Approach

Fig. 9-23 shows a block diagram of a ''general-purpose'' descrambler. It works by taking apart the scrambled signal and then using those parts of the scrambled signal that are necessary to reconstruct (regenerate) the original components of the composite video signal. This descrambler system should work on some of the satellite systems currently in use, but it is a composite taken from various sources and, in no way, is guaranteed to work optimally on any par-

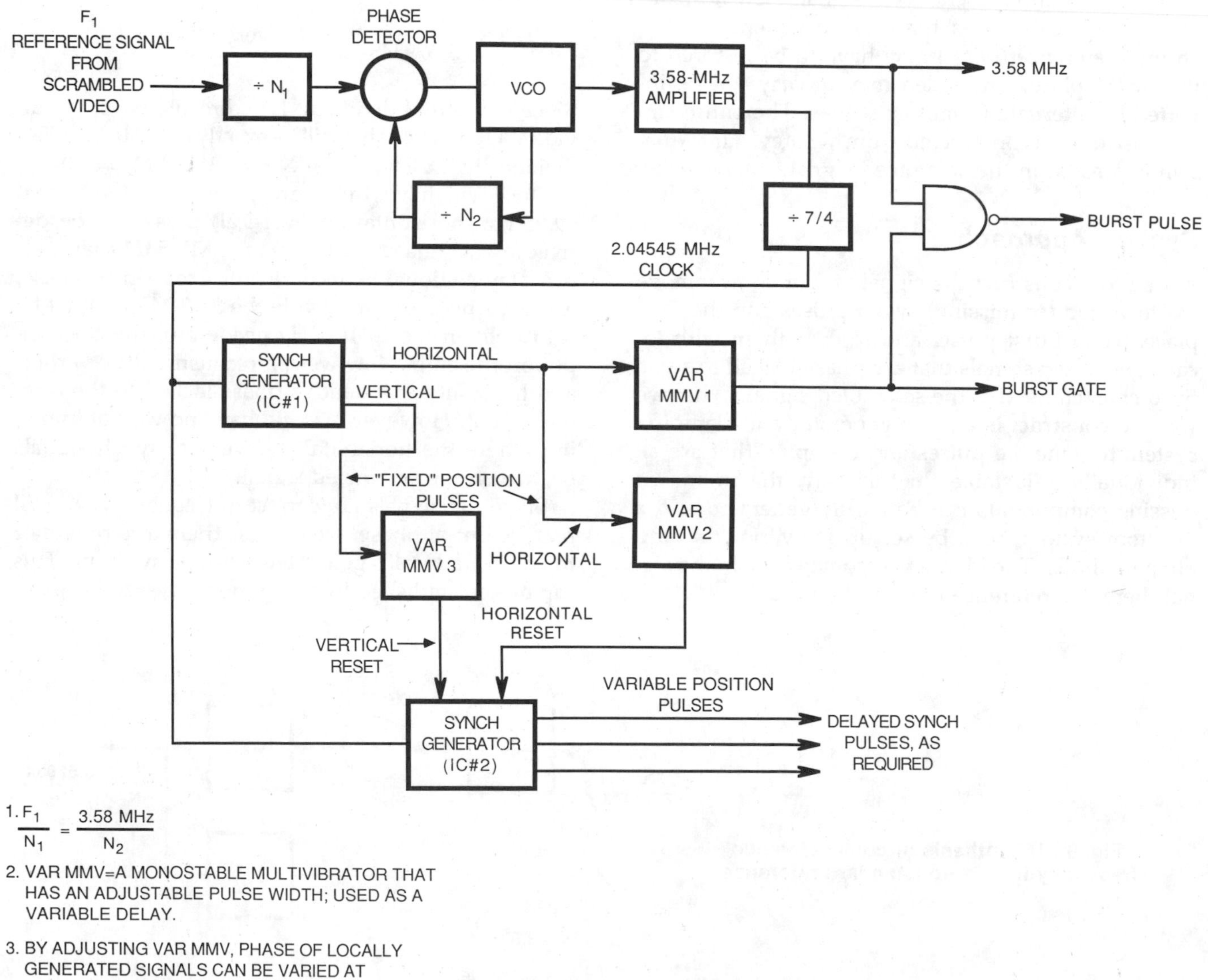

Fig. 9-22. Synch pulse synthesis.

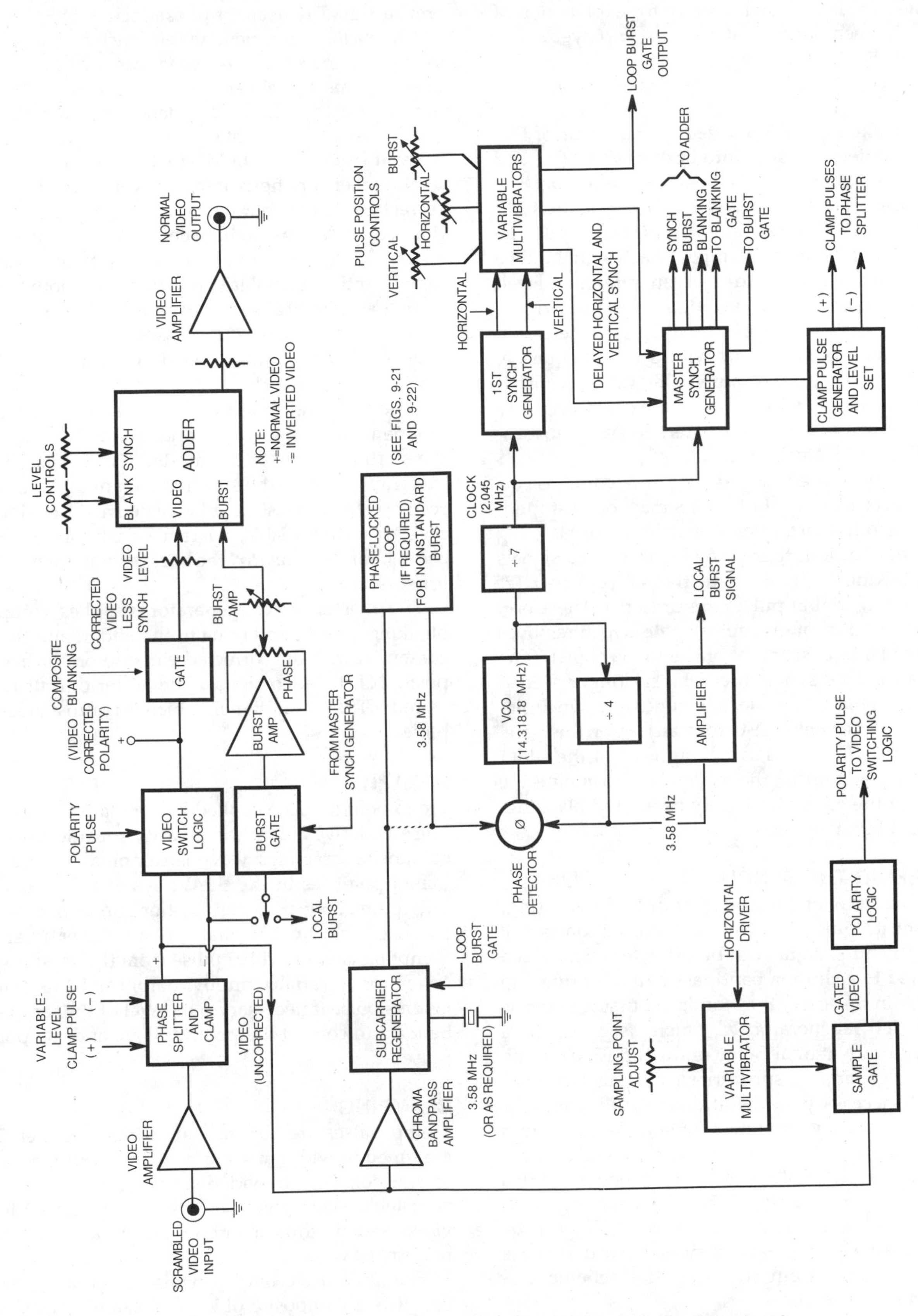

Fig. 9-23. Block diagram of a "general-purpose" descrambler circuit.

ticular system. It is shown merely for explanation of some of the techniques that may be employed.

Circuit Operation

The composite scrambled video at the input of Fig. 9-23 is amplified and split into both positive (+) and negative (−) polarities. The signals differ only in phase—otherwise they are mirror images. In addition, reference levels can be independently set. The two signals are fed into "clamps." Each signal can be independently clamped to a given reference level. This is so that flicker is avoided when polarity is changed. This video polarity may be switched from frame to frame randomly or with a fixed sequence. The two independently clamped video signals are fed into a video switch, which selects the correct polarity pulse. This will be discussed later. For now, just assume the pulse is available.

The output of the video switch now contains correctly polarized video but there may be distorted, missing, or otherwise altered synch in the blanking interval. In addition, there may be unwanted signals in the blanking interval (as in the VideoCipher II™ system, where digital pulses are present). Therefore, the signal is fed through another gate which removes the entire blanking segment of the video signal. (The original scrambled synch interval is no longer needed and will be replaced by a locally generated synch signal that is of normal NTSC format.) After this gate, the corrected video, less synch, appears at the adder input. The function of the adder is to combine the signal from the gate with locally generated blanking, synch, and burst signals.

RECOVERING THE SYNCH

How do we recover the synch and, if it is missing, regenerate it? Since the chroma information is still present, the burst signal can be extracted. Scrambled video is fed to a chroma bandpass amplifier tuned to 3.58 MHz; in this case, the standard burst frequency. A subcarrier regenerator IC, such as a Motorola MC1398P or a National Semiconductor TBA540, can be used to generate a subcarrier from this burst frequency. If necessary, a digital divider/PPL combination can be used to generate a 3.58-MHz signal from the burst signal. Of course, this is not needed if the burst signal is originally 3.58 MHz. We now use this frequency to generate the 15.734-kHz and the 59.94-Hz horizontal and vertical frequencies. This can be done by taking the 3.58-MHz signal and deriving a 2.045-MHz signal from it, using the scheme diagrammed in Fig. 9-21.

In the descrambler, the recovered 3.58-MHz reference signal is used to phase-lock a 14.31818-MHz VCO (*exactly* four times the reference frequency). Dividing 14.31818 by seven produces a 2.045-MHz frequency. This signal can be used to drive a National Semiconductor MM5231 synch generator IC. Two synch outputs, horizontal drive and vertical drive, are derived from the 2.045-MHz signal.

Now there are horizontal and vertical pulses of the correct frequency, but where are they with respect to the phase relationship between them and the original signal? What is needed is a means of obtaining independently adjustable synch phase for both the horizontal and vertical synch (framing). This is done by using a second MM5231 IC as a master synch generator. Whatever phase the derived horizontal and vertical synch may have, the pulses can be delayed in time by using monostable multivibrators that are independently adjustable, and, then, by controlling the starting points of the master synch generator. In this way, we place the synch and blanking pulses correctly on the reconstructed waveform. Also, since the positions are variable, system variations can be accommodated. Similarly, the burst signal can be moved if necessary.

The master synch generator supplies composite blanking, synch, and burst to the adder, and they are combined with the corrected video to produce a composite NTSC video signal. The adder output is a restored NTSC video signal, which is fed to a video amplifier.

POLARITY

Video polarity is controlled by a polarity pulse. This pulse is derived from the polarity sensing logic. The composite scrambled video is sampled at a predetermined point (as in the SSAVI system, for example). This point may be a certain horizontal line, a back porch of the video signal, or any other reference. Sampling is done with a pulse from the master synch generator. A variable multivibrator can be used to delay this pulse if necessary. The level of the sample can be used to control the polarity logic and the polarity pulse.

CLAMPING

Clamp pulses are obtained in a similar matter. They are timed in width and are positioned to clamp a reference point on the video signal to a voltage that is adjustable. This gives the correct DC levels to the video signal, thus avoiding flicker and unwanted brightness variations.

The system we have just discussed is "generic," i.e., it is a composite of several satellite system decoders that have been successfully used. It is a workable system and can be adapted to several different scrambling methods.

A "Universal" Decoder

We now present a block diagram, schematic, and the construction details of still another cable and satellite decoder. It regenerates video that has been scrambled by removing synch, inverting video, and moving the color burst to a nonstandard location. Oak Orion and VCII use this kind of scrambling, as do the SSAVI, gated synch, and similar types of cable-scrambling systems.

The circuit is "generic," i.e., not *exactly* optimized for any one mode of operation. It does work well, but it requires careful adjustment for each mode of operation. Some knowledge of video fundamentals is assumed on the part of the experimenter. This decoder can do *almost anything* that any other "box" which we have seen can do. However, it will *not* decode encrypted audio. As mentioned previously, no experimenter has at the time of this writing (June 1986) decoded VCII™ audio, as far as we know.

Block Diagram

The block diagram (Fig. 9-24) of this "universal" decoder shows basically how the decoder system works, and illustrates the signal processing and flow from input to output.

The video input, terminated by a 75-ohm resistance, is coupled through a 100-μF capacitor (refer to Fig. 9-25) and a 2.2K resistor to the input of IC1, an LM733. The LM733 is a differential video amplifier. Normal and inverted video appears on pins 7 and 8, depending on the input polarity.

In order to correct information removed by the scrambling process, a DC clamper is required. This prevents the video from flickering when video scene changes take place. A good DC clamp requires a low-impedance source feeding a high-impedance load. The LM733 video amplifier has a low (20 ohms) output impedance, and the following LM318 op amp has a high (> 10,000 ohm) input impedance.

Both IC2 and IC3 are LM318 operational amplifiers that are set for proper gain, bandwidth, and frequency adjustment, and having high-impedance noninverting inputs. Clamping occurs at pin 3 of the LM318s with transistors Q1 and Q2 being the clamping switches. IC18 (Fig. 9-26) provides a clamp pulse through 560-ohm resistors to the bases of transistors Q1 and Q2 (Fig. 9-25). The clamp position control at IC17 (Fig. 9-26) is used to adjust the timing of the clamp pulse. The DC level of the clamp is adjusted by the clamp level controls.

Both normal and inverted video are present at the outputs (pin 6) of IC2 and IC3. These outputs are gain, frequency, and DC level matched. The video from IC2 and IC3 are identical—except for polarity.

IC4 is used as a 3PDT analog switch, capable of switching video at high speeds, with section C of the

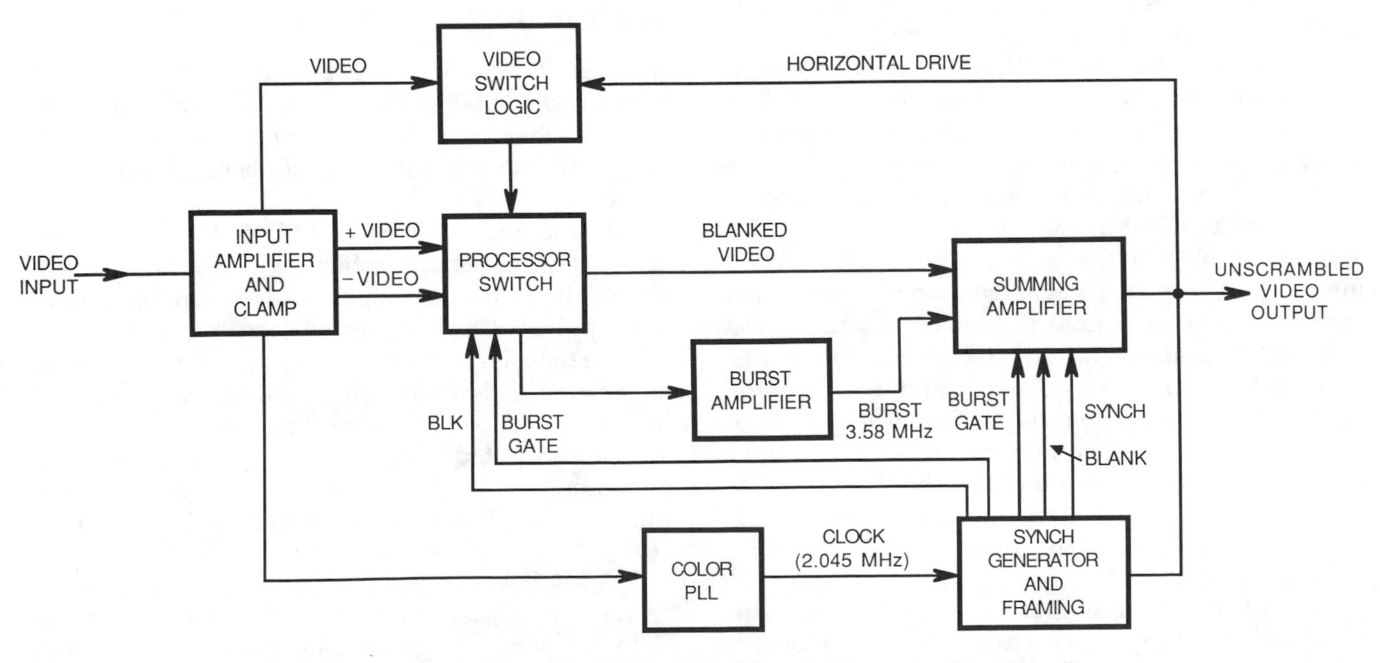

Fig. 9-24. Block diagram of a "universal" satellite and cable decoder.

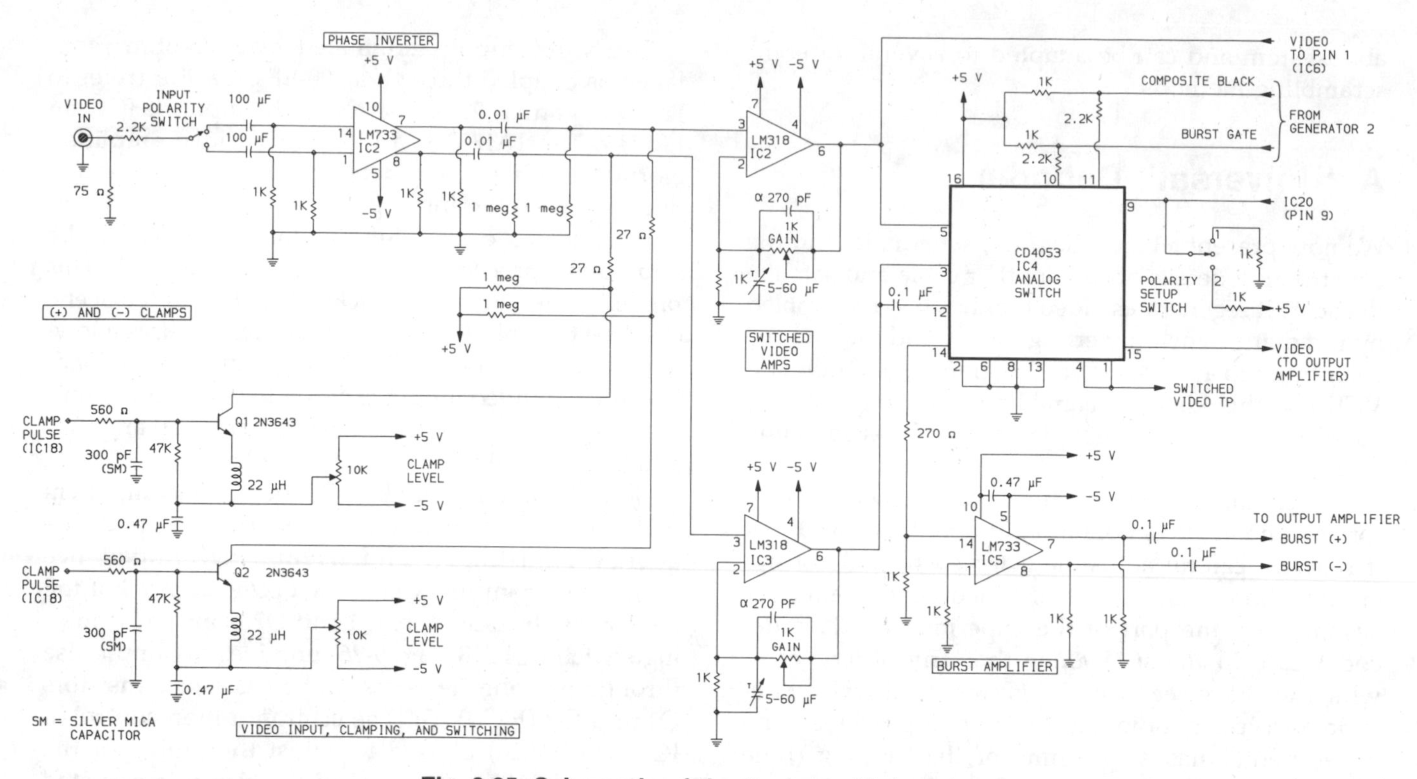

Fig. 9-25. Schematic of the "universal" decoder.

switch being used to select the correct video polarity. Pins 3 and 5 are inputs to the switch; pin 4 is the output. The control voltage for switch section C comes from pin 9 of IC20. A manually operated toggle switch, with the common terminal connected to pin 9 of IC4 is shown. This is a center-off switch, which will always be OFF for normal operation. The two ON positions (1 and 2) are for setup.

The signal at pin 4 of IC4 always has the correct-polarity video signal during normal operation. The signal at pin 4 still has the information in horizontal and vertical blanking intervals. Video from pin 4 goes to one input of video switch section B, while the other input to section B, pin 2, is grounded.

The control input for section B (pin 10) is blanking information from synch generator No. 2. This switch section returns to ground in horizontal and vertical intervals. This switch action removes the data pulses present between the scan lines and between the frames. The B switch output (pin 15) is blanked video. This video has no synch information, color burst, or data signals. Pin 12, one input to the A switch section, couples video from pin 6 of IC3 via a 0.1-μF capacitor to one input. The other input, pin 13, is grounded.

Pin 11 is the control input to switch section A and is fed with a burst gate pulse from generator 2. Switch A clamps everything to ground expect the color burst. Color burst appears on pin 14 and is coupled to pin 14 of IC5, the color burst amplifier. The burst outputs of IC5, an LM733, are pins 7 and 8 with (+) and (−) burst phasing, respectively.

PLL Circuit

A system is shown in Fig. 9-27 that can lock to the color burst component of the video signal. The video can be either normal or inverted, clamped or unclamped. Also, the video DC reference level does not matter.

Integrated circuit IC8, an LM318, functions as an active chroma bandpass filter. Video from IC2 (not shown) is coupled through two 270-ohm resistors to the input of IC8. The output from pin 6 of IC2 is chroma (or color) information. There is a chroma level adjustment so that the correct chroma level can be fed to IC9, an MC1398 color-processing IC. A ''free run'' switch, connected to pin 5 of the MC1398, is used to temporarily remove the chroma so that the oscillator can run free. This will be covered in the set-up instructions.

The MC1398 integrated circuit (IC9) is a color processor. This IC takes the chroma input as a reference from the loop burst gate IC14 buffers (Fig. 9-26), and produces a continual 3.58-MHz CW output at pin 13,

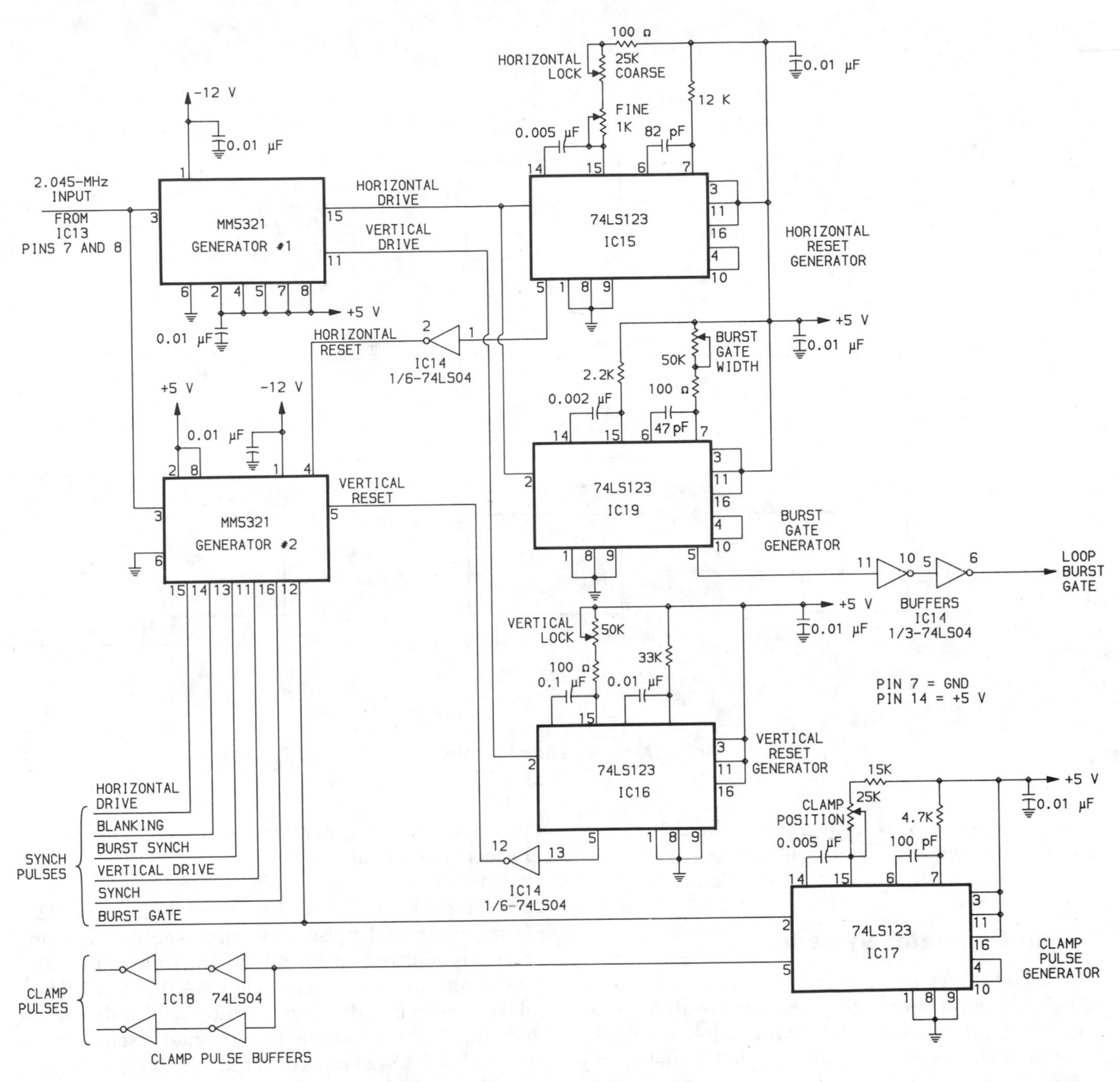

Fig. 9-26. A synch generator system.

phase-locked to the burst component of the chroma input signal. Then, IC10, an LM733, is used as a 3.58-MHz chroma amplifier. The 3.58-MHz (+) and (−) phases appear at pins 7 and 8. The (+) phase is used as a reference input to IC11.

IC11, an NE564 phase-locked loop IC, is designed for a free-running frequency of about 14.3 MHz by the adjustment of a 3–13 pF trimmer capacitor and the LOOP GAIN control. The 14.3-MHz signal is then fed to IC12, a 74LS74, which divides-by-2 to obtain 7.16 MHz. The 7.16-MHz signal is again divided-by-2 to obtain 3.58 MHz. This 3.58-MHz signal is fed to pin 3 of IC11, the NE564 PLL. The 14.3-MHz oscillator is phase-locked to the burst of incoming video in normal operation. When the loop locks, the exact frequency will be four times 3.579545 MHz, or 14.3181818 MHz.

The 14.31818-MHz signal is fed to pin 14 of IC13, or 74LS90 IC. IC13 divides-by-7 to supply a 2.04545-MHz signal, phase-locked to color burst, to clock the

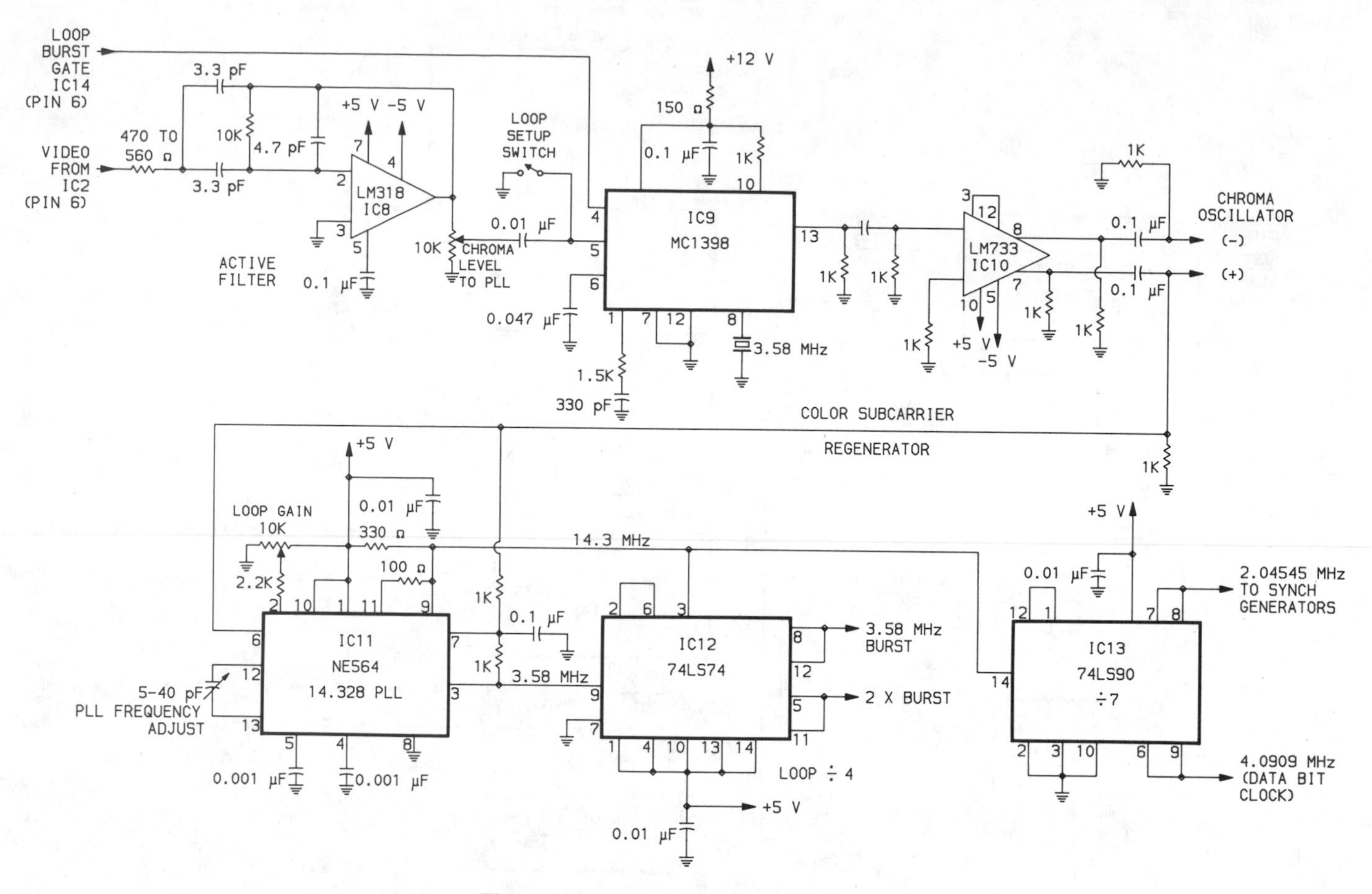

Fig. 9-27. Schematic of a color PPL.

two synch generator chips (generator 1 and 2), which are MM5321 synch-generator IC devices.

Synch Generator System

The circuit shown in Fig. 9-26 produces the necessary synch and pulse information needed to feed the input processor, the video switch, and the output amplifier, so a standard NTSC composite video signal is obtained.

The 2.04545-MHz clock signal from IC13 is fed to pin 3 of generator 2. This generator is an LSI MOS chip and provides all synch functions. These synch functions are identical to those used in standard NTSC nonscrambled video signals.

Generator 1 is a lock-up generator for resetting generator 2. Remember, this system is locking to color information. After lock occurs, generator 1 does not know exactly where the horizontal and vertical intervals start. Therefore, horizontal and vertical drive pulses from generator 1, which is locked to color information, are used to generate adjustable reset pulses for synch generator 2. This provides an adjustable horizontal reset.

The horizontal drive from pin 15 of synch generator 1 is fed to pin 2 of IC15 and IC19. IC15, a 74LS123, is the horizontal pulse generator. The system horizontal is adjusted by the 20K control from pin 15. The pulse on pin 5 is inverted via IC14. IC14, a 74LS04 device, is an inverter (pins 1 and 2) which drives the synch generator 2 at pin 4 (horizontal reset).

IC19, a 74LS123 device, is the loop burst gate generator. Its pin 5 output is buffered by two sections of IC14 (via pins 11, 10, 5, and 6). The resistor at pin 11 is used as a circuit board jumper, and is not critical. The buffered burst gate pulse from pin 6 of IC14 is routed to pin 4 of IC9 (Fig. 9-27). The burst gate width can be set by the 50K trim potentiometer connected to pin 7, IC19.

The vertical reset pulse generator is IC16, a 74LS123 device. The vertical drive from pin 11, generator 1, is connected to pin 2 (IC16), and a reset pulse appears at pin 5 of IC16. The pin 5 output is inverted by a section of IC14 (pins 12 and 13). The system vertical lock is adjusted by a 50K potentiometer con-

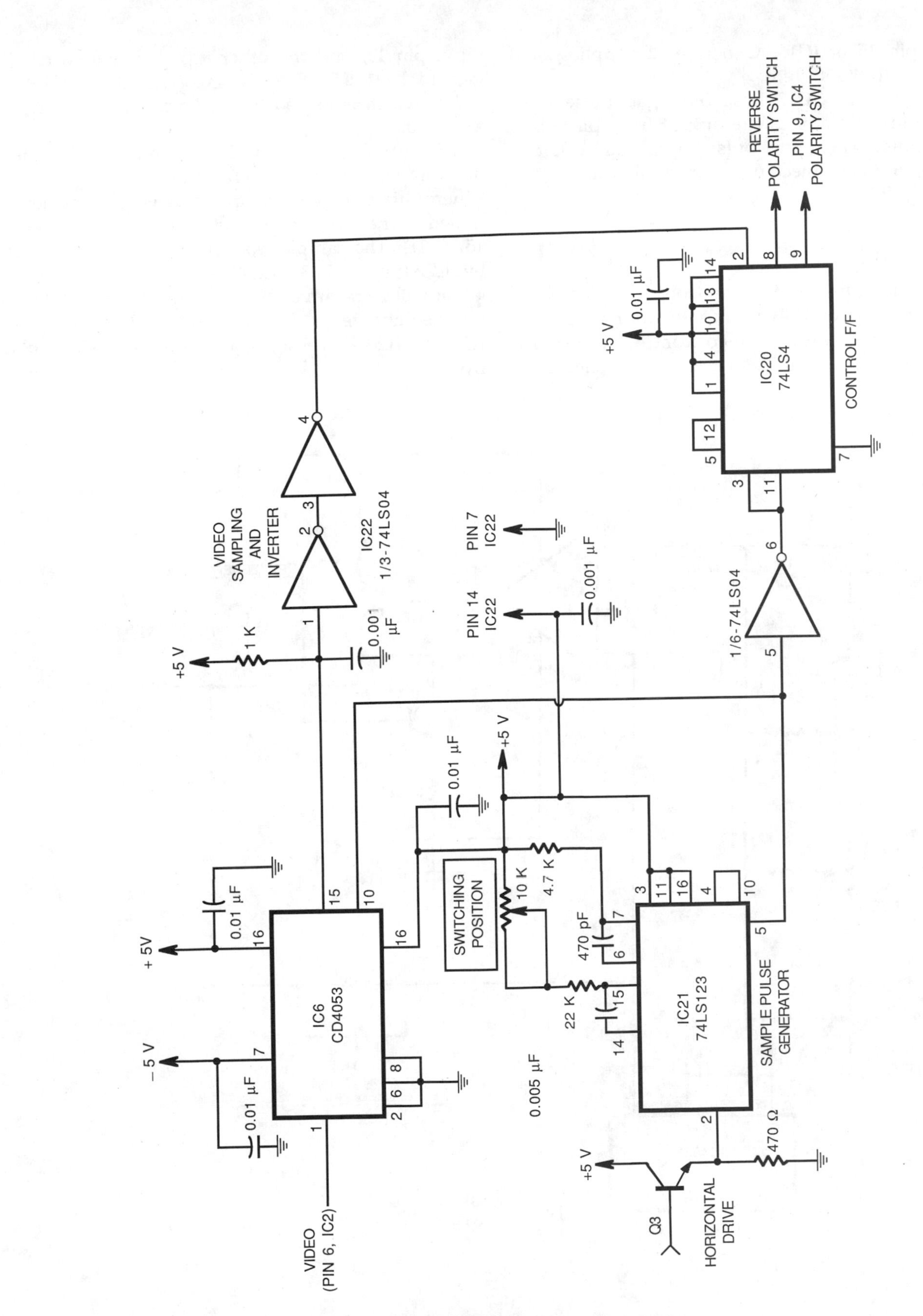

Fig. 9-28. Video polarity switch logic.

nected to pin 15 of IC16. Generator 2 supplies all synch pulses to the system.

A 74LS123 device (IC17) has a burst gate pulse fed to pin 2 from generator 2; the output from pin 5 is the clamp pulse. This clamp pulse is buffered by IC18 (a 74LS04), and is connected to clamp transistors Q1 and Q2 (Fig. 9-25).

Video Polarity

The circuit shown in Fig. 9-28 controls the video polarity switch (part of IC4) to supply normal or inverted video, as necessary. Video from IC2 is fed to pin 1 of IC6. IC6, a CD4053 switch, is controlled by input pin 10, and the other input of this switch is ground. IC21, a 74LS123 device, generates the control pulse, which is locked to the horizontal drive from generator 2.

The output of IC6, pin 15, is a sample of the video or a sample blanking edge from the scrambled input. Where this sample or sampling takes place is determined by the 10K potentiometer switch setting (at pin 15, IC21). The sample from IC6 (pin 15) is buffered by IC22 (pins 1, 2, 3, and 4). A pulse appears on pin 4 that will be positive (on) when the receiver video is inverted and zero (off) when the receiver is normal. IC20, a 74LS74, is reset by the video sampling pulse from pin 4 of IC22.

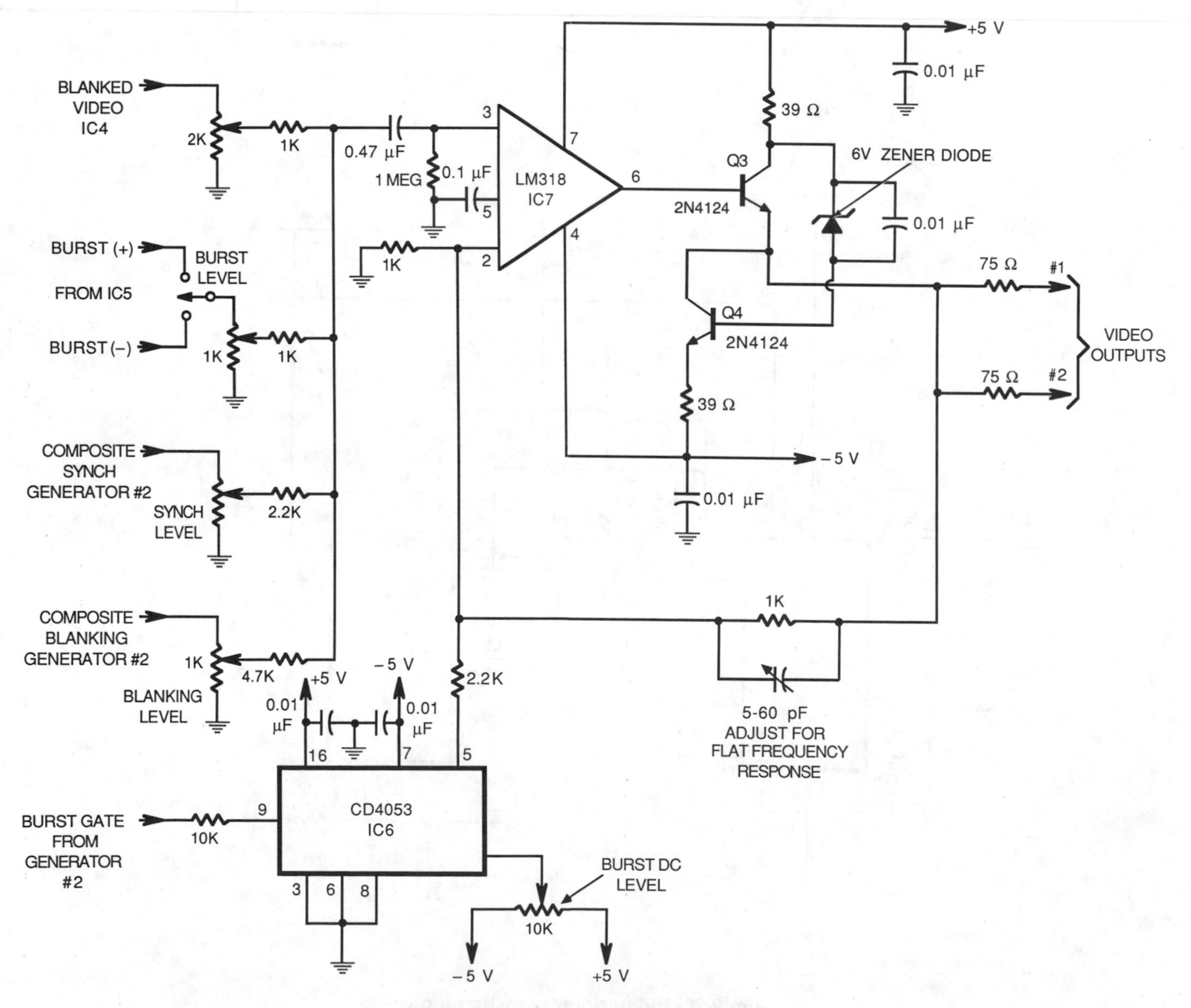

Fig. 9-29. Video output amplifier.

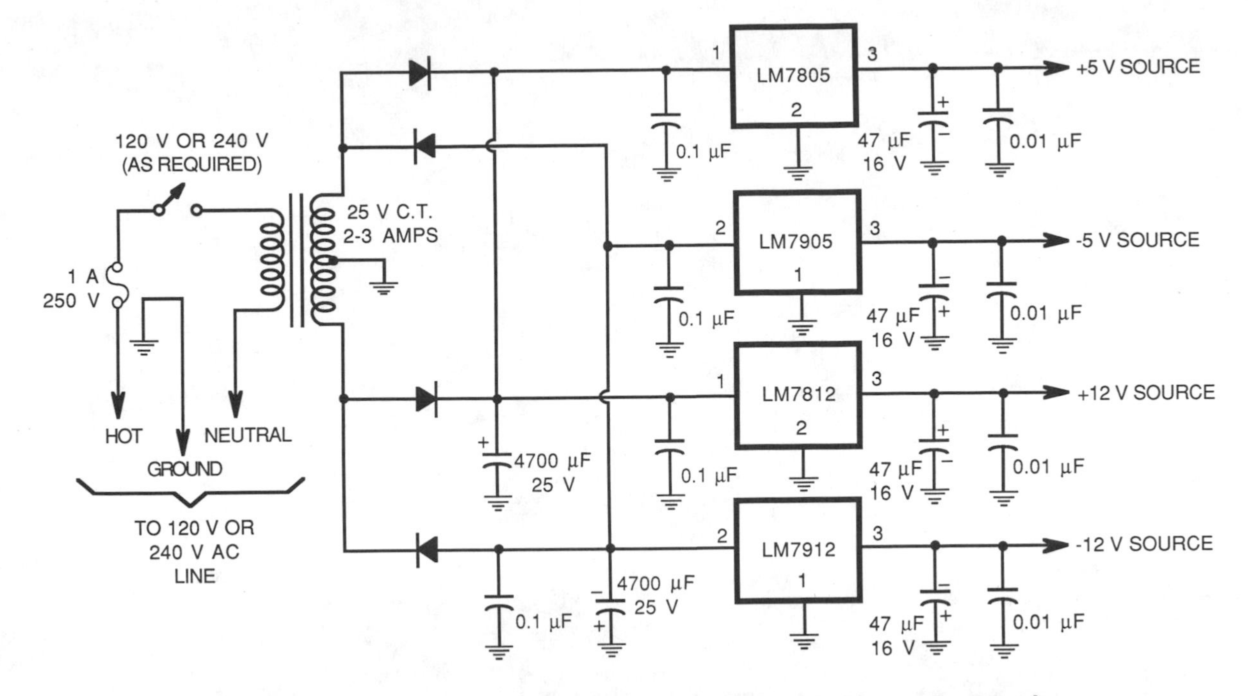

Fig. 9-30. Circuit for the "suggested" universal descrambler power supply.

The control pulse from pin 5 of IC21 is inverted by a section of IC22 (pins 5 and 6), and is used to clock the control flip-flop, IC20. When the video sample is positive (on) at the same time that the clock is changing from low to high, pin 9 of IC20 will be the correct voltage (high or low) to achieve the correct polarity (IC4).

Video Output

The circuit shown in Fig. 9-29 shows a resistor network made up of fixed and variable elements connected to the high-impedance input of an LM318 op amp, IC7.

Blanked video from IC4, a negative (−) burst from IC5, and composite synch from generator 2 (and composite blanking) are combined at the proper levels at pin 3 of IC7. Transistors Q3 and Q4 are low-impedance output drivers which buffer the output of IC7 (pin 6).

A section of IC6 is used to restore the correct DC level to the color burst. The burst gate signal from generator 2 is connected to control pin 9, with the output (pin 5) being routed to pin 2 of IC7 through a 2.2K resistor. The 10K burst DC level control is used to set the color burst to the correct DC position. Two undistorted video outputs are then available.

Power-Supply Considerations

Fig. 9-30 shows the power supply. It uses half-wave rectification, with output voltages of −12 V, +12 V, −5 V, and +5 V. The transformer is 12-volts, center tapped. A 2-ampere transformer is more than adequate.

All IC devices should be adequately heat sinked. They may run warm (but not too hot to touch).

Space limitations do not permit the inclusion of the component layouts and detailed tune-up and alignment procedures that are necessary for proper operation of this decoder. This data is included with the PC board set for this decoder, which is available from:

North Country Radio
P.O. Box 53
Wykagyl Station, New York 10804

Send SASE for more details.

10

Commercial Satellite Encryption Systems

VideoCipher II™

VideoCipher II™ is the current system being used by HBO and Cinemax; and it is seriously being considered by other satellite programmers. The system contains several capabilities not included in the previously described digital scrambling systems. (Fig. 10-1 shows the M/A-COM VIDEOCIPHER Model 2000E satellite descrambler.) Briefly, the VideoCipher II™ (Fig. 10-2) has the following features.

1. Video encryption. This is done by inverting the video polarity and moving the color burst level to a nonstandard position on the waveform.
2. An 88-bit data stream and a color burst (3.58 MHz), in place of a synch pulse, is inserted in the waveform. No 4.5-MHz sound carrier is needed.

The 88 data bits are used for 2-channel digital audio, program control and billing, synch information, security features, and as an auxiliary data channel.

The system is similar in principle to the previously described satellite system, which uses a 24-bit code for synch and audio, except that the 88-bit code of the VCII provides much more flexibility in the areas of audio, control, and security. The 88-bit data stream contains Data Encryption Standard (DES) encrypted audio data. The two audio channels are filtered and digitized. Each digital sample has a random binary sequence and error-coding bits added to it. These are generated by the DES algorithm. Audio bits encrypted this way appear completely random. The error-coding scheme enables the descrambler to correct all single-bit errors, while double-bit errors are corrected by interpolation. (For example, if one segment of a waveform known to be ''smooth'' has large impulses present, these are ignored, and a voltage that is midway between the previous sample and the next sample is instead inserted to smooth things out.) This improves the audio signal-to-noise ratio for all RF signal levels and ensures better quality. The descrambler must have the appropriate DES key to decrypt the audio. The lack of a 4.5-MHz sound carrier gets rid of the serious interference often seen on NTSC TV receivers (sound bars, beat interference, etc.), and allows the total transmitter power to be used for video. Up to a 2 dB of S/N improvement is claimed for the video system, at all signal levels.

Control and Customer Addressing

Security and control of the system are also encrypted. As previously mentioned, the DES keys are 72 quadrillion in number, and without the proper key, that many combinations must be tested by trial and error. A multilevel key organization and distribution method is used. Each descrambler has a unique address and a number of DES keys contained within it. (It is not, of course, constructed with an easily probed ROM or other such device.)

In order to receive scrambled programming for a given period (e.g., one month or one billing cycle), the individual descrambler must receive a message from the satellite containing the monthly key and service status (tiering, credit). This message is transmitted over the control channel to each individual descrambler, and is encrypted with that descrambler's key. A specific descrambler stores this information, and no other descrambler can use the information since it is encrypted with that descrambler's key.

Individual programs are encrypted with different keys. The program key is combined with program tiering, rating, and cost information, encrypted with the monthly key, and then broadcast over the control

Fig. 10-1. The M/A-COM VIDEOCIPHER Model 2000E satellite descrambler—for subscriber use. (Courtesy M/A-COM LINKABIT, Inc.)

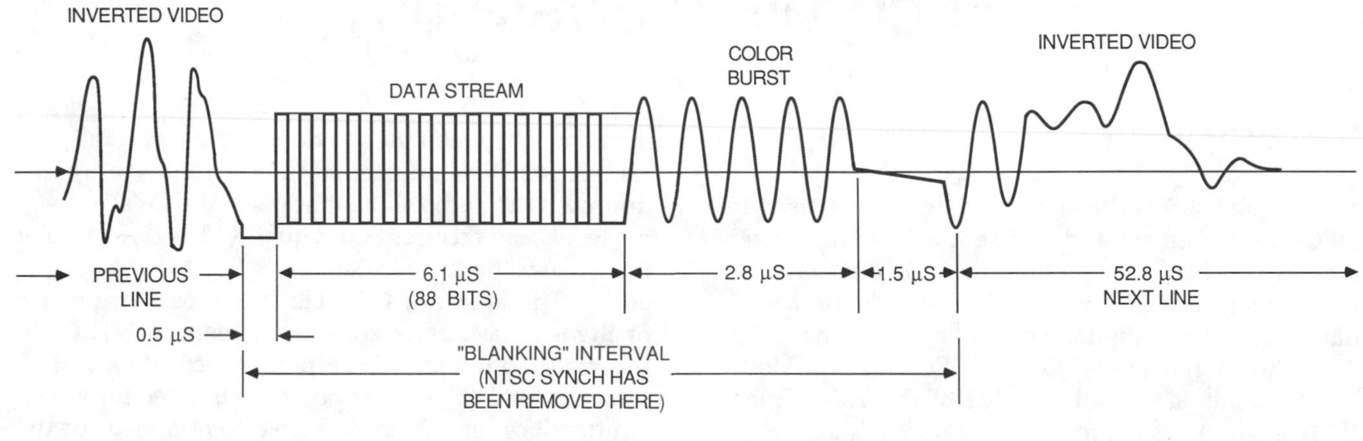

Fig. 10-2. VideoCipher™ horizontal format.

channel to all descramblers. Only descramblers having the monthly key can receive the program.

Monthly, before each billing period, messages are transmitted to authorized descramblers. By using a different key for each program, a single transmission can automatically authorize and de-authorize many descramblers at once, according to the program schedule. New subscribers can be authorized, and those who do not pay their bills can be turned off, and, in addition, subscribers requesting different programming can be activated or deactivated. All told, up to 600,000 descramblers per hour can be addressed. Control messages are error-checked and, if not received correctly, are ignored until the next message. Also, the descrambler has nonvolatile memory to guard against power failures.

A stolen descrambler can be deactivated, and it cannot be reused until activated from the satellite. Descrambler addresses and descrambler keys are DES encrypted. If they are discovered and decrypted, they can simply be replaced with others. Since each descrambler has many unique keys stored in it, no modification of the descramblers is required.

Other Features of VCII

Fifty-six tiers of programming, in any combination, can be accommodated using the VCII. This information is in the monthly key message. Therefore, each descrambler can only receive programming in tiers for which it has been authorized.

Using the tiering feature, the blacking out of programs in certain areas of the country can be accomplished simply by placing them in tiers not authorized for descramblers in that area. Each descrambler can be given coordinates based on the Postal ZIP code system. Then a programmer could simply select an area of coordinates to be blacked out and all the descramblers in that area would be deactivated for that program. Programs with ratings of PG, X, R, etc., could also be tiered, and the unauthorized viewing by children could thus be prevented.

Teletext, personal messages, dual language capability, and emergency messages can also be handled by the data system in the VCII. VideoCipher II™ may very well be the standard satellite decoder system of the future.

The VideoCipher II™ currently has a price of $395. Of course, this does not include monthly fees for programs. This fee question may determine the success or failure of the system, as well as the whole TVRO market's success or failure. Currently (June 1986), there are not too many descramblers in use among the general public.

B-MAC System

Scientific Atlanta has developed a system known as "B-MAC." This is a new transmission format (MAC stands for Multiplexed Analog Components) that makes use of time-division multiplex (TDM) of analog luminance and chrominance components. This system has several technical advantages. For satellite transmission methods, using FM, the NTSC signal spectrum does not make the best use of the FM channel from a signal-to-noise standpoint. A saturated color in a televised scene may over-deviate the transmitter. This causes "sparklers" to appear in these areas of the picture when low-cost satellite receivers, using less than full video bandwidth, are employed. Also, the S/N ratio is poorest for chroma due to the noise distribution in the FM channel. The B-MAC systems uses TDM techniques to circumvent this inherent difficulty.

Basic Theory

Time-division multiplex is the principle of sending portions of several information channels in a timed sequence. For example, a TDM system can send the information content of an NTSC signal in several time-sequenced groups. Synch can be sent, followed by the chrominance information, and, then, the luminance. The data can be stored, combined, and converted to video. In addition, multilevel data (security, addressing, stereo audio, teletex, etc.) can be sent along with the synch information. NTSC signals are frequency-division multiplex (FDM) in nature. All data are sent at the same time (chrominance, luminance, color burst, audio, etc.) on different frequencies; i.e., on the picture carrier, color subcarrier, and sound subcarrier. Since subcarriers can be eliminated in a TDM system, cross-talk and intermodulation between components is not a problem, since the components are not present at the same time.

Since a same rate of information transmission (the TV program) is required, the fact that the sequential transmission of the various components is necessary requires time compression of the components (Fig. 10-3). On a satellite channel, the bandwidth necessary to do this is available. *Time compression* is an exchange of time for bandwidth. For example, if we play a record twice as fast, all audio is doubled in frequency, and the record takes just half the time to play. This process is also used in tape duplicating to speed up production. It should be an obvious concept to anyone who has ever played a record or tape at a faster speed than the speed it was recorded at.

Another benefit of B-MAC is that since the chrominance signal is at baseband, the noise performance of the chroma channel is improved. By transmitting only one chroma component per line, either (R-Y) or (B-Y), chroma transmit time can be reduced 50%. This makes the storage of chrominance information necessary at the receiver, either digitally or using analog methods. The chrominance signals are filtered to restrict bandwidth to 2 MHz or less. The (R-Y) and (B-Y) components are band limited, and alternate data samples are discarded to reduce the data rate to about 7 samples per microsecond. A data rate of 14.31818 MHz is used, since it is better suited to NTSC translation (4 times 3.579545). For digital video, a clock rate of 13.5 MHz has been recommended for studio use of component-coded digital video. This, after dis-

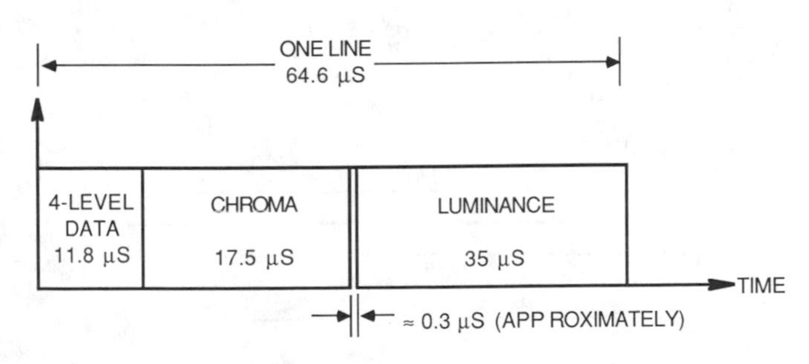

(A) B-MAC horizontal line.

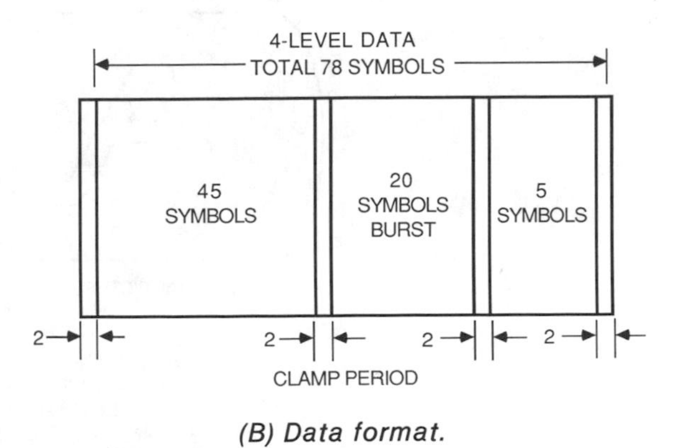

(B) Data format.

Fig. 10-3. Time compression of video components.

carding alternate samples, gives 7.16 samples per microsecond. The (R-Y) and (B-Y) samples are stored in 384-byte memory. By reading out the memory at 7.16 MHz × 3 = 21.48 MHz, the signals can be time compressed to 17.5 microseconds. Using a similar method, the luminance signal is compressed to 35 microseconds. This leaves about 11½ microseconds for the data.

The data rate used is 1.86 megabits per second for the data, audio, and synch. The data pulses are 2- or 4-level symbols during the blanking intervals. Actually 1.573 megabits/second are provided during the horizontal blanking intervals, and the six digital audio channels take 1.510 megabits per second. The remaining 62.5 kilobits are used for a utility data channel. This data channel is encrypted and controlled by the broadcaster for utilization by each user. Unused audio channels can be used as data channels. The six audio channels are Dolby®* digital audio and use 251.7 kilobits/second, including error coding. The frequency response is 20 Hz to 18 kHz at a 30-dB bandwidth. The audio channels can be encrypted and decrypted separately.

The vertical interval contains all the control data which is synchronized with the 4-level data in the horizontal lines. The first lines (1 through 8) carry control data for synch and for clock data recovery. The synch is only on one line in the vertical blanking interval and allows receiver lock-up at only 1-dB carrier-to-noise (c/n) ratio of the received signal. Lines 9 through 13 contain teletext information, with 40 ASCII characters per each line.

*Dolby is the registered trademark of Dolby Laboratories, Inc.

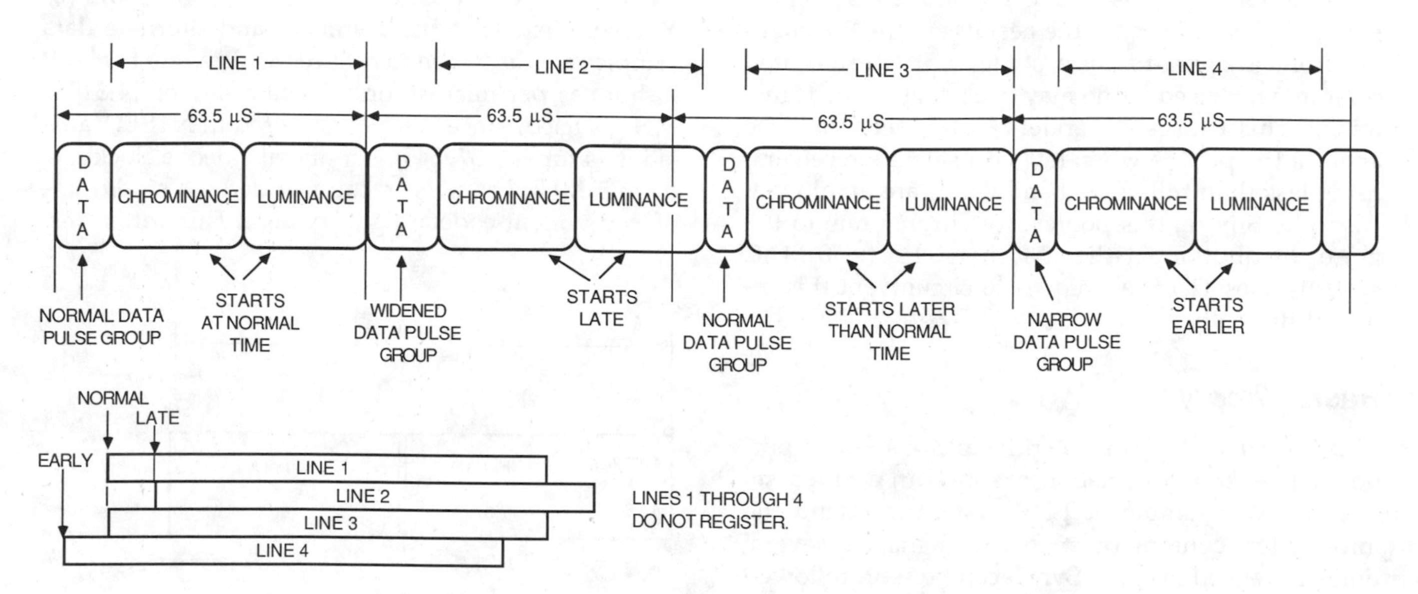

(A) Data, chrominance, and luminance pulses.

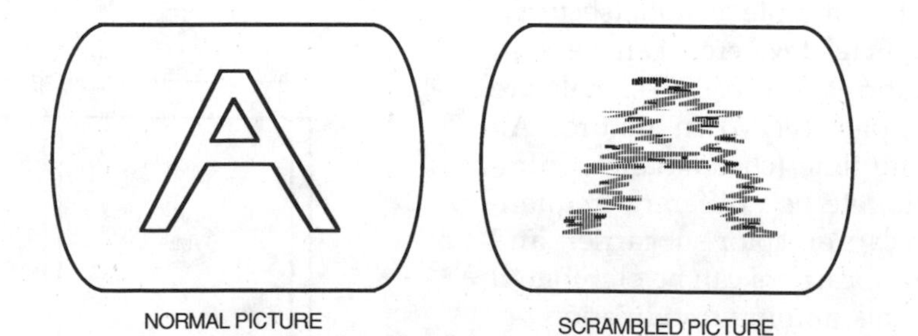

(B) Normal or scrambled picture.

Fig. 10-4. B-MAC video scrambling.

B-MAC Modes of Operation

The B-MAC system also provides for up to 256 million addresses, or about 1 million per hour, with redundancy. Decoders contain multiple addresses for independent programmers. Audio and data can be DES encrypted, with keys and codes changing four times per second.

Video scrambling is accomplished by a time-shifting process on each line. The width of the digital data packet can be varied, sending more data on one line than on the preceding line (or less). This either delays or advances the start of each line, producing a slewing of the individual lines (Fig. 10-4). No data is omitted—it is merely sent in advance or saved for later. This effectively scrambles the picture. Descrambling is done by reversing this process. Since merely the transmission time is shifted, no loss in picture quality occurs.

The B-MAC system has a technical edge on noise performance. The threshold of the system is defined as the c/n ratio at which the FM demodulator used in the system produces 100 ''clicks'' or cycle slips per second, with a 24-MHz BW IF system (± 1 dB). Above threshold, the B-MAC chrominance signal is 8.3 dB better than the B-NTSC. Audio performance on three of the six audio channels is essentially perfect at c/n ratios down to 6.7 dB, with a total loss at about 4.7 dB.

Conclusion

The B-MAC system has been selected in Australia for program distribution to TV stations and, later, to individual homesteads in remote communities. This system is an alternative to VCII™ and has technical merit with regard to color quality. It may find wide acceptance in the future, as well.

11

Satellite TV and Interference Considerations

Along with the growth of satellite TV, the use of satellites for various auxiliary purposes has become commonplace. It is feasible to use a few tens of watts and a moderate antenna system and establish reliable satellite communications. Nowadays, uplink terminals can easily be placed in a van or small truck-type vehicle, since high power is unnecessary and high antenna gains can be readily achieved at the microwave frequencies used for this purpose. Fig. 11-1* is a nomogram curve for showing the gain of a V_S size parabolic dish. Note that the gain is also proportional to frequency. These curves are theoretical and there is some loss due to incomplete feed illumination and inevitable parabolic surface errors.

> *Note:* The figures about to be given are illustrative only and may or may not be what is used or can be used in a practical situation. Note that program suppliers use generally larger dishes than does a home TVRO installation. Transmitter power is high enough to ensure reliability at all times. Therefore, powers considerably in excess of the figures quoted in this discussion could be used.

Examining the nomogram, you will note that at the satellite downlink frequency of 3.7–4.2 GHz, a 10-foot dish has a gain of about 40 dB and a beam width of a little less than 2°. At the uplink frequency, the gain would be about 2 dB greater and the beam width would be slightly narrower. The example shown in Fig. 11-1 is for a frequency of 3000 MHz, a gain of 32 dB, and a reflector diameter of 6 feet.

---

*From *Reference Data for Engineers: Radio, Electronics, Computer, and Communications,* Howard W. Sams & Co., Indianapolis, IN, p. 33-22.

Calculations

If we know the receiver antenna gain, receiver bandwidth, and noise figure, and also know the transmitter power and transmitter antenna gain, we can theoretically calculate the range of a radio or TV link, and the S/N ratio at the receiver, by using a few simple equations.

The equation for free-space path loss between two points is:

$$\text{dB path loss} = 37 + 20 \log_{10} f_{MHz} + 20 \log_{10} D \text{ miles}$$

As an example, at 6 GHz, the path loss of a geosynchronous satellite orbiting 22,500 miles above the earth is:

$$\begin{aligned} \text{Path loss} &= 37 + 20 \log_{10} (f_{MHz}) \\ &\qquad + 20 \log_{10} (\text{Distance in miles}) \\ &= 37 + 20 \log_{10} (6000) + 20 \log_{10} (22500) \\ &= 37 + 75.56 + 87.04 \\ &= 199.6 \text{ dB (approximately)} \end{aligned}$$

At 4 GHz, it would be slightly less, or

$$\begin{aligned} \text{Path loss} &= 37 + 20 \log_{10} (4000) + 20 \log_{10} (22500) \\ &= 37 + 72.04 + 87.04 \\ &= 196.1 \text{ dB (approximately)} \end{aligned}$$

Let us calculate a typical satellite downlink. Taking typical figures, assume an average effective radiated power (ERP) of 36 dBw (dBw = decibels with respect to 1 watt). The effective transmitter radiated power is 3981 watts (or about 4000 watts) which is 36 dBw. If the satellite transmitter power is 8 watts, or 9 dBw, the antenna gain would be 36 − 9, or 27 dB. This is arrived at by subtracting the value of the TX power

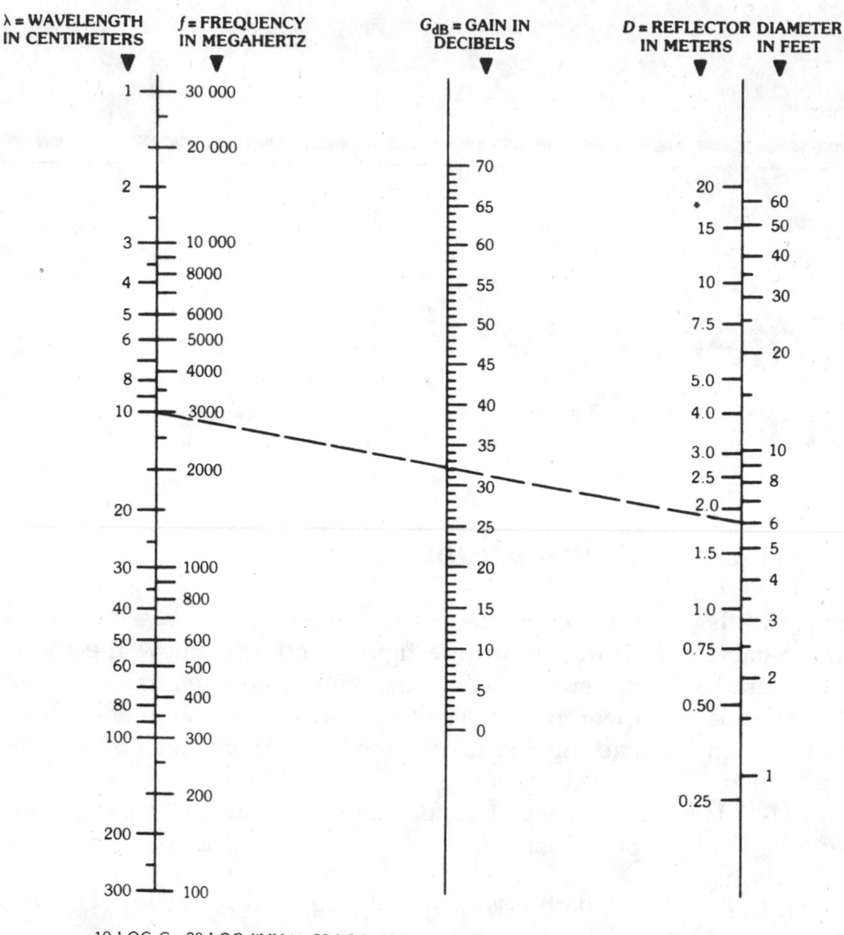

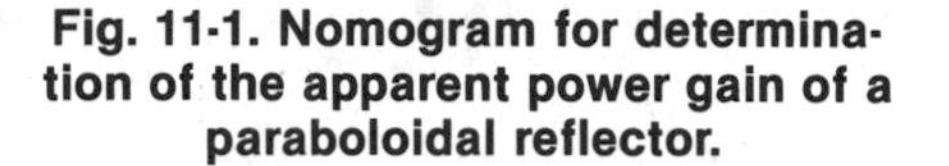

Fig. 11-1. Nomogram for determination of the apparent power gain of a paraboloidal reflector.

in dB from the ERP in dB. Assume a receiver bandwidth of 30 MHz, a 10-foot dish, and a 1.5 dB (LNA + losses) noise figure. We can calculate the received signal (Fig. 11-2*) using the formula Received Power = ERP − Path Loss + antenna gain.

$$\text{Received Power (at 4 GHz)} = +36 - 196.1 + 40 = -120.1 \text{ dBw}$$

Since 0 dBw = +30 dBm (dB referred to 1 milliwatt),

Received Power (dBm) = −90.1 dBm (0 dBm = 223 mV into 50 ohms)

Now, since,

$$\text{dB} = 10 \log_{10} \frac{P2}{P1} \left(\text{or } 20 \log_{10} \frac{V2}{V1}\right)$$

into a 50-ohm system, with 223 mV as a reference (where, again, 0 dBm = 223 mV into 50 ohms),

$$90.1 = 20 \log_{10} \frac{0.223}{(V_{received}{}^{**})}$$

$$4.505 = \log_{10} \frac{0.223}{V_{received}}$$

$$V_{received} = 6.97 \text{ microvolts (into 50 ohms)}$$

The receiver sensitivity, for a 30-MHz BW and NF = 1.5 dB, is

$$\begin{aligned} \text{RCVR SENS} &= -174 + \log_{10} BW_{Hz} + NF_{dB} \\ &= -174 + \log_{10}(30 \times 10^6) + 1.5 \\ &= -174 + 74.8 + 1.5 \\ &= -97.72 \text{ dBm (Note: This is a 0-dB S/N ratio.)} \end{aligned}$$

*From *Reference Data for Engineers: Radio, Electronics, Computer, and Communications,* Howard W. Sams & Co., Indianapolis, IN, p. 33-21.

**$V_{received}$ = received signal

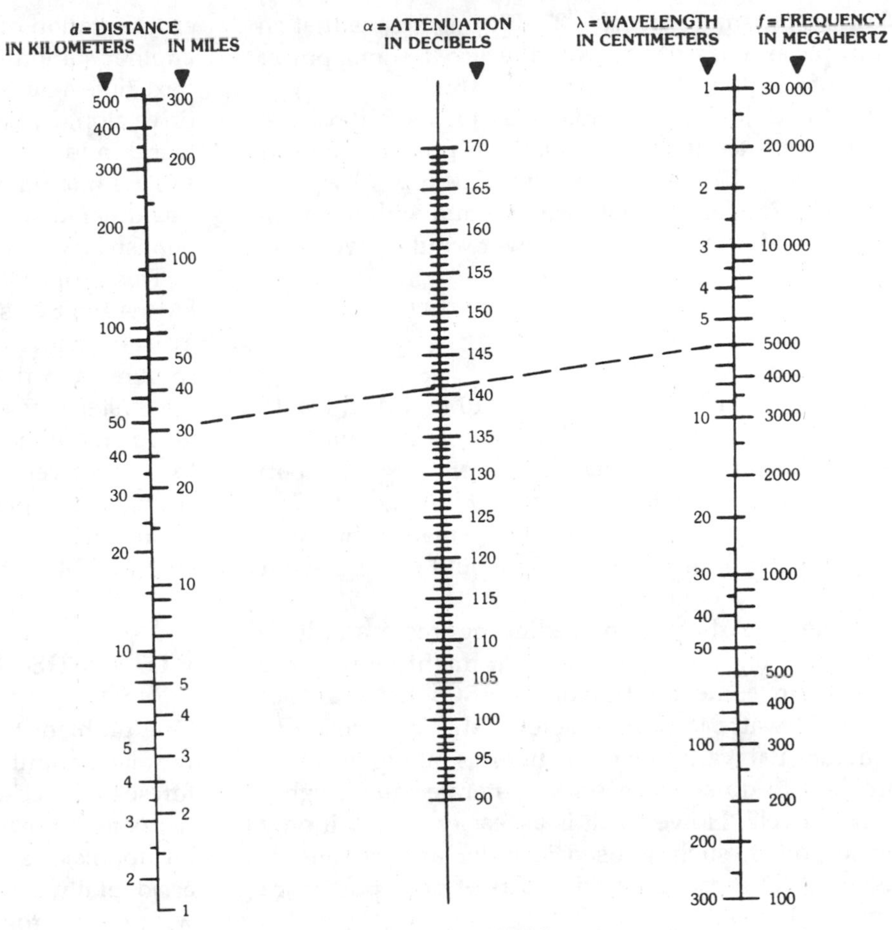

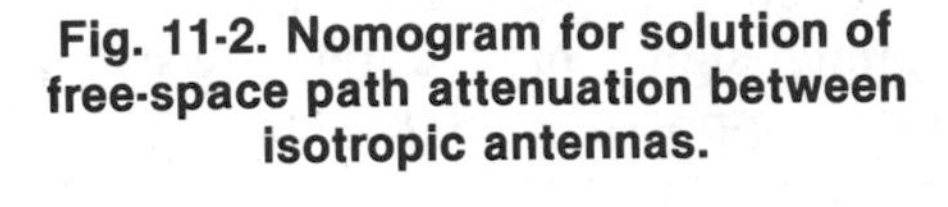

Fig. 11-2. Nomogram for solution of free-space path attenuation between isotropic antennas.

where -174 = noise power (KTB), in dBm per Hz at 25°C, and where (for KTB)

K = Boltzmann's constant (1.38×10^{-16} erg/°C),
T = Temperature in Kelvins,
B = Bandwidth in Hz.

Since our received signal is -90.1 dBm and our receiver sensitivity is -97.7 dBm,

S/N ratio = Received signal − Receiver Sensitivity
$= (-90.1) - (-97.7) = +7.6$ dB

This is marginal. In this case, while a picture would be present, a larger (12 foot) dish would give another 2 dB of gain and a much improved picture.

The preceding illustration was done not to bewilder the reader, but to show what we are dealing with in a satellite TV system.

Practical Considerations

A typical TVRO (3.7 to 4.2 GHz) receiver needs to have better than approximately an 8-dB S/N ratio for good performance, due to the "threshold effect" encountered in an FM system. For purposes of illustration, we will assume a threshold of 8 dB for a TVRO receiver. Much less than 8-dB S/N causes things to rapidly go downhill, and another 2 dB would be better—say, up to 10-dB S/N (sometimes called c/n*), will result in a noticeable improvement in the video S/N (less "snow" or "sparklies").

Let's take the case of our hypothetical satellite that we used for the last calculation. Assume a receiving antenna gain of 27 dB (two identical antennas would be carried aboard the hypothetical satellite in this case), and assume a satellite receiver with an NF of 1.5 dB and a 36-MHz IF bandwidth. Let us further assume that we want at least a 10-db S/N (worst case), and, typically, 16-dB S/N under normal operation. The required received power would be:

Received Power (dBm)
$= -174 + \log_{10} BW_{Hz} + NF_{dB} + S/N_{dB}$
$= -174 + 75.56 + 1.5 + 16$
$= -80.94$ dBm (call this -81 dBm)

*c/n = carrier to noise ratio.

With an antenna gain of +27 dB, this means that an isotropic antenna (0 dB) would receive a signal power level of (−81) + (−27), or −108 dBm.

The path loss was calculated as 199.6 dB at 6 GHz. Therefore, the effective transmitted power would be (199.6) + (−108) = +91.6 dBm (or +61.6 dBw). Assuming a modest 10-foot dish antenna with a gain of about 42 dB, the transmitter power would have to be:

$$\text{TX power} = (61.6) - (42) = 19.6 \text{ dBw}$$

This is equal to 91.2 watts.

In other words, 91 watts into a 10-foot dish would do the job in fine style as an uplink, and would give several dBs as a safety margin. A very large (16 foot) dish would require less power (about 40 watts). One would not probably use anything smaller than a 6-foot dish for home TVRO. This would require about 280 watts of TX power.

Remember that the preceding figures are only illustrative, but are typical of what might be used for uplinks in some applications. (HBO, for example, uses 750 watts into a 10-meter dish.) It is costly to generate 100 watts at 6 GHz, but it is no big technical problem to do so, or even to generate much higher power levels. However, it is an easier matter if only pulse power, such as used in radar applications, is required. In these cases, kilowatts of peak power are commonly used.

Capture Effect

A characteristic of FM systems is a result that is called the "capture effect." This is the property of an FM system to receive only the stronger of two signals—suppressing the weaker of the two. In the case of an AM system (as used in conventional TV broadcasting for video, and in AM broadcasting and CB radio), a signal that is 20 dB weaker than the desired signal can cause a noticeable, annoying interference. In an FM system (assuming we are above threshold), the weaker signal will be suppressed to the point where it is inaudible. In fact, this is true even when the signals are only 3 decibels different in amplitude. A performance figure given for high-fidelity FM receivers, called the "capture ratio," is the numerical value that measures this effect.

The *capture ratio* is generally given as that ratio of two received signals in which the stronger signal suppresses the weaker one by so many decibels (usually 20 or 30 dB). For good high-fidelity FM receivers, the capture ratio is usually under 2 dB, with 1 dB often claimed. If you want a demonstration of this, tune in an FM station on a radio channel that is also used by another station which is somewhat further away, the next time you are driving in a remote area. As you drive along, you will note that at times one station is heard, and then the other, but seldom, if ever, both at the same time. With an AM radio, there would be a wide area where both stations are heard simultaneously.

This property is due to the effect of *limiting* in the FM receiver IF system. The excellent limiting that is available with modern IC circuitry makes for excellent capture ratio figures in modern FM receivers. Since the systems used for satellite TV use FM for both video and audio, a satellite transponder will respond to a (received) pirate signal only a few decibels stronger, completely suppressing the legitimate program carrier. This is, again, an inherent property of both an FM system and the satellite transponder.

Problems with Interference

One problem that has arisen, which anyone with a normal amount of brains and imagination could have foreseen, is accidental or deliberate interference. It has happened, for example, that during the setting up of an uplink at a remote site, the wrong satellite was accidentally accessed. This form of interference can wipe out a programmer's entire uplink signal.

A technique that is used in mobile 2-way radio repeaters to avoid this problem of deliberate or accidental access is the placing of a sub-audible tone on the carrier. This tone is detected at the repeater, and the repeater will be activated only if this tone is present. It is a rather crude system, but widely used. A better approach uses a digital code system, such as Motorola's Digital Private Line. However, *once the door is opened by the desired signal, a stronger competing signal can take over the repeater, even if it is uncoded, until the repeater is caused to shut down* because the repeater circuitry either sees an improper or no access code. (This effectively prevents the desired signal from activating the repeater.) We can prevent unauthorized access, but not interference in this case, since the repeater has already been accessed and is operational at the time that the undesirable signal appears.

Captain Midnight

Recently (April 28, 1986), HBO experienced deliberate interference with its uplink (Fig. 11-3). Undoubtedly, a lot of people are upset at HBO's scrambling policy and the chronic lack of VCII decoders (and their

Fig. 11-3. Diagram of Captain Midnight's raid.

HBO

CAPTAIN MIDNIGHT

CABLE SYSTEM RECEIVER

PRIVATE SUBSCRIBER

1. HBO TRANSMITS SIGNAL TO SATELLITE THAT RELAYS IT.

2. CAPTAIN MIDNIGHT SENT A STRONGER SIGNAL TO THE SAME TRANSPONDER-HBO's CHANNEL ON THE SATELLITE-AND ON THE SAME FREQUENCY AS HBO.

3. STRONGER SIGNAL REPLACED THE ONE SENT BY HBO AND IT WAS RELAYED TO ALL CABLE OPERATORS AND PRIVATE SATELLITE SUBSCRIBERS.

high cost), and one individual, calling himself ''Captain Midnight,'' decided to complain by transmitting a printed message which apparently was in protest of HBO's scrambling its TV signals. Of course, this interference made national headlines. HBO glibly said that they will simply increase power. But this may not be the answer.

As it turned out, the interference experienced by HBO came from a satellite uplink station in Florida. The operator voluntarily surrendered (in July 1986) and pleaded guilty, due to fear of his being found out anyway.

Due to the short duration of any interference, it is almost impossible to trace. Interference cannot easily be detected electronically, since it is rather difficult to trace RF interference that has occurred in the past and is no longer present. Detection will have to be done via legal and/or police methods. But, the FCC is severely understaffed and, unless political pressure is applied, probably can and will do little.

Locating a "Pirate"

Let us consider the possibility of using search aircraft to find a ''satellite pirate.'' Assuming a 3° antenna bandwidth being used by a pirate (a lousy antenna; it is probably more like 2°) and assuming detectable signal-pattern side lobes out to 10° at 60,000 feet altitude, a pirate's interfering RF radiation could be detected over a circular area of only about 3.14 square miles (Fig. 11-4). For a 10% probability of detection, we would need an aircraft equipped with search equipment every 31 square miles. This is ridiculous.

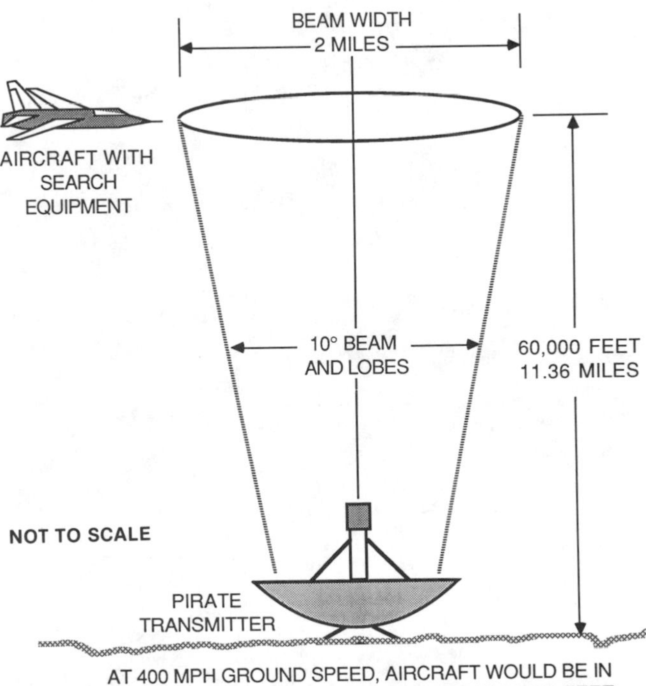

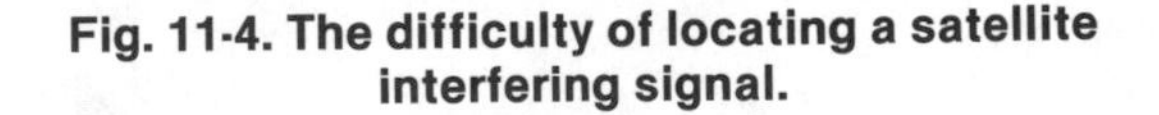

Fig. 11-4. The difficulty of locating a satellite interfering signal.

Satellites would do better in a search operation, but do we have them to do this? And furthermore, who really cares but those individuals who work for HBO and/or the FCC, and whose jobs may depend on finding this pirate. Since we do not know where in the United States to look, and since any interference will probably exist only for a few minutes, the chance of finding a ''pirate'' are approximately zero, unless someone who knows him or her spills the beans.

A Nasty but Real Possibility

What would happen if a video pirate used an old radar transmitter to only interfere with HBO's data stream? (This is 88 data bits between horizontal lines, containing audio and control channels, and replacing normal synch.) Conceivably, the encrypted data would come down from the satellite, radically altered and causing havoc among subscribers, ruining audio, possibly deactivating authorized VCII decoders, and destroying billing data. This caper could disrupt HBO operations all over the country, and subscribers would soon lose patience at missing paid-for shows and events.

And if there were *several* pirates doing this, what would happen? Even if the data were not interfered with, the picture already has been. An error-correcting code scheme can only deal with so much interference, and can correct just so many bit errors. This could cause many more expenditures on a far more sophisticated encoding/decoding system (funny that no one thought of this).

We were amused at the statement that appeared in the *New York Times* stating that $60,000 worth of equipment was necessary to cause this interference. The person who made this statement has probably never heard of surplus parts, basement ingenuity, and teenage ''hackers''—not to mention all the surplus government radar gear floating around that is for sale to experimenters.

What Do We Do?

The possibility of deliberate interference to a satellite is the weak link in the satellite communications chain, and this has been obvious to the military every since satellites have existed. The military have methods to deal with this interference problem, but will HBO and the other service programmers be able to afford the expense? Remember, the competition of unscrambled programmers' events and other video entertainment forms, such as VCRs and the better terrestrial programming, may limit the economic feasibility of the very sophisticated pay-TV security systems. And, consider this, will prospective viewers want to invest in $400 decoders that may soon be obsolete? The next few months (and years) will be very interesting, indeed, for the TVRO industry.

12

Data Encryption Standard (DES)

In order to answer the question of "What is the DES, anyway?" we have included excerpts from the Data Encryption Standard in this book. We regretfully cannot, due to space limitations, include the entire DES. However, we have picked out the basic "meat" from it and have included this information. The reader with a computer science or mathematical interest will do well to obtain FIPS 46, 74, and 81 from the U.S. Government Printing Office. This will give the complete story. In any event, the rest of this chapter and the excerpts contained therein will basically show the reader what the DES algorithm is.

The DES algorithm is a mathematical device, not an IC chip, computer system, or other piece of hardware. Included in Chapter 13 is a data sheet on a processor IC that implements this mathematical operation in hardware (the AT&T T7000A Digital Encryption Processor). The authors have heard the buzzword "DES Encrypted" many times and, yet, few individuals seem to know what it *really* means. Therefore, this chapter was included. We hope it answers many questions. It will probably generate many questions, as well.

DES Algorithm—Its Development

The computer age brought with it computer usage in banking and the financial institutions. Inevitably, computer crime came along with it. There arose the problem that, with sufficient knowledge and a computer terminal, one could transfer funds into his own account, make credit card purchases on someone else's card, or even get money from a cash-dispensing machine.

IBM quickly realized this and, in the late 1960s, set up a research group to develop a suitable cipher code to protect data. In 1971, a code named *Lucifer* was developed. It was sold to Lloyds of London for use with an IBM-developed cash-dispensing system.

Lucifer

Lucifer was successful but it had some weaknesses. IBM then spent about three years refining and strengthening Lucifer. The code was analyzed over and over by experts in cryptology. It withstood sophisticated cryptoanalytical attacks and, by 1974, it was ready to market.

Around the same time, the National Bureau of Standards (NBS), which, since 1965, was responsible for developing standards for the purchase of computer equipment by the Federal government, initiated a study of computer security. The NBS saw a need for an encryption method, and solicited for a suitable encryption algorithm. This was done in May, 1973, and August, 1974. The algorithm was to be for the storage and transmission of unclassified data.

In response to this solicitation, IBM submitted its Lucifer cipher. This cipher consisted of an extremely complex algorithm embedded in an IC structure. Basically, the cipher key goes into a series of eight "S" boxes—complex mathematical formulas that encrypt and decrypt data, with the appropriate key. The initial Lucifer cipher had a 128-bit key. Before it submitted the cipher to NBS, IBM shortened it by removing more than half the key.

NSA Participation

The National Security Agency (NSA), however, had taken an enormous interest in Project Lucifer. It had lent IBM a hand in the development process and had helped to develop the S-box structures. NSA was now getting concerned. They were, for the first time, experiencing competition within their own country.

For years, NSA had been dependent on international data communications. It monitored data communications, such as Middle East oil transactions and messages, and the financial and trade transactions from Latin American, Europe, and the Far East. Also, military and diplomatic intelligence (encrypted using crude techniques) were picked up and deciphered by NSA. Thus, much information about Communist countries was obtained from nonCommunist countries. Now, the development of an economical, highly secure, data-encryption device threatened to cause NSA serious trouble. Also, outside researchers might stumble across some of NSA's methods.

Problems and Changes

Meetings of NSA and IBM resulted in an agreement by IBM to reduce its key from 128 bits to 56 bits, and to classify certain details about their selection of the eight "S" boxes for the cipher.

The National Bureau of Standards passed this cipher to NSA for analysis. The National Security Agency certified the algorithm as "free" of any mathematical or statistical weaknesses and recommended it as the best candidate for the National Data Encryption Standard (DES). This suggestion was met with criticism. Was the cipher just long enough to prevent corporate eavesdroppers from penetrating it, and just short enough for NSA's code breakers? Was there a mathematical trick (classified) that would enable NSA to quickly break the code?

The Agency had been tinkering with the critical "S" boxes, and it had therefore insisted that certain details were to be classified. The reason cited for this was simple. Since the DES would be commercially available and would be sold abroad as well, NSA would be hanging itself by permitting the foreign use of an unbreakable cipher. The weaknesses designed into the cipher would still allow the agency to penetrate every communications channel and data bank using the DES. The code breakers at NSA wanted to be sure the NSA could break the cipher. As a result, bureaucratic compromise was reached. The S-box part of the cipher was strengthened, and the key, which was dependent on the users of the code, was weakened.

Computer experts argued, however, that it would be possible to build a computer using a million special "search chips" that could test a million possible solutions per second, and, therefore, in 72,000 seconds (20 hours), all possible combinations could be tried. There would be a 50% probability that just 10 hours of trial-time would break the code (with 56 bits, there are 2^{56} combinations). Such a computer would cost about $20 million to build and, prorated over five years, this would mean about $10,000 a day. If a 24-hour effort would be used, each code would average about $5,000 to break. As technology brought the costs down, these figures could be divided by a factor of 10 or 100.

The Original Lucifer

What if the 128-bit key, the original Lucifer, had been submitted for consideration? There are now 2^{128} solutions. This is equal to 34.03×10^{37}, or 34 followed by 37 zeros. This number is astronomical and incomprehensible to most people. If one trillion solutions per second could be tested, it would take a mere 34×10^{25} seconds, or about 1.08×10^{19} years. This is a rather long time. The known universe has been in existence, it is thought, about 2.6×10^{10} (26 billion) years. Therefore, IBM's Lucifer code, at present, is probably unbreakable.

DES Becomes Accepted

On June 15, 1977, the Data Encryption Standard (DES) became the official civilian cipher of the government. It is now widely used, with one of the users being HBO with its VideoCipher II™ system. With increases in computer speeds, new technologies, and lower costs, cipher security will slowly disappear. Some authorities give it five years, some ten, but very few any more than that. The advent of VideoCipher II™ has focused even more attention on the DES and, sooner or later, it will be defeated by someone. By that time, however, new scrambling methods will probably replace VideoCipher II™.

Excerpts from the Data Encryption Standard

Explanation

The Data Encryption Standard (DES)* specifies an algorithm to be implemented in electronic hardware devices and used for the cryptographic protection of computer data. The publications concerning this standard provide a complete description of a mathematical algorithm for encrypting (enciphering) and decrypting (deciphering) binary-coded information.

* See Federal Information Processing Standards (FIPS) Publication 46. Copies are for sale by the National Technical Information Service, U.S. Department of Commerce, Springfield, VA 22161. Order by FIPS PUB number and title.

Encrypting data converts the data to an unintelligible form called a *cipher.* Decrypting a cipher converts the data back to its original form. The algorithm described in this standard specifies both enciphering and deciphering operations which are based on a binary number called a key. The key consists of 64 binary digits (''0s'' or ''1s'') of which 56 bits are used directly by the algorithm and 8 bits are used for error detection.

Binary-coded data may be cryptographically protected using the DES algorithm in conjunction with a key. The key is generated in such a way that each of the 56 bits used directly by the algorithm are random and the 8 error-detecting bits are set to make the parity of each 8-bit byte of the key odd, i.e., there is an odd number of ''1s'' in each 8-bit byte. Each member of a group of authorized users of encrypted computer data must have the key that was used to encipher the data in order to use the data. This key, held by each member in common, is used to decipher any data received in cipher form from other members of the group. The encryption algorithm specified in this standard is commonly known among those using the standard. The unique key chosen for use in a particular application makes the results of encrypting data, using the algorithm, unique. Selection of a different key causes the cipher, which is produced for any given set of inputs, to be different. The cryptographic security of the data depends on the security provided for the key that is used to encipher and decipher the data.

Data can be recovered from a cipher only by using the exactly same key that was used to encipher it. Unauthorized recipients of the cipher, who know the algorithm but do not have the correct key, cannot derive the original data algorithmically. However, anyone who does have the key and the algorithm can easily decipher the cipher and obtain the original data. A standard algorithm, which is based on a secure key, thus provides a basis for exchanging encrypted computer data, by issuing the key that is used to encipher it only to those authorized to have the data. Additional Federal Information Processing Standards (FIPS) guidelines for implementing and using the DES are being developed and will be published by NBS.

Alternative Modes of Using the DES

The ''Guidelines for Implementing and Using the NBS Data Encryption Standard,'' FIPS Publication 74, describes two different modes for using the algorithm described in this standard. Blocks of data containing 64 bits may be directly entered into the device where 64-bit cipher blocks are generated under control of the key. This is called the *Electronic Codebook (ECB) mode.*

Alternatively, the device may be used as a binary stream generator to produce statistically random binary bits, which are then combined with the clear (unencrypted) data (1 to 64 bits) using an ''Exclusive-OR'' logic operation. In order to assure that the enciphering device and the deciphering device are synchronized, their inputs are always set to the previous 64 bits of cipher that were transmitted or received. This second mode of using the encryption algorithm is called the *Cipher Feedback (CFB) mode.*

The Electronic Codebook mode generates blocks of 64 cipher bits. The Cipher Feedback mode generates a cipher having the same number of bits as the plain text. Each block of cipher is independent of all others when the Electronic Codebook mode is used, while each byte (group of bits) of cipher depends on the previous 64 cipher bits when the Cipher Feedback mode is used.

The cryptographic algorithm specified in this standard transforms a 64-bit binary value into a unique 64-bit binary value based on a 56-bit variable. If the complete 64-bit input is used (i.e., none of the input bits should be predetermined from block to block) and if the 56-bit variable is randomly chosen, no technique other than that of trying all the possible keys, using a known input and output for the DES, will guarantee finding the chosen key. As there are over 70,000,000,000,000,000 (70 quadrillion) possible keys of 56 bits, the feasibility of deriving a particular key in this way is extremely unlikely in typical ''threat'' environments. Moreover, if the key is changed frequently, the risk of this event happening is greatly diminished. However, users should be aware that it is theoretically possible to derive the key in fewer trials (with a correspondingly lower probability of success depending on the number of keys tried), and should be cautioned to change the key as often as practical. Users must change the key and must provide it a high level of protection in order to minimize the potential risks of its unauthorized computation or acquisition. The feasibility of computing the correct key may change with advances in technology.

Data Encryption Methods

Basic Methods

Encryption is the transformation of data from its original intelligible form to an unintelligible cipher form. Two basic transformations may be used: *permutation* and *substitution.* Permutation changes the order of the individual symbols comprising the data. In a substitution transformation, the symbols themselves are replaced by other symbols. During permutation, the symbols retain their identities but lose their positions. During substitution, the symbols retain their positions but lose their original identities.

The set of rules for a particular transformation is expressed in an algorithm. Basic transformations may be combined to form a complex transformation. In a computer system, the symbols of the data are groups of one or more binary digits (1s and 0s) called bits. A group of bits is called a byte. In computer applications, the encryption transformation of permutation reorders the bits of the data. The encryption transformation of substitution replaces one bit with another or one byte with another.

Block Ciphers

A cipher that is produced by simultaneously transforming a group of message bits into a group of cipher bits is called a *block cipher.* In general, the groups are the same size.

Product Ciphers

Combining the basic transformations of permutation and substitution produces a complex transformation termed a *product cipher.* If permutation and substitution operations are applied to a block of data, the resulting cipher is called a *block product cipher.*

Data Encryption Algorithm

Introduction

The algorithm is designed to encipher and decipher blocks of data consisting of 64 bits under control of a 64-bit key. Deciphering must be accomplished by using the same key that was used for enciphering, but with the schedule of addressing the key bits altered so that the deciphering process is the reverse of the enciphering process.

A block to be enciphered is subjected to an initial permutation, IP, and then to a complex key-dependent computation, and, finally, to a permutation which is the inverse of the initial permutation (IP^{-1}). The key-dependent computation can be defined simply, in terms of a function "f," called the cipher function, and a function "KS," called the key schedule. A description of the computation is given first, along with details as to how the algorithm is used for encipherment. Next, the use of the algorithm for decipherment is described. Finally, a definition of the cipher function "f" is given in terms of the primitive functions, which are called the selection functions "S_i," and the permutation function "P." The primitive functions S_i, P, and KS of the algorithm are contained in the Appendix of FIPS Publication 46.

The following notation is convenient: Given two blocks (L and R) of bits, LR denotes the block consisting of the bits of L followed by the bits of R. Since concatenation is associative, $B_1 B_2 \ldots B_8$, for example, denotes the block consisting of the bits of B_1 followed by the bits of $B_2 \ldots$ followed by the bits of B_8.

Enciphering

A sketch of the enciphering computation is given in Fig. 12-1. The following information is given more clearly and accurately in FIPS Publications 46 and 74. It is quoted here for informational purposes only.

The 64 bits of the input block to be enciphered (Fig. 12-1) are first subjected to the following permutation, called the initial permutation, *IP*:

IP

58	50	42	34	26	18	10	2
60	52	44	36	28	20	12	4
62	54	46	38	30	22	14	6
64	56	48	40	32	24	16	8
57	49	41	33	25	17	9	1
59	51	43	35	27	19	11	3
61	53	45	37	29	21	13	5
63	55	47	39	31	23	15	7

That is, the permuted input has bit 58 of the input as its first bit, bit 50 as its second bit, and so on, with bit 7 as its last bit. The permuted input block is then the input to the complex key-dependent computation described below. The output of that computation, called the preoutput, is then subjected to the following permutation, IP^{-1}, *which is the inverse of the initial permutation:*

IP^{-1}

40	8	48	16	56	24	64	32
39	7	47	15	55	23	63	31
38	6	46	14	54	22	62	30
37	5	45	13	53	21	61	29
36	4	44	12	52	20	60	28
35	3	43	11	51	19	59	27
34	2	42	10	50	18	58	26
33	1	41	9	49	17	57	25

That is, the output of the algorithm has bit 40 of the preoutput block as its first bit, bit 8 as its second bit,

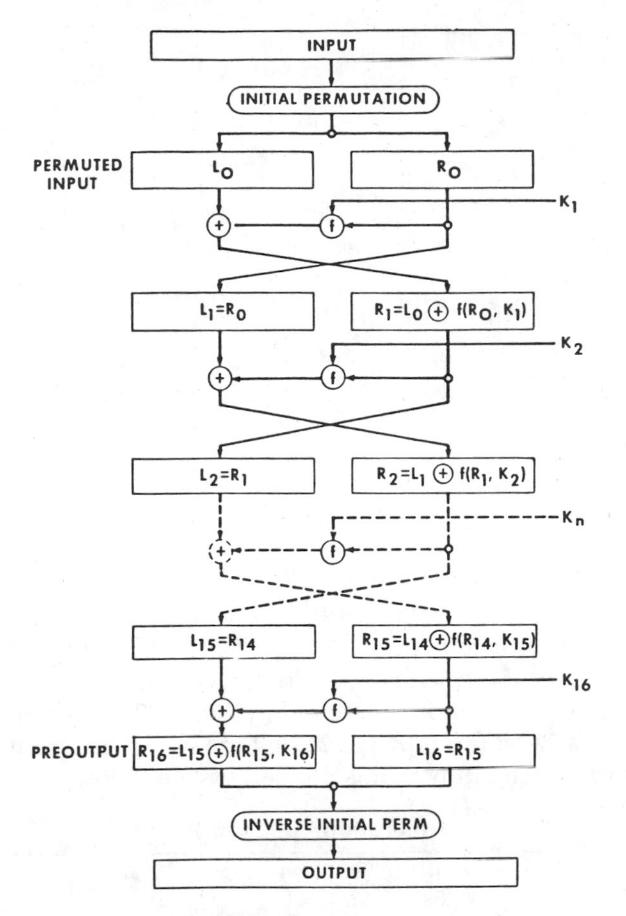

Fig. 12-1. Enciphering computation.

and so on, until bit 25 of the preoutput block is the last bit of the output.

The computation that uses the permuted input block as its input to produce the preoutput block consists, but for a final interchange of blocks, of 16 iterations of a calculation that is described below in terms of the cipher function *f*, which operates on two blocks (one of 32 bits and one of 48 bits) and produces a block of 32 bits.

Let the 64 bits of the input block to an iteration consist of a 32-bit block *L*, followed by a 32-bit block *R*. Using the notation defined in the introduction, the input block is then *LR*.

Let *K* be a block of 48 bits chosen from the 64-bit key. Then, the output *L'R'*, of an iteration with input *LR*, is defined by:

$$L' = R \qquad R' = L \oplus f(R,K) \qquad \text{(Eq. 12-1)}$$

where ⊕ denotes bit-by-bit addition modulo 2.

As remarked before, the input of the first iteration of the calculation is the permuted input block. If *L'R'* is the output of the sixteenth iteration, then *R'L'* is the preoutput block. At each iteration, a different block *K* of key bits is chosen from the 64-bit key designated by *KEY*.

With more notation, we can describe the iterations of the computation in more detail. Let *KS* be a function which takes an integer *n* in the range from 1 to 16 and a 64-bit block *KEY* as input, and yields as output, a 48-bit block K_n which is a permuted selection of bits from *KEY*. Thus,

$$K_n = KS\,(n,KEY) \qquad \text{(Eq. 12-2)}$$

with K_n determined by the bits in 48 distinct bit positions of *KEY*. *KS* is called the key schedule because the block *K*, used in the *n*'th iteration of Equation 12-1, is the block K_n determined by Equation 12-2.

As before, let the permuted input block be *LR*. Finally, let L_o and R_o be, respectively, *L* and *R*, and let L_n and R_n be, respectively, *L'* and *R'* of Equation 12-1 when *L* and *R* are, respectively, L_{n-1} and R_{n-1} and *K* is K_n; that is, when *n* is in the range from 1 to 16,

$$L_n = R_{n-1} \qquad R_n = L_{n-1} \oplus f(R_{n-1}, K_n) \qquad \text{(Eq. 12-3)}$$

The preoutput block is then $R_{16}\,L_{16}$.

The key schedule *KS* of the algorithm is described in detail in the next section (Primitive Functions). The key schedule produces the 16 K_n which are required for the algorithm.

Deciphering

The permutation IP^{-1} applied to the preoutput block is the inverse of the initial permutation, *IP*, applied to the input. Further, from Equation 12-1, it follows that:

$$R = L' \qquad L = R' \oplus f(L', K) \qquad \text{(Eq. 12-4)}$$

Consequently, to *decipher*, it is only necessary to apply the *very same algorithm to an enciphered message block*, taking care that at each iteration of the computation, *the same block of key bits, K, is used* during decipherment as was used during the encipherment of the block. Using the notation of the previous section, this can be expressed by the equations:

$$R_{n-1} = L_n \qquad L_{n-1} = R_n \oplus f(L_n, K_n) \qquad \text{(Eq. 12-5)}$$

where $R_{16}L_{16}$ is the permuted input block for the deciphering calculation, L_0R_0 is the preoutput block. That is, for the decipherment calculation, with $R_{16}L_{16}$

as the permuted input, K_{16} is used in the first iteration, K_{15} in the second, and so on, with K_1 used in the sixteenth iteration.

The Cipher Function f

A sketch of the calculation of *f(R, K)* is given in Fig. 12-2. Let *E* denote a function which takes a block of 32 bits as input and yields a block of 48 bits as output. Also, let *E* be such that the 48 bits of its output, written as 8 blocks of 6 bits each, are obtained by selecting the bits in its inputs in order, according to the following:

E Bit-Selection Table

32	1	2	3	4	5
4	5	6	7	8	9
8	9	10	11	12	13
12	13	14	15	16	17
16	17	18	19	20	21
20	21	22	23	24	25
24	25	26	27	28	29
28	29	30	31	32	1

Thus, the first three bits of *E(R)* are the bits in positions 32, 1, and 2 of *R* while the last two bits of *E(R)* are the bits in positions 32 and 1.

Each of the unique selection functions, S_1, S_2, . . . ,S_8, takes a 6-bit block as input and yields a 4-bit block as output. This is illustrated by using a table containing the recommended S_1, such as the following:

Selection Function S_1

Row	Column Number															
No.	0	1	2	3	4	5	6	7	8	9	10	11	12	13	14	15
0	14	4	13	1	2	15	11	8	3	10	6	12	5	9	0	7
1	0	15	7	4	14	2	13	1	10	6	12	11	9	5	3	8
2	4	1	14	8	13	6	2	11	15	12	9	7	3	10	5	0
3	15	12	8	2	4	9	1	7	5	11	3	14	10	0	6	13

If S_1 is the function defined in this table and *B* is a block of 6 bits, then $S1_1(B)$ is determined as follows: The first and last bits of *B* represent, in base 2, a number in the range 0 to 3. Let that number be *i*. The middle 4 bits of *B* represent, in base 2, a number in the range 0 to 15. Let that number be *j*. In the table, look up the number in the *i*'th row and *j*'th column. It is a number in the range of 0 to 15 and is uniquely represented by a 4-bit block. That block is the output $S_1(B)$ of S_1 for the input *B*. For example, for input 011101, the row is *01*, (that is row 1), and the column is determined by 1101, that is column 13. In row 1, column 13 appears as a 5, so that the output is 0101.

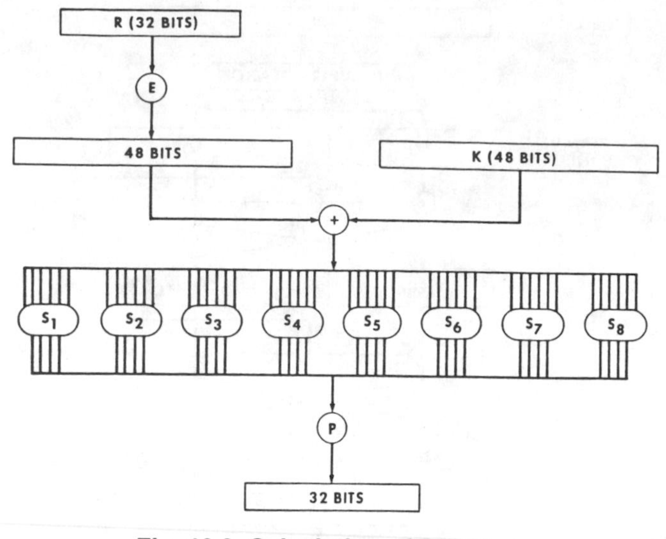

Fig. 12-2. Calculation of *f (R,K)*.

Selection functions S_1, S_2, . . . , S_8 of the algorithm appear in the next section (Primitive Functions).

The permutation function, *P*, yields a 32-bit output from a 32-bit input by permuting the bits of the input block. Such a function is defined by the following:

Permutation Function P

16	7	20	21
29	12	28	17
1	15	23	26
5	18	31	10
2	8	24	14
32	27	3	9
19	13	30	6
22	11	4	25

The output *P(L)* for the function *P*, defined by this table, is obtained from the input *L* by taking the sixteenth bit of *L* as the first bit of *P(L)*, the seventh bit as the second bit of *P(L)*, and so on until the twenty-fifth bit of *L* is taken as the thirty-second bit of *P(L)*. The permutation function, *P*, of the algorithm is repeated in the next section.

Now let S_1, . . . , S_8 be eight distinct selection functions, let *P* be the permutation function, and let *E* be the function defined above. To define *f(R, K)*, we first define B_1, . . . , B_8 to be blocks of 6 bits each for which

$$B_1B_2 \ldots B_8 = K \oplus E(R) \qquad \text{(Eq. 12-6)}$$

The block, *f(R, K)*, is then defined to be

$$P(S_1(B_1)S_2(B_2 \ldots S_8(B_8)) \qquad \text{(Eq. 12-7)}$$

Thus, $K \oplus E(R)$ is first divided into the 8 blocks as indicated in Equation 12-6. Then, each B_i is taken as an input to S_i and the 8 blocks $S_1(B_1)$, $S_2(B_1)$, . . . , $S_8(B_8)$ of 4 bits each are consolidated into a single block of 32 bits which forms the input to P. The output Equation 12-7 is then the output of the function f for the inputs R and K.

Primitive Functions for the Data Encryption Algorithm

The choice of the primitive functions KS, $S_1, \ldots, S_8$, and P is critical to the strength of an encipherment resulting from the algorithm. Specified now is a recommended set of functions describing $S_1, \ldots, S_8$, and P in the same way that they are described in the algorithm. For the interpretation of the tables describing these functions, see the discussion in the body of the algorithm.

The primitive functions, $S_1, \ldots, S_8$, are:

S_1

14	4	13	1	2	15	11	8	3	10	6	12	5	9	0	7
0	15	7	4	14	2	13	1	10	6	12	11	9	5	3	8
4	1	14	8	13	6	2	11	15	12	9	7	3	10	5	0
15	12	8	2	4	9	1	7	5	11	3	14	10	0	6	13

S_2

15	1	8	14	6	11	3	4	9	7	2	13	12	0	5	10
3	13	4	7	15	2	8	14	12	0	1	10	6	9	11	5
0	14	7	11	10	4	13	1	5	8	12	6	9	3	2	15
13	8	10	1	3	15	4	2	11	6	7	12	0	5	14	9

S_3

10	0	9	14	6	3	15	5	1	13	12	7	11	4	2	8
13	7	0	9	3	4	6	10	2	8	5	14	12	11	15	1
13	6	4	9	8	15	3	0	11	1	2	12	5	10	14	7
1	10	13	0	6	9	8	7	4	15	14	3	11	5	2	12

S_4

7	13	14	3	0	6	9	10	1	2	8	5	11	12	4	15
13	8	11	5	6	15	0	3	4	7	2	12	1	10	14	9
10	6	9	0	12	11	7	13	15	1	3	14	5	2	8	4
3	15	0	6	10	1	13	8	9	4	5	11	12	7	2	14

S_5

2	12	4	1	7	10	11	6	8	5	3	15	13	0	14	9
14	11	2	12	4	7	13	1	5	0	15	10	3	9	8	6
4	2	1	11	10	13	7	8	15	9	12	5	6	3	0	14
11	8	12	7	1	14	2	13	6	15	0	9	10	4	5	3

S_6

12	1	10	15	9	2	6	8	0	13	3	4	14	7	5	11
10	15	4	2	7	12	9	5	6	1	13	14	0	11	3	8
9	14	15	5	2	8	12	3	7	0	4	10	1	13	11	6
4	3	2	12	9	5	15	10	11	14	1	4	6	0	8	13

S_7

4	11	2	14	15	0	8	13	3	12	9	7	5	10	6	1
13	0	11	7	4	9	1	10	14	3	5	12	2	15	8	6
1	4	11	13	12	3	7	14	10	15	6	8	0	5	9	2
6	11	13	8	1	4	10	7	9	5	0	15	14	2	3	12

S_8

13	2	8	4	6	15	11	1	10	9	3	14	5	0	12	7
1	15	13	8	10	3	7	4	12	5	6	11	0	14	9	2
7	11	4	1	9	12	14	2	0	6	10	13	15	3	5	8
2	1	14	7	4	10	8	13	15	12	9	0	3	5	6	11

The primitive function, P, is:

16	7	20	21
29	12	28	17
1	15	23	26
5	18	31	10
2	8	24	14
32	27	3	9
19	13	30	6
22	11	4	25

Recall that K_n, for $1 \leq n \leq 16$, is the block of 48 bits in Equation 12-2 of the algorithm. Hence, to describe *KS*, it is sufficient to describe the calculation of K_n from *KEY* for $n = 1, 2, \ldots, 16$. That calculation is illustrated in Fig. 12-3. To complete the definition of *KS*, it is therefore sufficient to describe the two permuted choices, as well as the schedule of left shifts. One bit in each 8-bit byte of the *KEY* may be utilized for error detection in key generation, distribution, and storage. Bits 8, 16, . . . , 64 are for use in assuring that each byte is of odd parity.

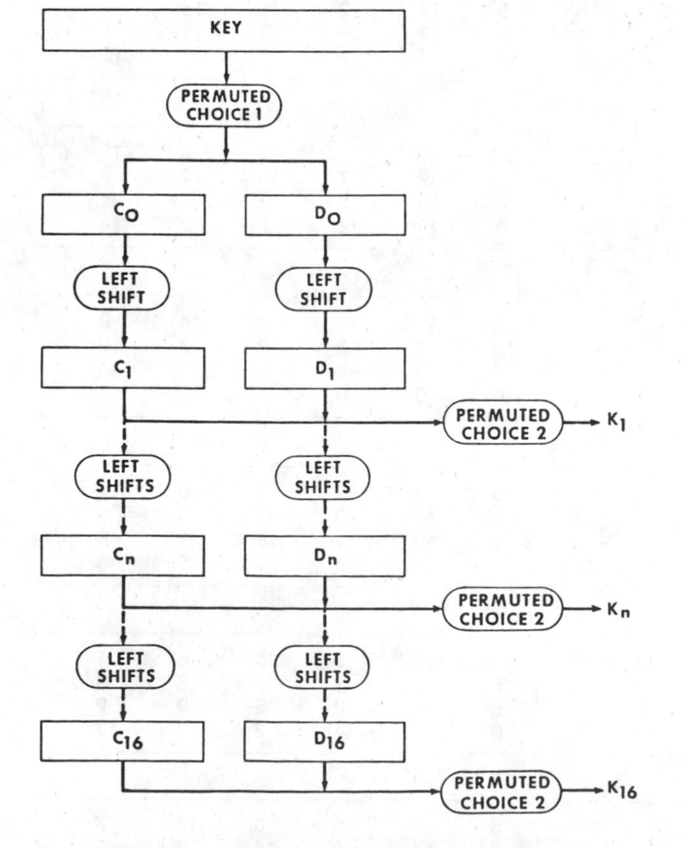

Fig. 12-3. *KEY* schedule calculation.

Permuted choice 1 is determined by the following:

Permuted Choice 1 (PC-1)

57	49	41	33	25	17	9
1	58	50	42	34	26	18
10	2	59	51	43	35	27
19	11	3	60	52	44	36
63	55	47	39	31	23	15
7	62	54	46	38	30	22
14	6	61	53	45	37	29
21	13	5	28	20	12	4

The table has been divided into two parts, with the first part determining how the bits of C_0 are chosen, and the second part determining how the bits of D_0 are chosen. The bits of *KEY* are numbered 1 through 64. The bits of C_0 are, respectively, bits 57, 49, 41, . . . , 44, and 36 of *KEY*, with the bits of D_0 being bits 63, 55, 47, . . . , 12, and 4 of *KEY*.

With C_0 and D_0 defined, we now define how the blocks C_n and D_n are obtained from the blocks C_{n-1} and D_{n-1}, respectively, for $n = 1, 2, \ldots, 16$. That is accomplished by adhering to the following schedule of left shifts of the individual blocks:

Iteration Number	Number of Left Shifts
1	1
2	1
3	2
4	2
5	2
6	2
7	2
8	2
9	1
10	2
11	2
12	2
13	2
14	2
15	2
16	1

For example, C_3 and D_3 are obtained from C_2 and D_2, respectively, by two left shifts, and C_{16} and D_{16} are obtained from C_{15} and D_{15}, respectively, by one left shift. In all cases, a single left shift is meant as a rotation of the bits one place to the left, so that, after one left shift, the bits in the 28 positions are the bits that were previously in positions 2, 3, . . . , 28, and 1.

Permuted choice 2 is determined by the following:

Permuted Choice 2 (PC-2)

14	17	11	24	1	5
3	28	15	6	21	10
23	19	12	4	26	8
16	7	27	20	13	2
41	52	31	37	47	55
30	40	51	45	33	48
44	49	39	56	34	53
46	42	50	36	29	32

Therefore, the first bit of K_n is the fourteenth bit of C_nD_n, the second bit is the seventeenth, and so on, with the forty-seventh bit being the twenty-ninth, and the forty-eighth bit being the thirty-second.

Recirculating Block Product Cipher

A *block product cipher* may be constructed by using a permutation operation and a substitution operation alternately and by recirculating the output of one pair of operations back into the input for some number of iterations. Each iteration is called a *round.* A cipher produced in this way is termed a *recirculating block product cipher.* If a recirculating block product cipher is properly constructed with an unknown key, then the alteration of a single bit of the plaintext block will unpredictably alter each bit of the ciphertext block. Altering a bit of the ciphertext will also result in an unpredictable change to the plaintext block after decryption.

Characteristics of the DES Algorithm

The *DES algorithm* is a recirculating, 64-bit, block product cipher whose security is based on a secret key. DES keys are 64-bit binary vectors consisting of 56 independent information bits and 8 parity bits. The parity bits are reserved for error-detection purposes and are not used by the encryption algorithm. The 56 information bits are used by the enciphering and deciphering operations and are referred to as the *active key.* Active keys are generated (selected at random from all possible keys) by each group of authorized users of a particular computer system or set of data. Each user should understand that the key must be protected and that any compromise of the key will compromise all data and resources protected by that key.

In the encryption computation, the 64-bit data input is divided into two halves, with each consisting of 32 bits. One half is used as input to a complex nonlinear function, and the result is Exclusive-OR'ed to the other half (Fig. 12-4). After one iteration, or round, the two halves of the data are swapped and the operation is performed again. The DES algorithm uses 16 rounds to produce a recirculating block product cipher. The cipher produced by the algorithm displays no correlation to the input. Every bit of the output depends on every bit of the input and on every bit of the active key.

The security provided by the DES algorithm is based on the fact that, if the key is unknown, an unauthorized recipient of encrypted data, knowing some of the matching input data, must perform an unacceptable effort to decipher other encrypted data or recover the key. Even having all but one bit of the key correct does not result in intelligible data.

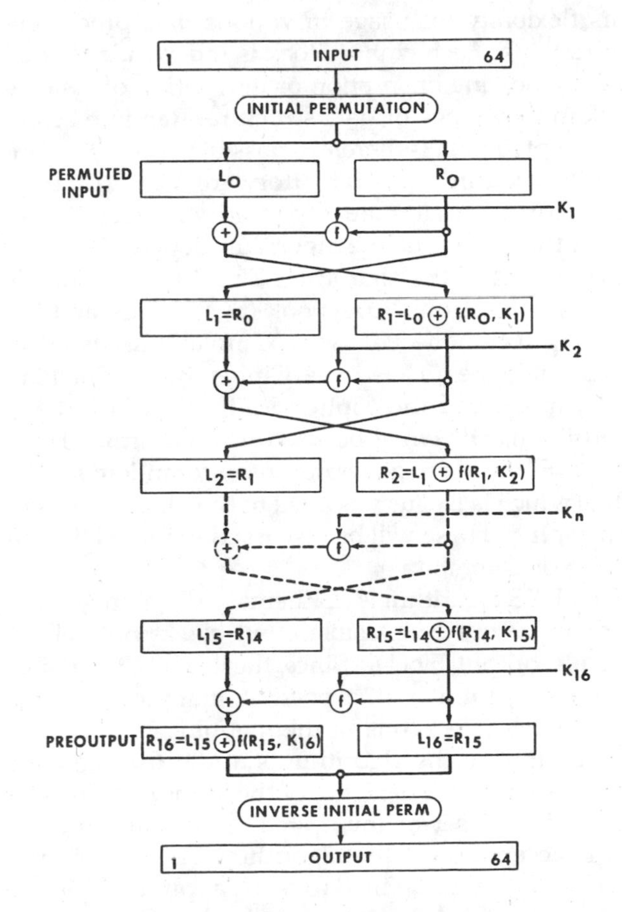

Fig. 12-4. Enciphering computation in the Electronic Codebook (ECB) mode.

The only known way of obtaining the key with certainty is by obtaining matched ciphertext and plaintext and, then, exhaustively testing the keys by enciphering the known plaintext with each key and comparing the result with the known ciphertext. Since 56 independent bits are used in a DES key, 2^{56} such tests are required to guarantee finding a particular key. The expected number of tests needed to recover the correct key is 2^{55}. At one microsecond per test, 1142 years would be required. Under certain conditions (not only knowing matched plaintext and ciphertext but also the complement of the plaintext and the resulting ciphertext), the expected effort could be reduced to 571 years. The possibility of 2^{56} keys (approximately 70 quadrillion) makes the guessing or computing of any particular key very unlikely, given that the guidelines for generating and protecting a key provided in this publication are followed. Of course, one can always reduce the time required to exhaust any cryptoalgorithm by having several devices working in parallel; time is reduced but initial expenses are increased.

An important characteristic of the DES algorithm is its flexibility for usage in various data-processing applications. Each cipher block is independent of all others, allowing encryption or decryption of a single block in a message or data structure. Random access to encrypted data is therefore possible. The algorithm may be used in this straightforward way to form a block cipher or, alternatively, used with chaining in which the output of the algorithm depends on previous results of the algorithm. The first technique is called the Electronic Codebook (ECB) mode and the chaining technique has two examples (discussed in these guidelines) called the Cipher Block Chaining (CBC) mode and the Cipher Feedback (CFB) mode. In addition, DES may be used in the Output Feedback (OFB) mode to generate a pseudorandom stream of bits which is Exclusive-OR'ed to the plaintext bits to form cipher. These will be discussed in the section on Modes of Operation.

The DES algorithm is mathematically a one-to-one mapping of the 2^{64} possible input blocks onto all 2^{64} possible output blocks. Since there are 2^{56} possible active keys, there are 2^{56} possible mappings. Selecting one key selects one of the mappings.

The input to the algorithm is under the complete specification of the designer of the cryptographic system and the user of the system. Any pattern of 64 bits is acceptable to the algorithm. The format of a data block may be defined for each application. In the ECB mode, the subfields of each block may be defined to include one or more of the following: a block sequence number, the block sequence number of the last block received from the transmitter, error-detecting/-correcting codes, control information, date and time information, user or terminal authentication information, or a field in which random data is placed to ensure that identical data fields in different input blocks will result in different cipher blocks. It is recommended that no more than 16 bits be used for known constant values. For example, the same 32-bit terminal identification value should not be used in every block. If it is desired that data blocks in the ECB mode display a sequence dependency, a portion of the last sent or last received block may be incorporated into the block, either as a subfield or Exclusive-OR'ed to the block itself.

The DES algorithm is composed of two parts: the enciphering (encryption) operation and the deciphering (decryption) operation. The algorithms are functionally identical except that the selected portion of the key used for rounds 1, 2, . . . , 16 during the encryption operation are used in the order 16, 15, . . . , 1 for the decryption operation. The algorithm uses two 28-bit registers called C and D to hold the 56-bit active key. The key schedule of the algorithm circularly shifts the C and D registers independently, *left for encryption* and *right for decryption*. If the bits of the C register are all zeros or all ones (after Permuted Choice 1 is applied to the key), and the bits of the D register are all zeros or all ones, then decryption is identical to encryption. This occurs for four known keys: 0101010101010101, FEFEFEFEFEFEFEFE, 1F1F1F1F0E0E0E0E, and E0E0E0E0F1F1F1F1. (Note that the parity bits of the key are set so that each 8-bit byte has odd parity.) It is likely that, in all other cases, data encrypted twice with the same key will not result in plaintext (the original, intelligible data form). This characteristic is beneficial in some data-processing applications in that several levels of encipherment can be utilized in a computer network even though some of the keys used could be the same. If an algorithm is its own inverse, then an even number of encryptions under the same key will result in plaintext.

There are certain keys such that for each key, K, there exists a key, K', for which encryption with K is identical to decryption with K', and vice versa. K and K' are called dual keys. Keys with duals were found by examining the equations which must hold in order for two keys to have reversed key schedules. Keys having duals are keys which produce all zeros, all ones, or alternating zero-one patterns in the C and D register after Permuted Choice 1 has operated on the key. These keys are:

KEY	DUAL
1. E001E001F101F101	01E001E001F101F1
2. FE1FFE1FFE0EFE0E	1FFE1FFE0EFE0EFE
3. E01FE01FF10EF10E	1FE01FE00EF10EF1
4. 01FE01FE01FE01FE	FE01FE01FE01FE01
5. 011F011F010E010E	1F011F010E010E01
6. E0FEE0FEF1FEF1FE	FEE0FEE0FEF1FEF1
7. 0101010101010101	0101010101010101
8. FEFEFEFEFEFEFEFE	FEFEFEFEFEFEFEFE
9. E0E0E0E0F1F1F1F1	E0E0E0E0F1F1F1F1
10. 1F1F1F1F0E0E0E0E	1F1F1F1F0E0E0E0E

The first 6 keys have duals different than themselves, hence each is both a key and a dual giving 12 keys with duals. The last four keys equal their duals, and are called self-dual keys. These are the four previously discussed keys for which double encryption equals no encryption, i.e., the identity mapping. The dual of a key (which has a dual) is formed by dividing the key into two halves of eight hexadecimal characters each and circular shifting each half by two characters. No other keys are known to exist which have duals.

Data may be decrypted first and then encrypted (rather than encrypted and then decrypted) and will result in plaintext. Plaintext may be encrypted several times and then decrypted the same number of times with the same key and will result in plaintext. Similarly, data may be encrypted successively by different keys and decrypted successively by the same keys to produce the original data, if the decryption operations are performed in the proper (inverse) order. If $D_1(E_1(P)) = P$ is read "Encrypting plaintext with Key 1 and then decrypting the result with Key 1 yields the plaintext," then the following are true:

1. $E_1(D_1(P)) = P$
2. $E_1(E_1(P)) = P$ for self-dual keys
3. $D_1(D_1(E_1(E_1(P)))) = P$
4. $E_1(E_1(D_1(D_1(P)))) = P$
5. $D_1(D_2(E_2(E_1(P)))) = P$
6. $D_1(D_2(\ldots (D_j(E_j \ldots (E_2(E_1(P) \ldots) = P$
7. $E_1(E_2(\ldots (E_j(D_j \ldots (D_2 (D_1(P) \ldots) = P$
8. $E_2(E_1(P)) = P$ for dual keys
9. $D_2(D_1(P)) = P$ for dual keys

but, in general, the following is not true:

10. $D_2(D_1(E_2(E_1(P)))) = P$.

Modes of Operation

The DES algorithm specifies a mathematical transformation of a 64-bit input block to a 64-bit output block using a key. Specific examples of this transformation are given in the NBS Special Publication 500-20[5]. $E_K(I) = O$ and $D_K(O) = I$ are read "Enciphering the input I using key K results in output O" and "Deciphering the output O using key K results in input I." Given the same I and K, the same O always results. Likewise, given the same O and K, the same I results.

If the input at time t is called I_t, then the output is called O_t. A sequence of input blocks to the DES may be denoted as $I_1, I_2, I_3, \ldots, I_n$. The outputs are similarly denoted as $O_1, O_2, O_3, \ldots, O_n$.

The DES specifies only the functions E and D. Other considerations will define the input and how the output is used. Many different possibilities exist but the application generally dictates which ones are feasible. In order to provide compatibility between devices which are able to communicate, four modes of operation are used. These are specified in FIPS PUB 81.

The Electronic Codebook (ECB) Mode

The simplest mode of operation, the Electronic Codebook (ECB), is the DES algorithm specified in FIPS PUB 46. The ECB mode is shown in Figs. 12-1 through 12-4. In the ECB mode of operation, the algorithm is independent of time and is called a memoryless system. Given the same data and the same key, the resultant cipher will always be the same. This characteristic should be considered when designing a cryptographic system using the ECB mode. The output block O_t is not dependent on any of the previous inputs, $I_1, I_2, \ldots, I_{t-1}$. It is important to note that the full 64 bits of the O_t must be available in order to obtain the original input I_t.

A general guideline for using the DES in this mode is that all possible inputs should be allowed and used whenever possible. Since the security of the data in this mode is based on the number of inputs in the Codebook, this number should be maximized whenever possible. In particular, this mode should never be used for enciphering single characters (e.g., enciphering 8-bit ASCII characters by entering them in a fixed 8-bit position and filling the other 56 bits with a fixed number). In this mode, 2^{64} inputs are possible, and as large a subset as feasible should be used. Random information should be used to pad small blocks and the random information discarded when the block is deciphered.

Data should be entered into the input register so that the first character of input appears on the left, the second character to the right of it, etc., and the last character on the far right. Using shift-register technology, the characters should be entered on the right and be shifted left until the register is full. Similarly, the output of the DES should be taken from left to right when being transmitted or stored in the character serial mode. Using shift-register technology, the characters should exit from the left and the register shifted left until the register is empty.

The Cipher Block Chaining (CBC) Mode

A method of using the DES algorithm, in which the blocks of cipher are chained together, is called the Cipher Block Chaining (CBC) mode. Fig. 12-5 demonstrates how the CBC mode is used to encrypt a message. The input to the DES at time *t* is defined to be the Exclusive-OR (represented by ⊕) of the data at time *t* and the cipher at time $t-1$. The cipher at time *0* is defined to be a quantity called the *initialization vector*, or IV. The CBC mode requires complete blocks of 64 bits until the final block is to be enciphered.

The final (terminal) data block of a message or record may not contain exactly 64 bits when processing in the CBC mode. When this occurs, either the terminal block must be padded to 64 bits or the terminal block must be enciphered in a way that yields the same number of bits as the input. The first technique is called *padding* and the second is called *truncation*.

When a sequence of characters is being enciphered and the terminal block contains less than the maximum number of characters (e.g., eight in the case of 8-bit characters), then padding may be used to format the final input block in the following way. Suppose "P" padding characters are needed to fill out the block. If P equals one, the character representing the number one should be put in the last byte position. If P is greater than one, the character representing the number P should be put in the last byte and zeros should be put in the remaining $P-1$ byte positions (Fig. 12-5). In most coding schemes, the last three bits of the character representing a digit are the same as the binary representation of the digit (e.g., the ASCII representation of the character 4 is a hexadecimal 34). One bit may either be used in the header block of a message packet to signify a padded message (i.e., that the final block of the packet is padded) or some other method must be devised.

Truncation may be used in the CBC mode when the number of cipher bits must be the same as the number of input bits. It may be necessary that an enciphered tape contain the same number of records and the same number of characters per record as the unenciphered tape. This requirement also occurs in some message-switching systems in which the record length is fixed. In these cases, the following method can be used to encipher the terminal block which does not contain 64 bits.

The short terminal block is enciphered by encrypting the previous cipher block in the ECB mode and

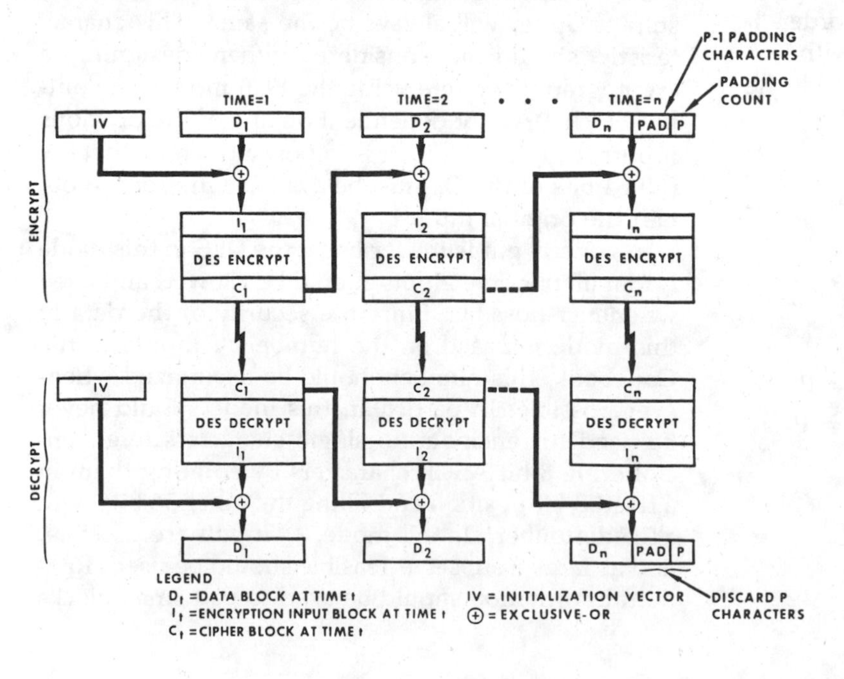

Fig. 12-5. Cipher Block Chaining (CBC) mode—with terminal block padding.

Exclusive-OR'ing the result to the terminal data block (Fig. 12-6). The receiver must detect the short cipher block and perform the same operation; i.e., encrypt the previous complete cipher block and perform the Exclusive-OR operation to obtain the original plaintext block. If a short terminal block contains B bits, then the leftmost B bits of the enciphered cipher block are used. This technique normally provides adequate security for the final block, but it should be noted that if the last B bits of plaintext are known to an active wiretapper, he or she may alter the last B bits of cipher so that they will decrypt to any desired plaintext. This is because, if only the last B bits are altered, the same value will be Exclusive-OR'ed to the short cipher block upon decryption.

One or more bit errors within a single cipher block will affect the decryption of two blocks (the block in which the error occurs and the succeeding block). If the errors occur in the tth cipher block, then each bit of the tth plaintext block will have an average error rate of 50%. The $(t+1)^{st}$ plaintext block will have only those bits in error which correspond directly to the cipher bits in error, and the $(t+2)^{nd}$ plaintext block will be correctly decrypted. Thus, the CBC mode synchronizes itself one block after the error.

The Cipher Feedback (CFB) Mode

The Cipher Feedback (CFB) mode of operation may be used in applications which require chaining to prevent substitution or where blocks of 64 bits cannot be used efficiently. Most computer data that are to be transmitted or stored are coded in 6- to 8-bit codes. FIPS PUB 1 [9] requires the use of the 7-bit ASCII code for interchange. In many communications protocols, the units of data are bits or characters rather than block. The Cipher Feedback mode of using the DES satisfies a requirement for encrypting data elements of length K where $1 \leq K \leq 64$.

The CFB mode of operation is shown in Fig. 12-7. The input to the DES algorithm is not the data itself but rather the previous 64 bits of cipher. The first encryption uses an initialization vector (IV) as its I_0 input. In the CFB mode, both the transmitter and the receiver of data use only the encryption operation of the DES. The output at time t is the 64-bit block O_t. The cipher at time t is produced by Exclusive-OR'ing the K bits of plaintext, P_t, to the leftmost K bits of O_t. This cipher, C_t, is transmitted and also is entered on the right side of the input register after the previous input is shifted K-bit positions to the left. The new input is used for the next encipherment.

A 64-bit IV is generated at time 0 and put into the input register. From that time on, the cipher text will depend on this initial input. In order to fill the receiver's input register, one of two events must occur:

1. The receiver must independently generate the identical initial fill.
2. The transmitter must transmit sufficient data to fill the receiver's input register.

A guideline is that the transmitter generates a pseudorandom number (48 to 64 bits) and transmits it as the IV. The transmitter and the receiver shall use this number (with the high-order bits of the 64-bit DES input padded with "0" bits if necessary) as the 64-bit IV. Using a higher number of bits provides higher

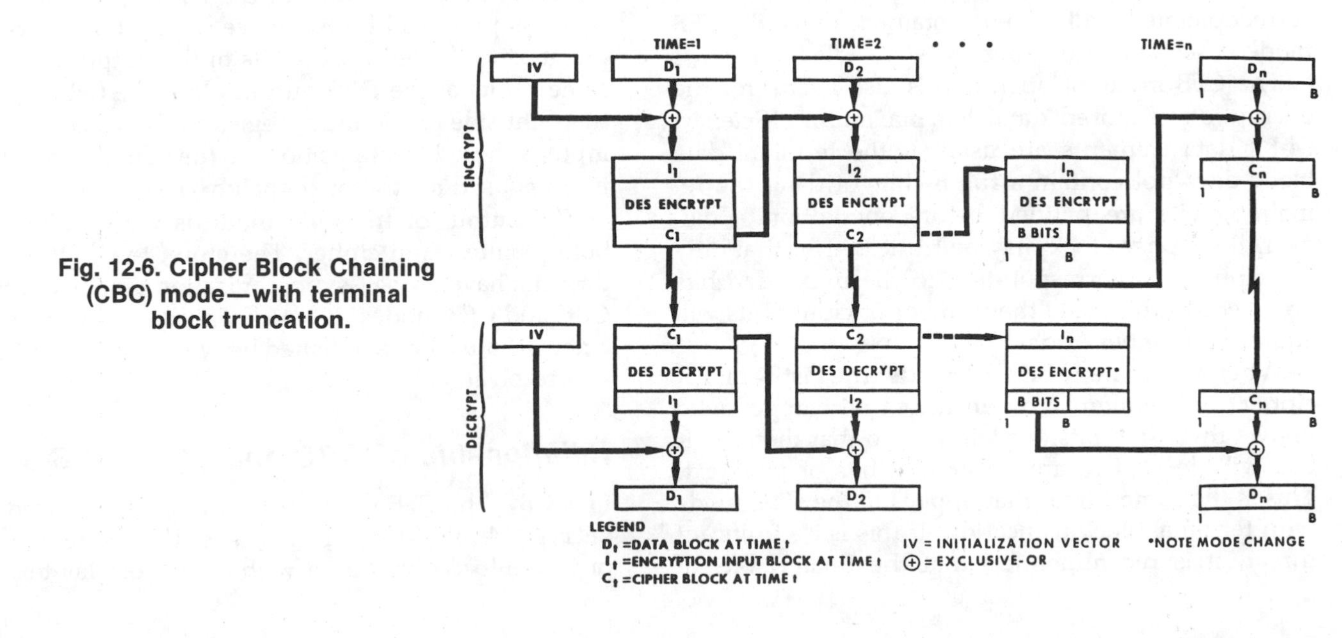

Fig. 12-6. Cipher Block Chaining (CBC) mode—with terminal block truncation.

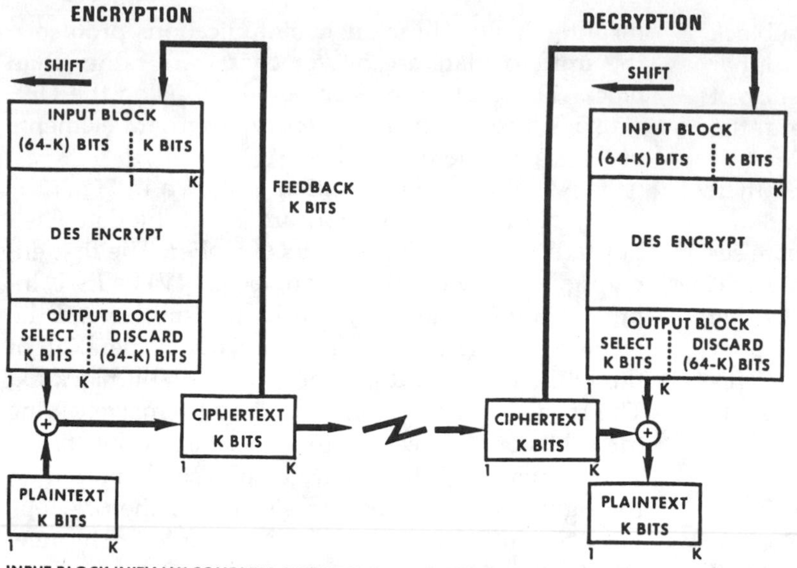

Fig. 12-7. K-bit Cipher Feedback (CFB) mode.

security but it also results in higher transmission overhead. It is desirable that no two messages enciphered with the same key use the same IV. The DES may be used as a pseudorandom number generator to generate the IV. Start–stop (asynchronous) communications devices should transmit the IV as characters with appropriate start–stop bits appended.

In the CFB mode, errors within a K-bit unit of cipher will affect the decryption of the garbled cipher and, also, the decryption of succeeding cipher until the bits in error have been shifted out of the DES input block. The first affected K-bit unit of plaintext will be garbled in exactly those places where the cipher is in error. Succeeding decrypted plaintext will have an average error rate of 50% until all errors have been shifted out of the input block. Assuming no additional errors are encountered during this time, the correct plaintext will then be obtained. Thus, the CFB mode is self-synchronizing.

The CFB mode of operation is also useful for the encryption of stored data. For maximum efficiency, 64-bit data elements are used. If the terminal data block does not contain a full 64 bits of data, the remaining bits are padded before encryption. However, the cipher block may be truncated so that only the cipher bits corresponding to the unpadded bits are used. In this case, the number of cipher bits will equal the number of data bits.

When using the K-bit CFB mode, the last K bits of cipher can be altered by an active wiretapper who knows the last K bits of plaintext, so that the final K bits will decrypt to any desired K bits of plaintext. This is the same threat that applies to the CBC mode with terminal block truncation. If this is a significant threat, it is recommended that the final K bits of plaintext be a function of the previous plaintext bits (i.e., a parity or sum check).

The Output Feedback (OFB) Mode

The Output Feedback (OFB) mode, like the CFB mode, operates on data units of length "K," where K is an integer from 1 to 64. However, the OFB mode does not chain cipher from one time to the next. A 1-bit error in cipher text causes only 1 bit of the decrypted plaintext to be in error. Therefore, this mode can be useful in applications where no error propagation is required.

Fig. 12-8 illustrates the OFB mode. The first encryption uses an initialization vector (IV) as its I_0 input, and both the transmitter and receiver use only the encryption operation of the DES. The cipher at time *t* is produced by Exclusive-OR'ing the K bits of plaintext to the leftmost K bits of the output O_t. The same K bits of the DES output block are fed back to the right side of the input register after the previous input is shifted K-bit positions to the left, and the new input is used for the next encipherment.

The output of the OFB mode is independent of both plaintext and cipher. Therefore, the OFB mode does not have the self-synchronization property of the CBC and CFB modes. If synchronization is lost, then a new IV must be established between the transmitter and receiver.

Relationship of CBC and 64-bit CFB

Like CBC, the CFB mode of operation can be used to encrypt 64-bit blocks. In this case, the entire 64 bits of O_t are Exclusive-OR'ed with 64 bits of plaintext at

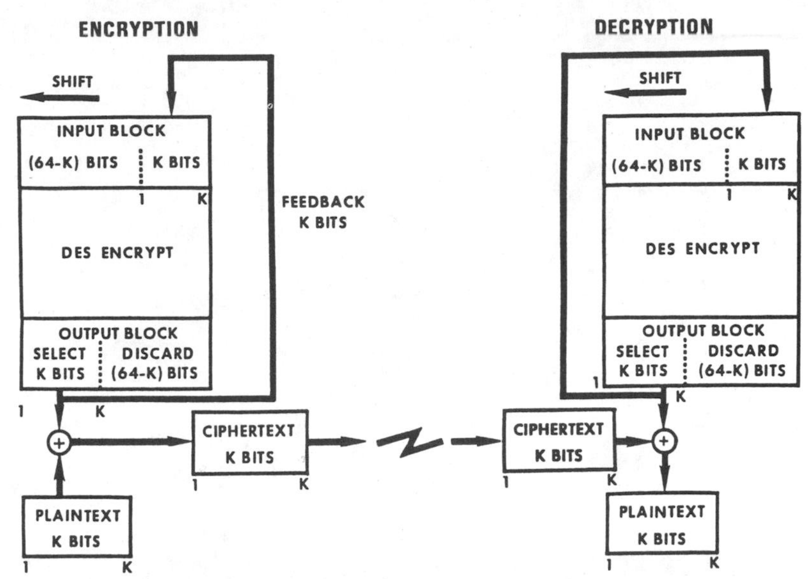

Fig. 12-8. K-bit Output Feedback (OFB) mode.

each encryption time. This is called the 64-bit CFB mode of operation.

Let M1 be a 64-bit CFB machine with key schedule, KS = (K_1,K_2, . . . ,K_{16}), on each of the 16 encryption rounds. (Fig. 12-3 shows the generation of a DES key schedule.) In CFB mode, the same schedule is also used for decryption. Let M2 be a CBC machine with a key schedule of KR = (K_{16},K_{15}, . . . ,K_1) for encryption (i.e., the DES decipher operation), and (K_1,K_2, . . . ,K_{16}) for decryption (i.e., the DES encipher operation). If M1 encrypts the 64-bit plaintext blocks P_1, P_2, and P_3 with initialization vector IV to form cipher C_1, C_2, and C_3, then M2 will encrypt P_3, P_2, and P_1 with initialization vector C_3 to form cipher C_2, C_1, IV. Similarly while M1 will decrypt C_1, C_2, and C_3 (using initialization vector IV) to P_1, P_2, and P_3, M2 will decrypt C_2, C_1, and IV (using initialization vector C_3) to P_3, P_2, and P_1. Thus, by reversing (IV,C_1,C_2,C_3) to (C_3,C_2,C_1,IV), we may decrypt cipher generated by M1 with M2.

To see that the above statements are true, let E[S](X) represent the encryption of X in the ECB mode using key schedule S, and let D[S](X) be the ECB decryption of X under schedule S. Note that S is the key schedule and not the key itself. Decryption uses the key schedule in the reverse order of encryption. Thus, E[KS](X) = D[KR](X). The encryption of P_1, P_2, and P_3 by M1, using IV, may be described by three equations.

$$P_1 \oplus E[KS](IV) = P_1 \oplus O_1 = C_1$$
$$P_2 \oplus E[KS](C_1) = P_2 \oplus O_2 = C_2$$
$$P_3 \oplus E[KS](C_2) = P_3 \oplus O_3 = C_3$$

O_1, O_2, and O_3 represent ECB encryption, with key schedule KS, of inputs IV, C_1, and C_2, respectively. The math symbol, $\oplus$, is a 64-bit Exclusive-OR operator. The encryption of P_3, P_2, and P_1 by M2, using C_3 as the initialization vector, may also be described by three equations.

$$E[KR]\ (P_3 \oplus C_3) = E[KR]\ (O_3) = D[KS]\ (O_3) = C_2$$
$$E[KR]\ (P_2 \oplus C_2) = E[KR]\ (O_2) = D[KS]\ (O_2) = C_1$$
$$E[KR]\ (P_1 \oplus C_1) = E[KR]\ (O_1) = D[KS]\ (O_1) = IV$$

By reversing the key schedules, the inputs, and the outputs, we have obtained equivalent machines. Similar equations may be derived for decryption, and the relationship holds for an arbitrary length stream of 64-bit plaintext blocks.

13

Manufacturers' Data Sheets

This chapter contains partial data specification sheets for some of the semiconductor devices used in descramblers. The devices included herein are useful in a large percentage of the scrambler/descrambler applications.

Space does not permit the inclusion of the complete data sheets or data sheets for all the semiconductor devices used in descramblers; thus, it is suggested the reader write to the various semiconductor and equipment manufacturers to obtain complete and up-to-date data sheets for the devices that are pertinent to their application.

Advance

DATA SHEET

T7000A Digital Encryption Processor

FEATURES

- Programmable DES Ciphering Modes
  - ☐ Electronic Codebook (ECB)
  - ☐ Cipher Block Chaining (CBC)
  - ☐ 1-, 8-, or 64-bit Cipher Feedback (CFB)
  - ☐ Output Feedback (OFB)
- Ciphering Rates of 235,000 Operations per Second for Any of the DES Modes. Data Throughput of 1.882M Bytes/s Using the Entire DES Output Block
- Programmable Multiple or Multiplexed Ciphering
- On-chip RAM and ROM Program Memory
- Flags Readable on the Data Bus or Independent Output Pins
- Four Sets of Key and Initial Value Registers
- Odd Parity Verification of Each Key Variable
- Separate Plain Text and Cipher Text Parallel (8-bit) Ports
- Separate Plain Text and Cipher Text Serial Ports
- Separate Serial Key Input Port
- ECB Program Available in ROM

DESCRIPTION

The T7000A Integrated Circuit is a programmable, Digital Encryption Processor (DEP) which provides a low-cost, high-security, cryptographic system for encrypting and decrypting digital signals. It is manufactured using CMOS technology, requires a single +5 volt supply, and is supplied in a 40-pin plastic DIP. It implements four Data Encryption Standard (DES) modes, and is capable of performing multiple encryption operations or multiplexed key and initial value ciphering. The DEP has been validated by the National Bureau of Standards to be in compliance with the Data Encryption Standard.

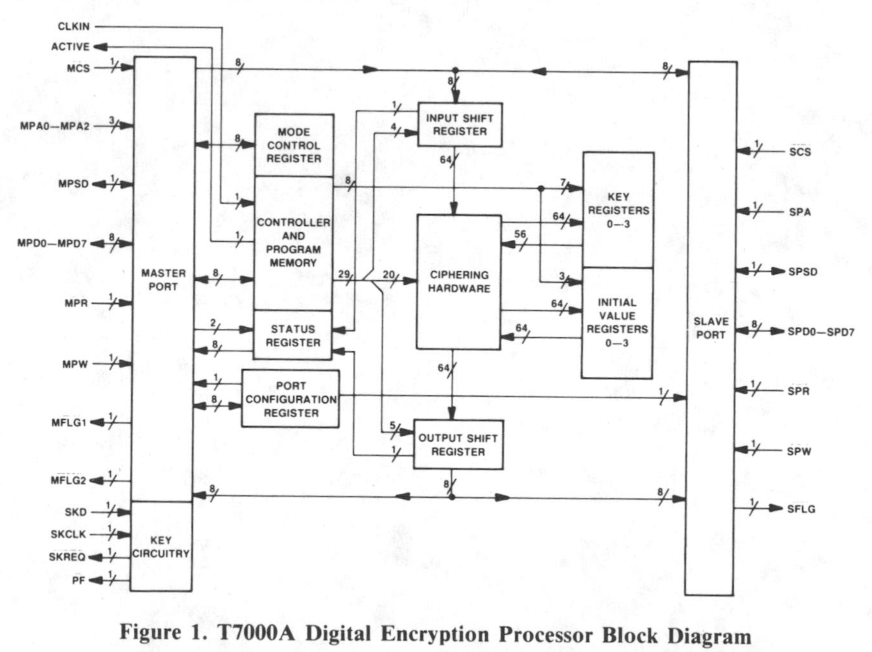

Figure 1. T7000A Digital Encryption Processor Block Diagram

Information contained on this document is advance and subject to change without notice.

August 1985

USER INFORMATION

Pin Descriptions

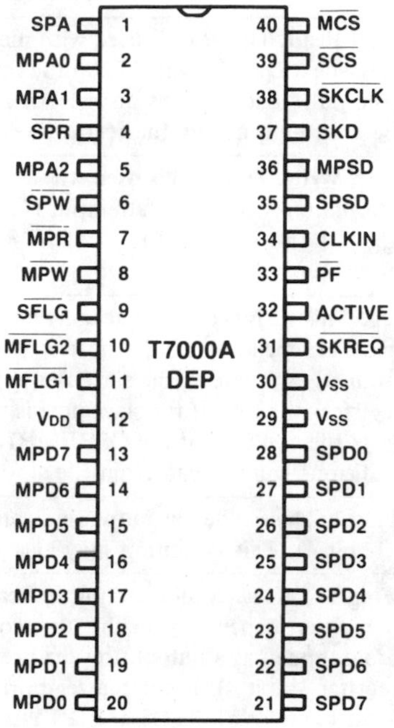

Symbol	Pin	Pin	Symbol
SPA	1	40	$\overline{MCS}$
MPA0	2	39	$\overline{SCS}$
MPA1	3	38	$\overline{SKCLK}$
$\overline{SPR}$	4	37	SKD
MPA2	5	36	MPSD
$\overline{SPW}$	6	35	SPSD
$\overline{MPR}$	7	34	CLKIN
$\overline{MPW}$	8	33	$\overline{PF}$
$\overline{SFLG}$	9	32	ACTIVE
$\overline{MFLG2}$	10	31	$\overline{SKREQ}$
$\overline{MFLG1}$	11	30	Vss
VDD	12	29	Vss
MPD7	13	28	SPD0
MPD6	14	27	SPD1
MPD5	15	26	SPD2
MPD4	16	25	SPD3
MPD3	17	24	SPD4
MPD2	18	23	SPD5
MPD1	19	22	SPD6
MPD0	20	21	SPD7

T7000A DEP

Figure 2. Pin Function Diagram

Table 1. Pin Descriptions

Pin	Symbol	Type	Name/Function
1	SPA	I	**Slave Port Address.** When high, the contents of the status register can be read, but not written, to the slave port data bus. When low, either the input shift register (ISR) or output shift register (OSR) is accessed, depending on the port configuration programmed.
2 3	MPA0 MPA1	I I	**Master Port Address Bits 0 and 1.** Used with MPA2 (pin 5) for internal register selection.
4	$\overline{SPR}$	I	**Slave Port Read.** This lead is used with the slave address lead to read from the output shift register (if the slave port is programmed as an output), or from the status register. Data is available on the slave port data bus following the falling edge of the pulse and remains on the bus as long as the $\overline{SPR}$ is low. The $\overline{SPW}$ lead should be held high during a read pulse.
5	MPA2	I	**Master Port Address Bit 2.** Used with MPA0 and MPA1 (pins 2 and 3) for internal register selection.
6	$\overline{SPW}$	I	**Slave Port Write.** This lead is used with the slave address lead to write to the input shift register if the slave port has been programmed as an input. The data input is latched on the rising edge of the write pulse. The $\overline{SPR}$ lead should be held high during the write pulse.

Pin	Symbol	Type	Name/Function
7	$\overline{\text{MPR}}$	I	**Master Port Read.** This lead is used with master port address bus to read one of the internal registers. Data is available on the master port data bus following the falling edge of the pulse and remains on the bus as long as $\overline{\text{MPR}}$ is low. The $\overline{\text{MPW}}$ lead should be held high during the read pulse.
8	$\overline{\text{MPW}}$	I	**Master Port Write.** This lead is used with the master port address bus to write to one of the internal registers. The data input is latched into the addressed register on the rising edge of the write pulse. The $\overline{\text{MPR}}$ lead should be held high during the write pulse.
9	$\overline{\text{SFLG}}$	O	**Slave Flag.** This active-low output indicates the status of either the input or output shift registers, depending on the port configuration programmed. If the slave port is programmed as an input, the slave flag reflects the contents of the ISRFULL flag (status register — bit 4). If the slave port is programmed as an output, then the slave flag reflects the contents of the OSREMPTY flag (status register — bit 5). Both of these conditions can be read from the status register.
10	$\overline{\text{MFLG2}}$	O	**Master Flag 2.** This active-low output indicates the status of the ISRFULL flag (status register — bit 4). This condition may also be read from the status register.
11	$\overline{\text{MFLG1}}$	O	**Master Flag 1.** This active-low output indicates the status of either the input or output shift register, depending on the port configuration programmed. If the master port is programmed as an input, this lead reflects the contents of the ISRFULL flag (status register — bit 4). If the master port is programmed as an output, this lead indicates the contents of the OSREMPTY flag (status register — bit 5). If the master port is programmed as both input and output, this lead indicates the contents of the OSRFULL flag and $\overline{\text{MFLG2}}$ indicates the contents of the ISRFULL flag. The status of the input and output shift register can also be read from the status register.
12	VDD	—	**+5 Volt Supply.**
13	MPD7	I/O	**Master Port Data Bit 7.** (Bidirectional, 8-bit Master Port I/O bus — pins 13–20)
14	MPD6	I/O	**Master Port Data Bit 6.**
15	MPD5	I/O	**Master Port Data Bit 5.**
16	MPD4	I/O	**Master Port Data Bit 4.**
17	MPD3	I/O	**Master Port Data Bit 3.**
18	MPD2	I/O	**Master Port Data Bit 2.**
19	MPD1	I/O	**Master Port Data Bit 1.**
20	MPD0	I/O	**Master Port Data Bit 0.**
21	SPD7	I/O	**Slave Port Data Bit 7.** (Bidirectional, 8-bit Slave Port I/O bus — pins 21–28)
22	SPD6	I/O	**Slave Port Data Bit 6.**
23	SPD5	I/O	**Slave Port Data Bit 5.**
24	SPD4	I/O	**Slave Port Data Bit 4.**
25	SPD3	I/O	**Slave Port Data Bit 3.**
26	SPD2	I/O	**Slave Port Data Bit 2.**
27	SPD1	I/O	**Slave Port Data Bit 1.**
28	SPD0	I/O	**Slave Port Data Bit 0.**
29	Vss	—	**Ground.**
30	Vss	—	**Ground.**

Table 1. Pin Descriptions (Continued)

(Reproduced with permission of AT&T)

Table 1. Pin Descriptions (Continued)			
Pin	**Symbol**	**Type**	**Name/Function**
31	$\overline{\text{SKREQ}}$	O	**Serial Key Request.** This active-low output indicates the DEP is expecting a key in put. This lead is active when IO Serial Act is programmed. The condition of this flag can be read from the status register.
32	ACTIVE	O	This active-high output flag is set by the microcode instruction IO ACT.
33	$\overline{\text{PF}}$	O	**Parity Fail.** When this output is low it indicates that one or more key input bytes had even parity. This flag is set on the 8th $\overline{\text{MPW}}$ pulse when the key is loaded through the parallel master port, and on the 64th SKCLK pulse when the key is loaded serially. The status of this flag can be read from the status register.
34	CLKIN	I	**Clock Input.** The clock signal input at this lead determines all internal timing. A microcode instruction is executed every two clock cycles. The master and slave ports' read and write signals are not required to be synchronous with this clock signal. The frequency range of this clock is 10 kHz to 2.5 MHz.
35	SPSD	I/O	**Slave Port Serial Data.** Depending on the programmed port configuration, this bidirectional lead is used to write data to the input shift register or read data from the output shift register. The first bit read or written is the most significant. When this port is selected by the port configuration register, the slave port signals $\overline{\text{SPW}}$, $\overline{\text{SPR}}$, and $\overline{\text{SFLG}}$ are used for control. This port may not be used to read or write to any of the other six registers.
36	MPSD	I/O	**Master Port Serial Data.** Depending on the programmed port configuration, this bidirectional lead is used to write data to the input shift register or read data from the output shift register. The first bit read or written is the most significant. When this port is selected by the port configuration register and master port address 0 is addressed, the master port signals $\overline{\text{MPW}}$, $\overline{\text{MPR}}$, $\overline{\text{MFLG1}}$, and $\overline{\text{MFLG2}}$ are used for control. This port may not be used to read or write to any of the other six ports.
37	SKD	I	**Serial Key Data.** This input port is used to load key variables serially. The data on this pin is latched into key memory on the falling edge of the serial key clock during the execution of a serial load key program. The key is entered with the most significant bit first and every 8th bit is treated as an odd parity bit. A parity failure will not prevent the 56-bit key from being loaded.
38	SKCLK	I	**Serial Key Clock.** This clock is used to latch key data into key memory. Data is latched on the falling edge of the clock. The key input circuitry is inhibited after the 64th clock is received.
39	$\overline{\text{SCS}}$	I	**Slave Chip Select.** This active-low input enables the slave port input and output leads. When high, all slave port outputs are placed in a high-impedance state. The $\overline{\text{SPW}}$, $\overline{\text{SPR}}$, SPSD, and SPD0 – SPD7 signals are affected.
40	$\overline{\text{MCS}}$	I	**Master Chip Select.** This active-low input enables the master port input and output leads. When high, all master port outputs are placed in a high-impedance state and all inputs are disabled. The $\overline{\text{MPW}}$, $\overline{\text{MPR}}$, MPSD, and MPD0 – MPD7 signals are affected.

T7000A Digital Encryption Processor

Overview

Figure 1 shows a block diagram of the DEP. There are three major sections: the ciphering hardware and peripheral circuitry, the controller and program memory, and the ports.

The **ciphering hardware** contains a high-speed hardware implementation of the National Bureau of Standards Data Encryption Algorithm (DEA), and the necessary hardware to configure the DES operating modes. (See Figure 3.)

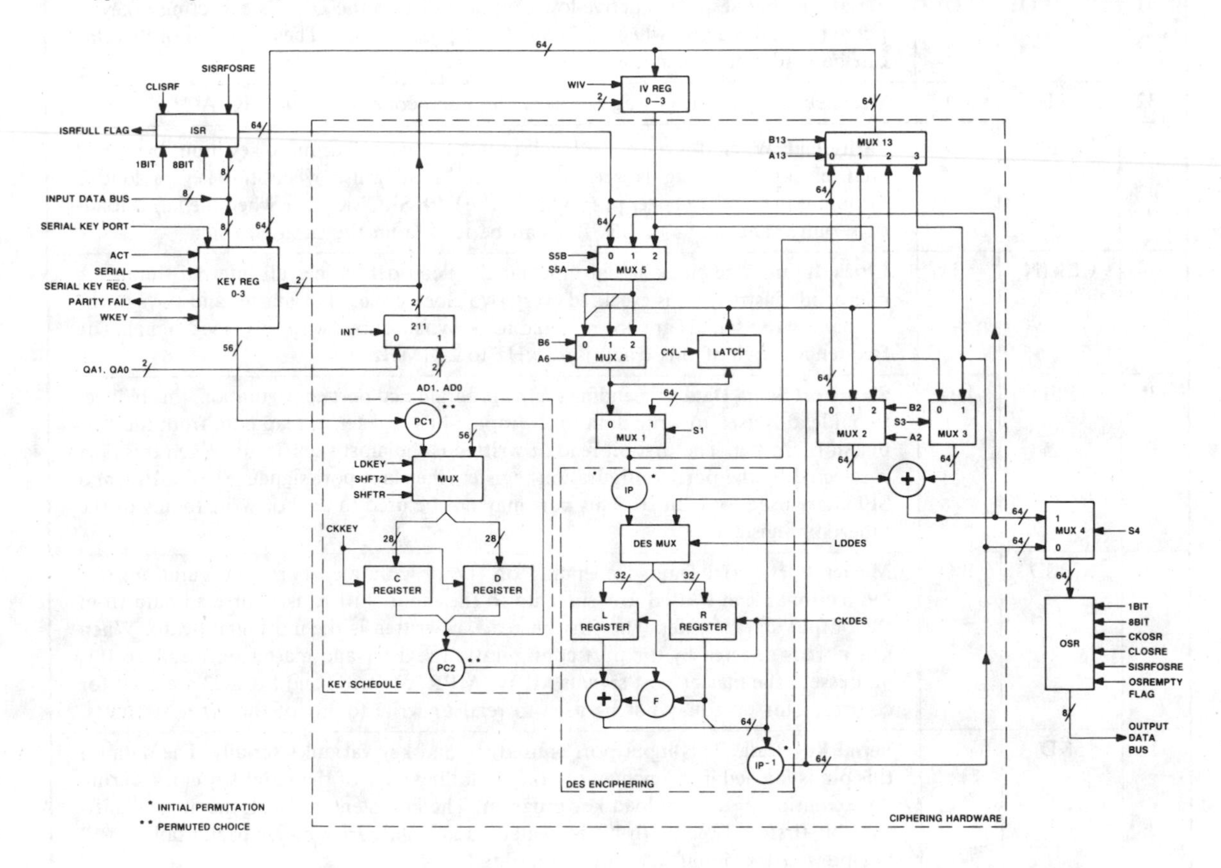

Figure 3. Ciphering Hardware Block Diagram

Both the key schedule and DES enciphering circuitry are part of the DEA algorithm. The remaining circuitry (seven multiplexers, an exclusive-OR gate, and a latch) is used for the DES operating modes. An input shift register (ISR), four key registers, four initial value registers, and an output shift register (OSR) support the ciphering hardware.

An internal **hardware controller** executes a 22-bit machine instruction every two clock cycles, thereby setting up the ciphering multiplexers and clocking the appropriate registers. Within the controller, a program counter is used to address the machine instruction stored in either RAM or ROM program memory. On-chip ROM (29 x 22 bits) contains a subroutine controlling the DES hardware, a load initial value program, a load key program, a serial load key program, and an ECB encrypt and decrypt program. These short programs are located at hex address 00 through 1C. User accessible on-chip RAM (32 x 22 bits) allows the user to tailor the ciphering operation to meet system requirements and eliminate external hardware. These ciphering programs must start at hex address 20 and may not exceed hex address 3F.

T7000A Digital Encryption Processor

Master and slave ports are provided so that the plain text and cipher text can be on separate buses. These ports have both serial and 8-bit parallel bidirectional data buses. When using the 8-bit parallel data bus, master or slave, the most significant data or key byte should be written/read first. In the serial mode, the most significant bit is written/read first.

Table 2. Register Assignments

Master Port (MP)		
MP Address	**Register**	**Size (Bytes)**
0 (Write)	Input Shift	8
0 (Read)	Output Shift	8
1	Status	1
2	Port Configuration	1
3	Mode Control	1
4	M1	1
5	M2	1
6	M3	1
Slave Port (SP)		
SP Address	**Register**	**Size (Bytes)**
0 (Write)	Input Shift	8
0 (Read)	Output Shift	8
1 (Read)	Status	1

Registers

Eight addressable, internal registers control device operation. Table 2 shows the register assignments for both the master and slave ports during either a read or write operation.

Both the **input and output shift registers** (master or slave port address 0) may be accessed from the Master Port Data bus (MPD), Master Port Serial Data pin (MPSD), Slave Port Data bus (SPD), or Slave Port Serial Data pin (SPSD). The Input Shift Register (ISR) is a 64-bit, write-only, shift register. The Output Shift Register (OSR) is a 64-bit, read-only, shift register. The port configuration register controls which port, master or slave, is associated with the input or output shift register. These shift registers are used to input and output data, and would not normally be accessed until the other registers are loaded.

If a parallel port is used, 8 bytes are read or written to empty or load these registers, except when the 1- or 8-bit Cipher Feedback (CFB) mode has been programmed. In these cases a single byte is expected. For 1-bit CFB, only the most significant bit of the byte is used.

If a serial port is used, 64 bits are read or written to empty or load these registers, except when the 1- or 8-bit CFB mode has been programmed. One bit is expected for 1-bit CFB and eight bits for 8-bit CFB.

The **status register** (master or slave port address 1) may be read or written from the master port data bus or read only from the slave port data bus. (see Figure 4.)

(Reproduced with Permission of AT&T)

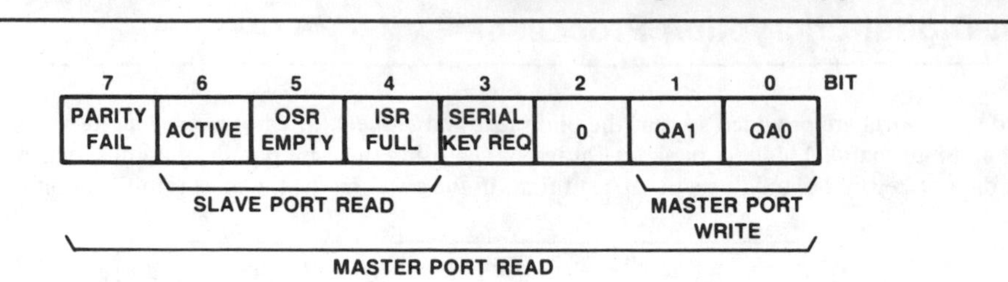

Figure 4. Status Register

Bits 1 and 0 (QA1, QA0) are read/write address lines that are used to select key and initial value register pairs 0-3 when the microcode instruction bit INT is not set. Key and initial value registers are matched sets; i.e., 00 selects key register 0 and initial value register 0 (see Table 3). The values are loaded into these registers by executing the appropriate program in ROM.

Table 3. Key and Initial Value Register Addresses

Bits		Key and Initial Value Register Number
QA1	**QA0**	
0	0	0
0	1	1
1	0	2
1	1	3

Bit 3 is a read-only, active-high, serial key request (SKREQ) flag. The complement of this flag ($\overline{\text{SKREQ}}$) is available at output pin 31. SKREQ is microcode-controlled and goes active when the SERIAL and ACT instructions are executed simultaneously.

Bit 4 is a read-only, active-high, input shift register full (ISRFULL) flag. The complement of this flag appears on an output pin; the specific pin ($\overline{\text{MFLG1}}$, $\overline{\text{MFLG2}}$, or $\overline{\text{SFLG}}$) is determined by the port configuration. An active signal indicates that the ISR is full and additional information written to that register is ignored. The ISRFULL flag is set automatically whenever the mode control register is written or after the microcode instruction SISRFOSRE is executed. It is cleared by microcode instruction CLISRF.

Bit 5 is a read-only, active-high, output shift register empty (OSREMPTY) flag. This flag's complement appears on an output pin; the specific pin ($\overline{\text{MFLG1}}$ or $\overline{\text{SFLG}}$) is determined by the port configuration. An active signal indicates that the OSR is empty and additional attempts to read that register are ignored. OSREMPTY is set automatically whenever the mode control register is written or after the microcode instruction SISRFOSRE is executed. It is cleared by the microcode instruction CLOSRE.

Bit 6 is a read-only, active-high, activity (ACTIVE) flag. The condition of this flag appears on output pin 32 (ACTIVE). It is set by the microcode instruction IO ACT and indicates processor activity. The ACTIVE flag has no effect on device operation.

Bit 7 is a read-only, active-high, parity fail flag. The complement of this flag is available at output pin 33 ($\overline{\text{PF}}$). This flag is latched whenever the WKEY instruction is executed. An active condition indicates that one or more of the key bytes entered had even parity. Device operation is not inhibited by the parity fail flag.

The **port configuration register** (master port address 2) is a read/write register accessible only through the master port data bus. Table 4 defines the possible port configurations and associated hex code for data encryption and decryption.

Table 4. Port Configuration (MP Address = 2)				
Port Type	Input	Output	Hex Code	
			Encrypt*	Decrypt*
parallel	MPD	SPD	04	84
parallel	SPD	MPD	11	91
parallel	MPD	MPD	01	81
serial	MPSD	SPSD	28	A8
serial	SPSD	MPSD	62	E2
parallel to serial	MPD	SPSD	08	88
serial to parallel	SPSD	MPD	61	E1

* The most significant bit in the hex code for the port configuration is an input flag. It is tested by the microcode mnemonic LT?. In the microcode for the standard modes given in this document, this bit is tested to determine the order in which the DES key schedule should be used (encrypt or decrypt).

The conditions indicated by the master and slave port flags are determined by the port configuration (see Table 5).

Table 5. Master and Slave Port Flag Conditions				
Port Configuration		Flag Condition		
Input	Output	$\overline{\text{MFLG1}}$	$\overline{\text{MFLG2}}$	$\overline{\text{SFLG}}$
MPD or MPSD	SPD or SPSD	ISRFULL	–	OSREMPTY
SPD or SPSD	MPD or MPSD	OSREMPTY	–	ISRFULL
MPD	MPD	OSREMPTY	ISRFULL	–

Bit 7 of the port configuration register is an input flag which is tested by microcode instruction LT?. This bit may be used to indicate the order in which the key schedule is used (encrypt or decrypt) or as a general purpose conditional jump.

The **mode register** (master port address 3) is a read/write register accessible only through the master port data bus. This register is used to address on-chip memory for read/write operations and to begin program execution. Only the six least significant bits are used in this register.

To run a microcode program, write the starting address for the set of instructions to be executed into the mode register. On the next instruction cycle this address is loaded into a program counter and execution begins.

To read/write the program memory, the address of the instruction is loaded into the mode control register and one of the three hex bytes (M1, M2, or M3) which make up an instruction is read/written on a subsequent $\overline{\text{MPR}}/\overline{\text{MPW}}$ pulse. M1, M2, or M3 is selected using the masterport address bus.

The **M1, M2,** and **M3** registers (master port addresses 4-6, respectively) are accessible only through the master port data bus. These three bytes define a 22-bit microcode instruction stored in on-chip program memory. The two most significant bits of register M3 are not used.

DATA SHEET

T7001 Random Number Generator

FEATURES

- On-chip or External High Frequency Oscillator Source Option
- On-chip or External Jitter Oscillator Source Option
- Generation of a 536-Bit Random Number Available in 8-Bit Bytes
- Internal Verification of RNG Output on the Data Bus
- Data Ready and Alarm Output Flags Readable from the Data Bus or Independent Output Pins Allowing either Processor Interrupt or Processor Polled Configuration
- Internal 4-Bit Statistical "Run-up" Test with Programmable Limits (Elementary Randomness Check). External Access to Generated Statistics

DESCRIPTION

The T7001 Integrated Circuit is a Random Number Generator (RNG) that produces random bits based on the phase jitter of a free-running oscillator. The output data stream is truly random (not pseudo random). The T7001 IC is processed in CMOS technology, requires a single +5 volt supply, and is supplied in a 32-pin plastic DIP.

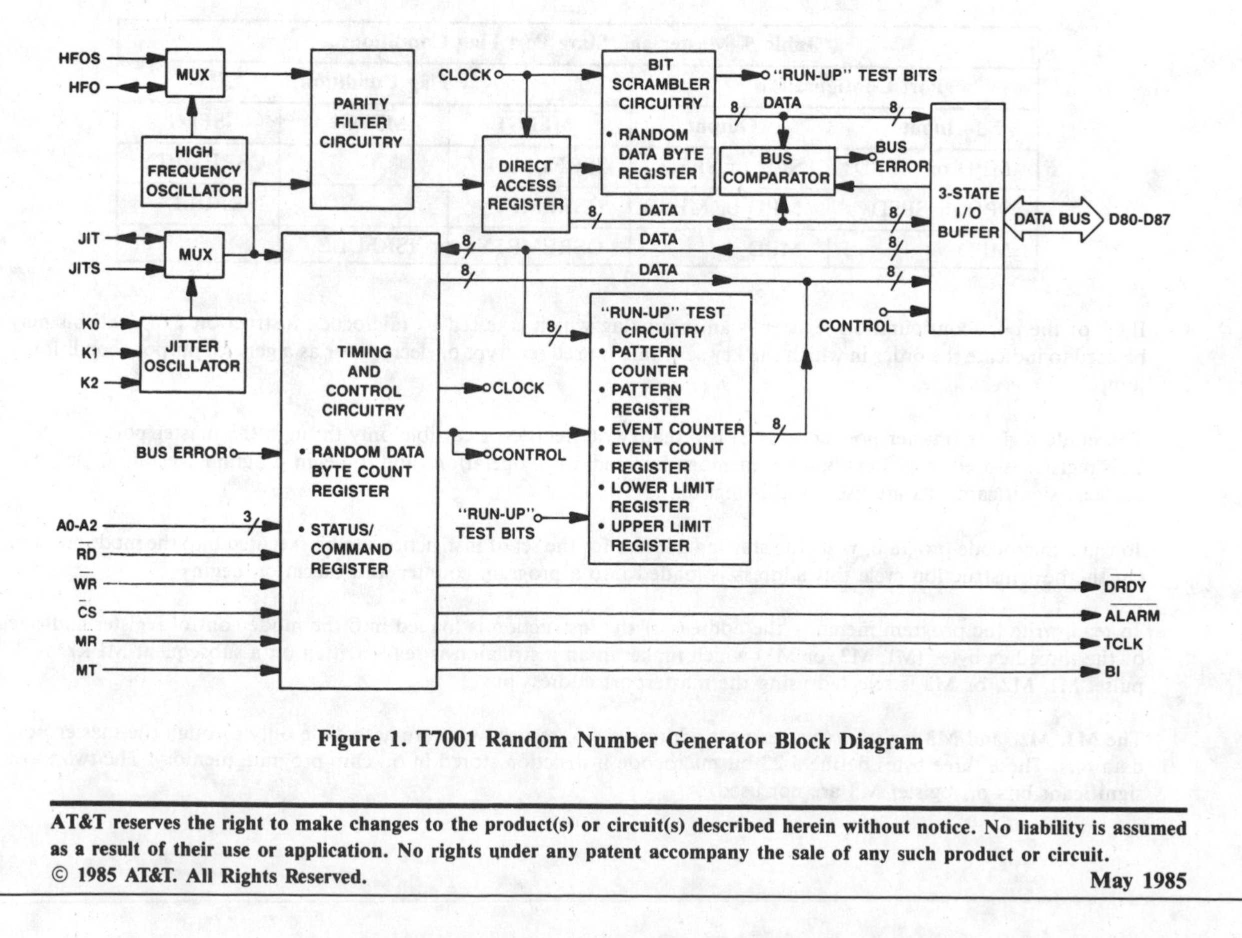

Figure 1. T7001 Random Number Generator Block Diagram

May 1985

(Reproduced with Permission of AT&T)

USER INFORMATION

Pin Descriptions

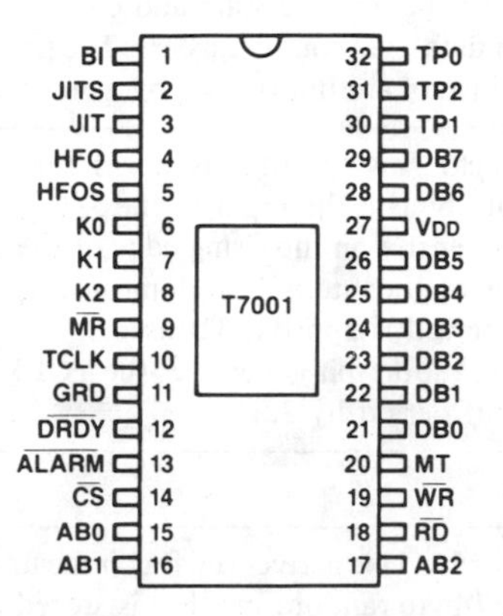

Figure 2. Pin Function Diagram

Table 1. Pin Descriptions

Pin	Symbol	Type	Name/Function
1	BI	O	**Test Pin.** Output of the positive-edge-triggered sampling D flip-flop which has the High Frequency Oscillator (HFO) as its data input and the jitter oscillator (JIT) as its clock. It may be used to verify that both HFO and JIT are working properly.
2	JITS	I	**Jitter Select.** Determines the jitter signal source for the device. When low, the internal jitter oscillator is used. When high, an external jitter oscillator signal is expected at input pin 3 (JIT).
3	JIT	I/O	**Jitter.** Output of the internal jitter oscillator when JITS (pin 2) is low. When JITS is high this pin is an input for an external jitter oscillator.
4	HFO	I/O	**High-Frequency Oscillator.** Output of the internal high-frequency oscillator when HFOS (pin 5) is low. This pin is the input for an external high-frequency oscillator signal when HFOS is high.
5	HFOS	I	**High-Frequency Oscillator Select.** Determines the high-frequency signal source for the device. When low, the internal high frequency oscillator (8 MHz) is selected. When high, an external oscillator signal is expected at pin 4 (HFO).
6 7 8	K0 K1 K2	I I I	A resistor (R) is connected between K0 and K1, and a capacitor (C) is connected between K0 and K2 to control the frequency of the on-chip jitter oscillator. The approximate frequency is determined by the equation: $f = \frac{1}{2.2\ RC}$

(Reproduced with Permission of AT&T)

Table 1. Pin Descriptions (Continued)

Pin	Symbol	Type	Name/Function
9	$\overline{\text{MR}}$	I	**Master Reset.** This active-low input resets the device when the chip select input (pin 14) is active. A master reset puts the command bits in the status/command register to the inactive state and clears the random data-byte, event count, and random data-byte count registers. The pattern register, and lower-limit and upper-limit registers are unaffected by a master reset.
10	TCLK	O	**Test Clock.** This output is used with the direct access register to monitor the random data byte at the input to the 536-bit shift register. Random data is latched into the register on the rising edge of the TCLK signal and can be read after this edge. Using the internal random-bit generator (with jitter osc. set to 1 kHz) the TCLK period is approx. 32 msec. Using a different oscillator frequency or an external random-bit generator, the TCLK period is computed by multiplying the JIT input period by 32.
11	GRD	–	**Ground.**
12	$\overline{\text{DRDY}}$	O	**Data Ready.** This active-low flag indicates that the "run-up" test was passed and that a 67-byte random number is stored in the random data-byte register. This flag may also be read from the status/command register. The $\overline{\text{DRDY}}$ flag goes inactive following the 67th $\overline{\text{RD}}$ pulse, when the random data-byte register (address 0) is addressed.
13	$\overline{\text{ALARM}}$	O	**Alarm.** This active-low flag indicates either a "run-up" test failure or a bus error. Both of these conditions may be read from the status/command register. This flag can only be cleared by a master reset.
14	$\overline{\text{CS}}$	I	**Chip Select.** This active-low input enables $\overline{\text{RD}}$ (pin 18), $\overline{\text{WR}}$ (pin 19), $\overline{\text{MR}}$ (pin 9), and the address bits (pin 15, pin 16, and pin 17). When inactive the data bus output buffers are held in a high-impedance state regardless of the state of any other input.
15	AB0	I	**Address Bus Bit 0.**
16	AB1	I	**Address Bus Bit 1.**
17	AB2	I	**Address Bus Bit 2.**
18	$\overline{\text{RD}}$	I	**Read.** This active-low input is used to read one of the eight internal registers. Data appears on the data bus following the falling edge of this signal and remains on the bus as long as $\overline{\text{RD}}$ is low. The $\overline{\text{WR}}$ lead (pin 19) should be held high (inactive) during a read operation.
19	$\overline{\text{WR}}$	I	**Write.** This active-low input is used to write to one of three internal registers. The data is latched into the addressed register on the rising edge of the write pulse. The $\overline{\text{RD}}$ lead (pin 18) should be held high (inactive) during a write operation.
20	MT	I	**Manufacture Test.** This pin must be grounded for device operation.
21	DB0	I/O	**Data Bus Bit 0.**
22	DB1	I/O	**Data Bus Bit 1.**
23	DB2	I/O	**Data Bus Bit 2.** Bidirectional
24	DB3	I/O	**Data Bus Bit 3.** 3-state I/O leads.
25	DB4	I/O	**Data Bus Bit 4.**
26	DB5	I/O	**Data Bus Bit 5.**

(Reproduced with Permission of AT&T)

Table 1. Pin Descriptions (Continued)			
Pin	Symbol	Type	Name/Function
27	V_{DD}	–	**+5 Volt Supply**
28	DB6	I/O	**Data Bus Bit 6.** Bidirectional 3-state I/O leads.
29	DB7	I/O	**Data Bus Bit 7.** Bidirectional 3-state I/O leads.
30	TP1	–	**Manufacture Test Point.** No Connection.
31	TP2	–	**Manufacture Test Point.** No Connection.
32	TP3	–	**Manufacture Test Point.** No Connection.

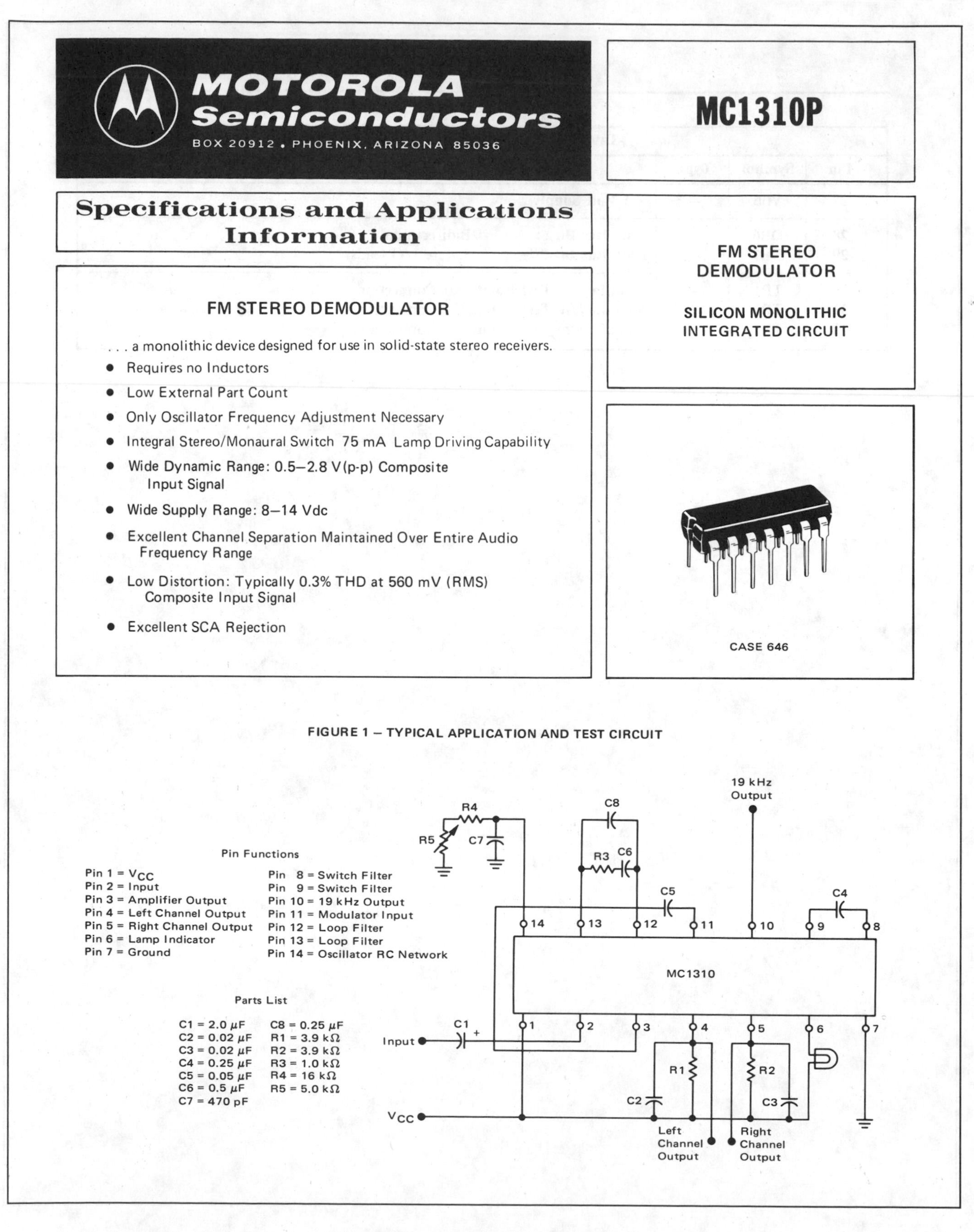

MOTOROLA Semiconductors

BOX 20912 • PHOENIX, ARIZONA 85036

MC1310P

Specifications and Applications Information

FM STEREO DEMODULATOR

. . . a monolithic device designed for use in solid-state stereo receivers.

- Requires no Inductors
- Low External Part Count
- Only Oscillator Frequency Adjustment Necessary
- Integral Stereo/Monaural Switch 75 mA Lamp Driving Capability
- Wide Dynamic Range: 0.5–2.8 V(p-p) Composite Input Signal
- Wide Supply Range: 8–14 Vdc
- Excellent Channel Separation Maintained Over Entire Audio Frequency Range
- Low Distortion: Typically 0.3% THD at 560 mV (RMS) Composite Input Signal
- Excellent SCA Rejection

FM STEREO DEMODULATOR

SILICON MONOLITHIC INTEGRATED CIRCUIT

CASE 646

FIGURE 1 – TYPICAL APPLICATION AND TEST CIRCUIT

Pin Functions

Pin 1 = V_{CC}	Pin 8 = Switch Filter
Pin 2 = Input	Pin 9 = Switch Filter
Pin 3 = Amplifier Output	Pin 10 = 19 kHz Output
Pin 4 = Left Channel Output	Pin 11 = Modulator Input
Pin 5 = Right Channel Output	Pin 12 = Loop Filter
Pin 6 = Lamp Indicator	Pin 13 = Loop Filter
Pin 7 = Ground	Pin 14 = Oscillator RC Network

Parts List

C1 = 2.0 μF	C8 = 0.25 μF
C2 = 0.02 μF	R1 = 3.9 kΩ
C3 = 0.02 μF	R2 = 3.9 kΩ
C4 = 0.25 μF	R3 = 1.0 kΩ
C5 = 0.05 μF	R4 = 16 kΩ
C6 = 0.5 μF	R5 = 5.0 kΩ
C7 = 470 pF	

(Courtesy Motorola Inc.)

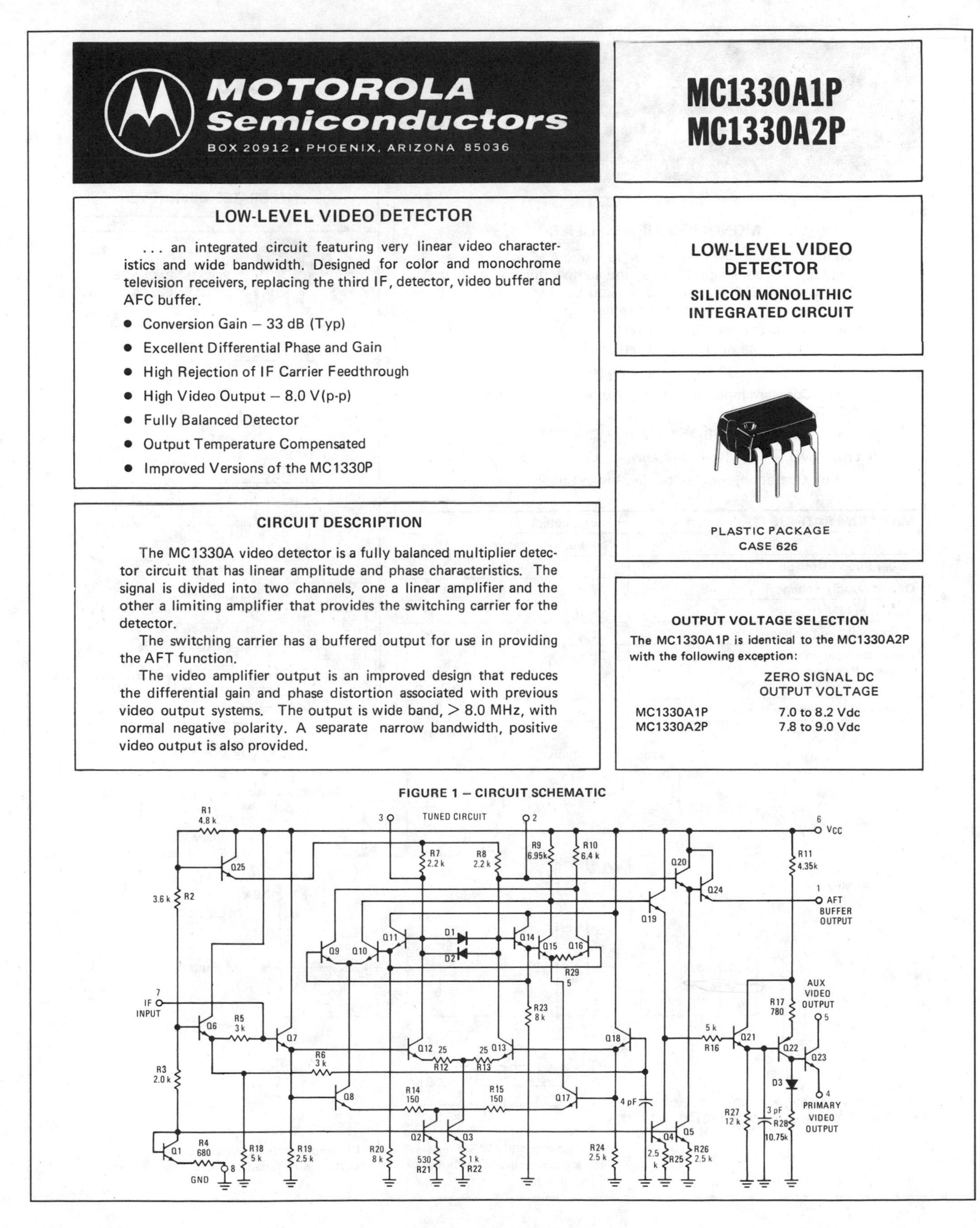

MOTOROLA Semiconductors

BOX 20912 • PHOENIX, ARIZONA 85036

MC1330A1P
MC1330A2P

LOW-LEVEL VIDEO DETECTOR

... an integrated circuit featuring very linear video characteristics and wide bandwidth. Designed for color and monochrome television receivers, replacing the third IF, detector, video buffer and AFC buffer.

- Conversion Gain – 33 dB (Typ)
- Excellent Differential Phase and Gain
- High Rejection of IF Carrier Feedthrough
- High Video Output – 8.0 V(p-p)
- Fully Balanced Detector
- Output Temperature Compensated
- Improved Versions of the MC1330P

LOW-LEVEL VIDEO DETECTOR

SILICON MONOLITHIC INTEGRATED CIRCUIT

PLASTIC PACKAGE
CASE 626

CIRCUIT DESCRIPTION

The MC1330A video detector is a fully balanced multiplier detector circuit that has linear amplitude and phase characteristics. The signal is divided into two channels, one a linear amplifier and the other a limiting amplifier that provides the switching carrier for the detector.

The switching carrier has a buffered output for use in providing the AFT function.

The video amplifier output is an improved design that reduces the differential gain and phase distortion associated with previous video output systems. The output is wide band, > 8.0 MHz, with normal negative polarity. A separate narrow bandwidth, positive video output is also provided.

OUTPUT VOLTAGE SELECTION

The MC1330A1P is identical to the MC1330A2P with the following exception:

	ZERO SIGNAL DC OUTPUT VOLTAGE
MC1330A1P	7.0 to 8.2 Vdc
MC1330A2P	7.8 to 9.0 Vdc

FIGURE 1 – CIRCUIT SCHEMATIC

(Courtesy Motorola Inc.)

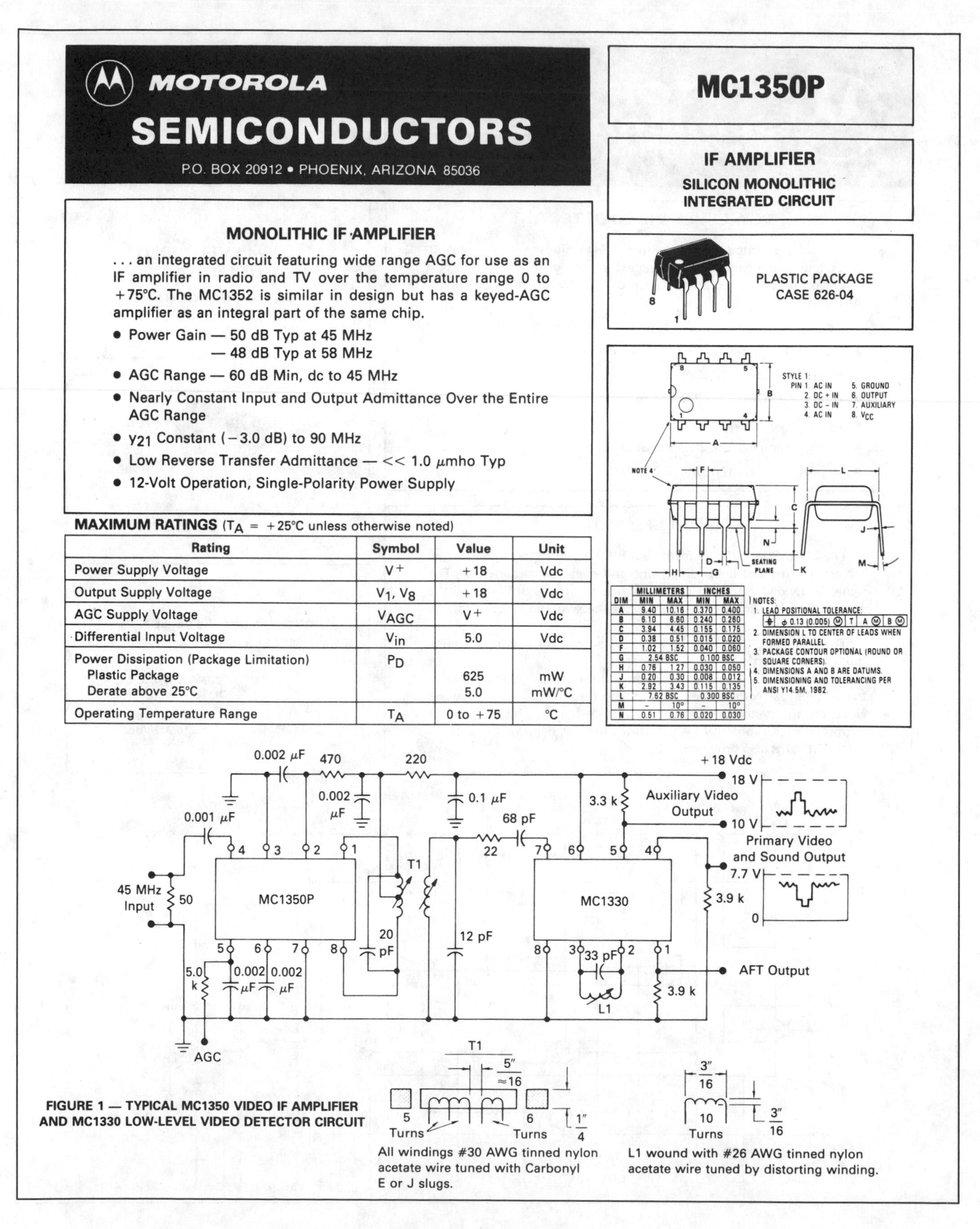

MOTOROLA

SEMICONDUCTORS

P.O. BOX 20912 • PHOENIX, ARIZONA 85036

MC1350P

IF AMPLIFIER

SILICON MONOLITHIC INTEGRATED CIRCUIT

MONOLITHIC IF AMPLIFIER

. . . an integrated circuit featuring wide range AGC for use as an IF amplifier in radio and TV over the temperature range 0 to +75°C. The MC1352 is similar in design but has a keyed-AGC amplifier as an integral part of the same chip.

- Power Gain — 50 dB Typ at 45 MHz
  — 48 dB Typ at 58 MHz
- AGC Range — 60 dB Min, dc to 45 MHz
- Nearly Constant Input and Output Admittance Over the Entire AGC Range
- y_{21} Constant (−3.0 dB) to 90 MHz
- Low Reverse Transfer Admittance — << 1.0 μmho Typ
- 12-Volt Operation, Single-Polarity Power Supply

PLASTIC PACKAGE
CASE 626-04

STYLE 1:
PIN 1. AC IN
2. DC + IN
3. DC − IN
4. AC IN
5. GROUND
6. OUTPUT
7. AUXILIARY
8. V_{CC}

MAXIMUM RATINGS (T_A = +25°C unless otherwise noted)

Rating	Symbol	Value	Unit
Power Supply Voltage	V^+	+18	Vdc
Output Supply Voltage	V_1, V_8	+18	Vdc
AGC Supply Voltage	V_{AGC}	V^+	Vdc
Differential Input Voltage	V_{in}	5.0	Vdc
Power Dissipation (Package Limitation) Plastic Package Derate above 25°C	P_D	 625 5.0	 mW mW/°C
Operating Temperature Range	T_A	0 to +75	°C

DIM	MILLIMETERS MIN	MILLIMETERS MAX	INCHES MIN	INCHES MAX
A	9.40	10.16	0.370	0.400
B	6.10	6.60	0.240	0.260
C	3.94	4.45	0.155	0.175
D	0.38	0.51	0.015	0.020
F	1.02	1.52	0.040	0.060
G	2.54 BSC		0.100 BSC	
H	0.76	1.27	0.030	0.050
J	0.20	0.30	0.008	0.012
K	2.92	3.43	0.115	0.135
L	7.62 BSC		0.300 BSC	
M	–	10°	–	10°
N	0.51	0.76	0.020	0.030

NOTES:
1. LEAD POSITIONAL TOLERANCE: ⌖ φ 0.13 (0.005) Ⓜ T A Ⓜ B Ⓜ
2. DIMENSION L TO CENTER OF LEADS WHEN FORMED PARALLEL.
3. PACKAGE CONTOUR OPTIONAL (ROUND OR SQUARE CORNERS).
4. DIMENSIONS A AND B ARE DATUMS.
5. DIMENSIONING AND TOLERANCING PER ANSI Y14.5M, 1982.

FIGURE 1 — TYPICAL MC1350 VIDEO IF AMPLIFIER AND MC1330 LOW-LEVEL VIDEO DETECTOR CIRCUIT

All windings #30 AWG tinned nylon acetate wire tuned with Carbonyl E or J slugs.

L1 wound with #26 AWG tinned nylon acetate wire tuned by distorting winding.

(Courtesy Motorola Inc.)

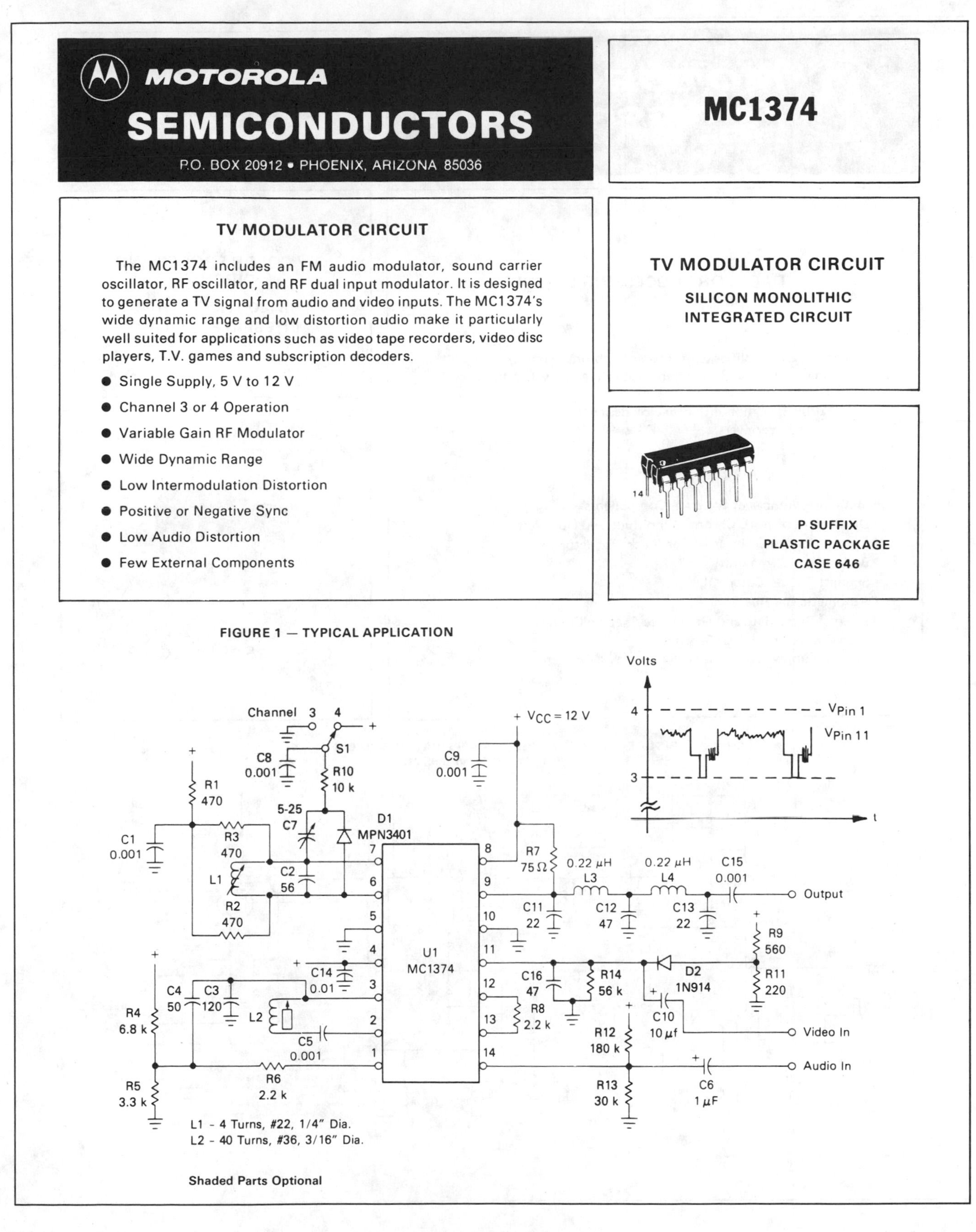

MOTOROLA

SEMICONDUCTORS

P.O. BOX 20912 • PHOENIX, ARIZONA 85036

MC1374

TV MODULATOR CIRCUIT

The MC1374 includes an FM audio modulator, sound carrier oscillator, RF oscillator, and RF dual input modulator. It is designed to generate a TV signal from audio and video inputs. The MC1374's wide dynamic range and low distortion audio make it particularly well suited for applications such as video tape recorders, video disc players, T.V. games and subscription decoders.

- Single Supply, 5 V to 12 V
- Channel 3 or 4 Operation
- Variable Gain RF Modulator
- Wide Dynamic Range
- Low Intermodulation Distortion
- Positive or Negative Sync
- Low Audio Distortion
- Few External Components

TV MODULATOR CIRCUIT

SILICON MONOLITHIC INTEGRATED CIRCUIT

P SUFFIX
PLASTIC PACKAGE
CASE 646

FIGURE 1 — TYPICAL APPLICATION

(Courtesy Motorola Inc.)

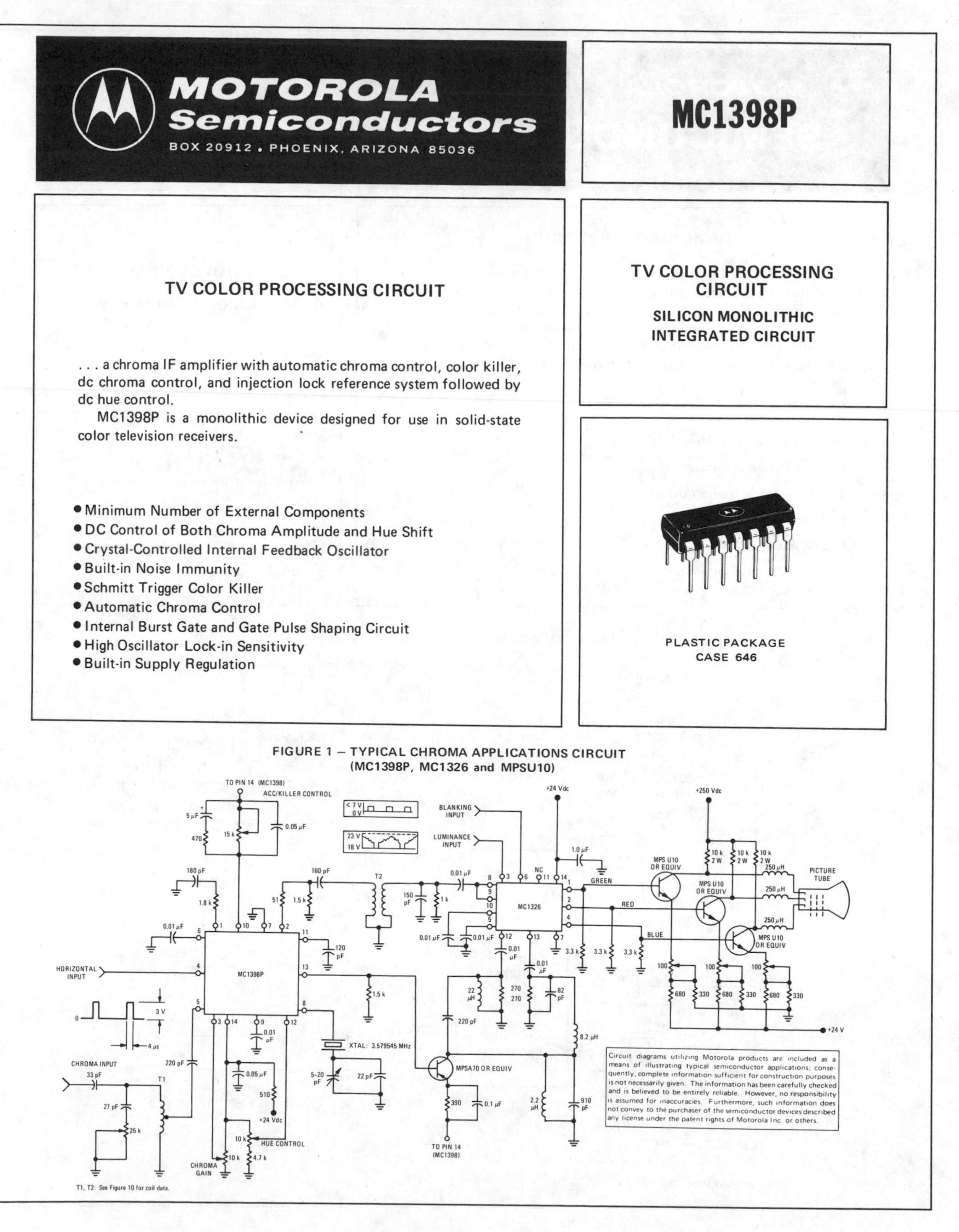

MOTOROLA Semiconductors

BOX 20912 • PHOENIX, ARIZONA 85036

MC1398P

TV COLOR PROCESSING CIRCUIT

. . . a chroma IF amplifier with automatic chroma control, color killer, dc chroma control, and injection lock reference system followed by dc hue control.

MC1398P is a monolithic device designed for use in solid-state color television receivers.

- Minimum Number of External Components
- DC Control of Both Chroma Amplitude and Hue Shift
- Crystal-Controlled Internal Feedback Oscillator
- Built-in Noise Immunity
- Schmitt Trigger Color Killer
- Automatic Chroma Control
- Internal Burst Gate and Gate Pulse Shaping Circuit
- High Oscillator Lock-in Sensitivity
- Built-in Supply Regulation

TV COLOR PROCESSING CIRCUIT

SILICON MONOLITHIC INTEGRATED CIRCUIT

PLASTIC PACKAGE
CASE 646

FIGURE 1 – TYPICAL CHROMA APPLICATIONS CIRCUIT (MC1398P, MC1326 and MPSU10)

(Courtesy Motorola Inc.)

FIGURE 2 – MC1398P TEST CIRCUIT

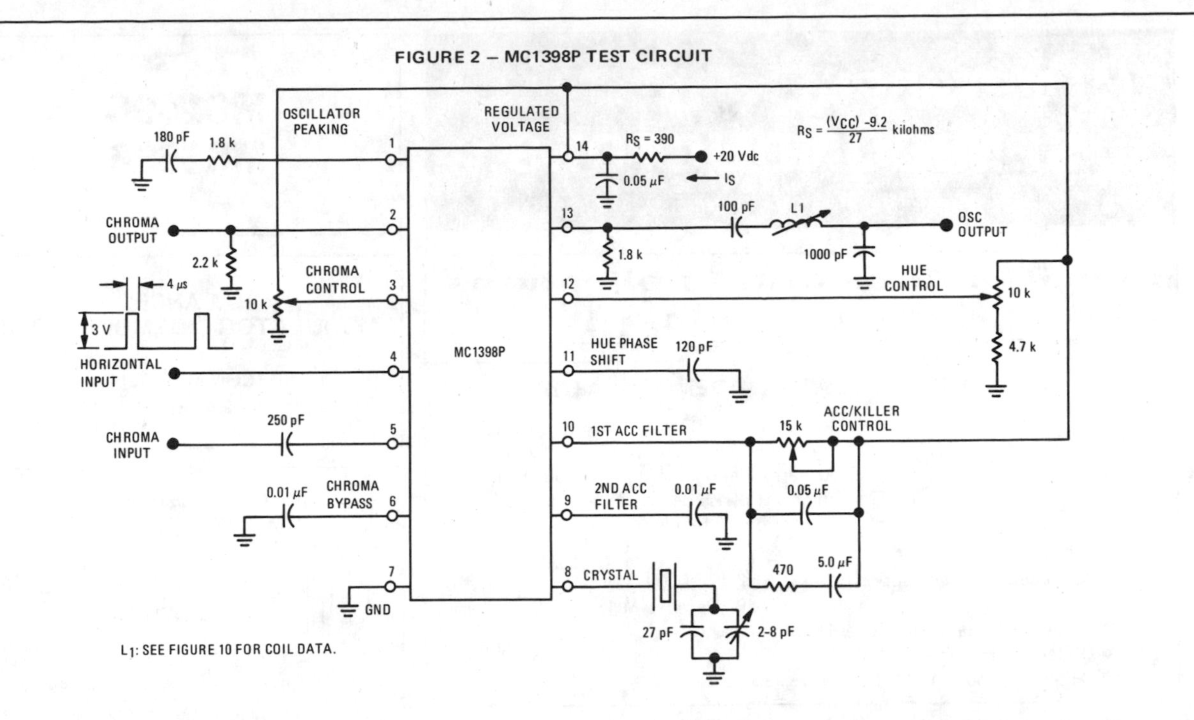

(Courtesy Motorola Inc.)

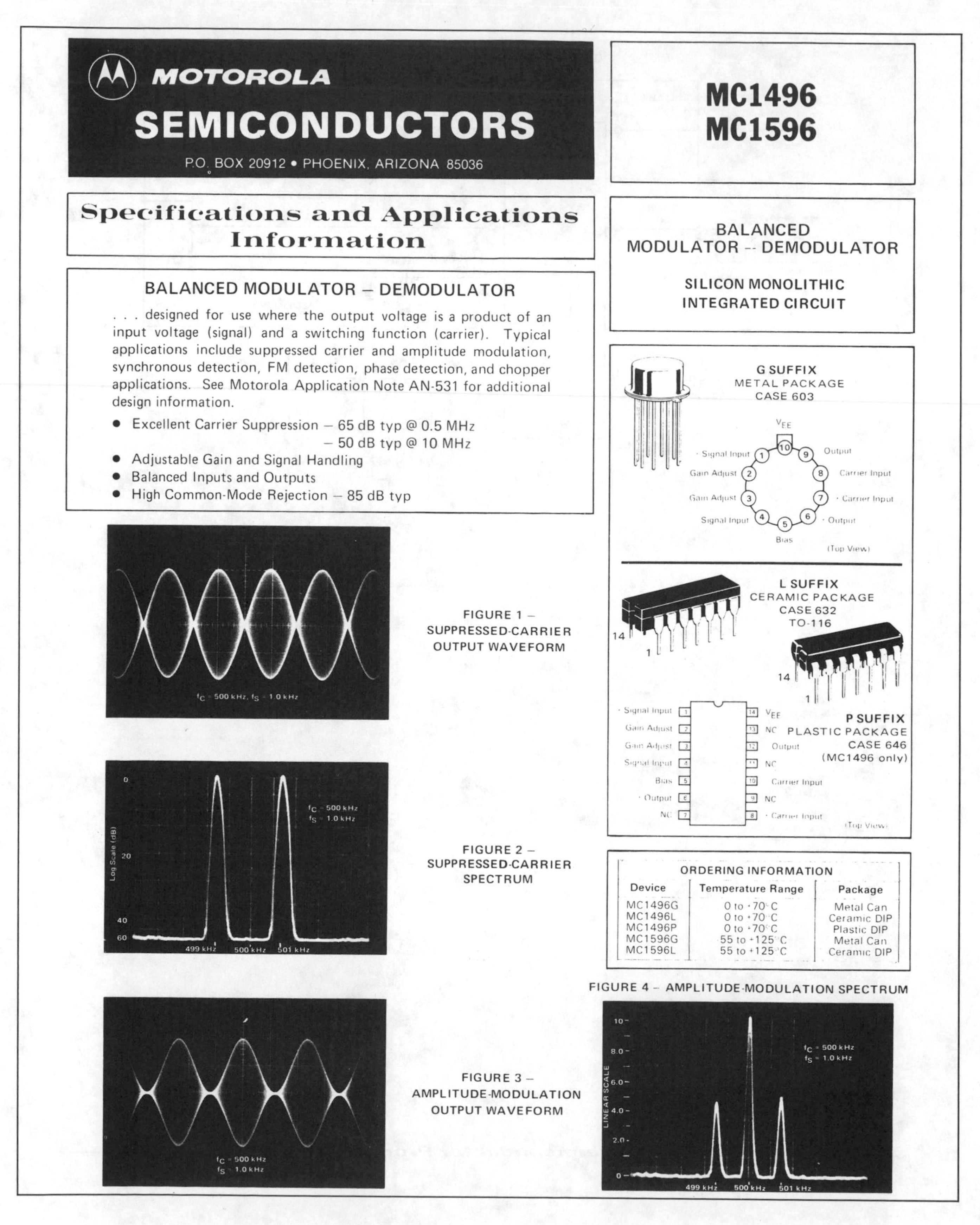

MOTOROLA
SEMICONDUCTORS
P.O. BOX 20912 • PHOENIX, ARIZONA 85036

MC1496
MC1596

Specifications and Applications Information

BALANCED MODULATOR – DEMODULATOR

. . . designed for use where the output voltage is a product of an input voltage (signal) and a switching function (carrier). Typical applications include suppressed carrier and amplitude modulation, synchronous detection, FM detection, phase detection, and chopper applications. See Motorola Application Note AN-531 for additional design information.

- Excellent Carrier Suppression – 65 dB typ @ 0.5 MHz
  – 50 dB typ @ 10 MHz
- Adjustable Gain and Signal Handling
- Balanced Inputs and Outputs
- High Common-Mode Rejection – 85 dB typ

BALANCED
MODULATOR – DEMODULATOR

SILICON MONOLITHIC
INTEGRATED CIRCUIT

G SUFFIX
METAL PACKAGE
CASE 603

L SUFFIX
CERAMIC PACKAGE
CASE 632
TO-116

P SUFFIX
PLASTIC PACKAGE
CASE 646
(MC1496 only)

ORDERING INFORMATION

Device	Temperature Range	Package
MC1496G	0 to +70°C	Metal Can
MC1496L	0 to +70°C	Ceramic DIP
MC1496P	0 to +70°C	Plastic DIP
MC1596G	55 to +125°C	Metal Can
MC1596L	55 to +125°C	Ceramic DIP

FIGURE 1 – SUPPRESSED-CARRIER OUTPUT WAVEFORM

FIGURE 2 – SUPPRESSED-CARRIER SPECTRUM

FIGURE 3 – AMPLITUDE-MODULATION OUTPUT WAVEFORM

FIGURE 4 – AMPLITUDE-MODULATION SPECTRUM

(Courtesy Motorola Inc.)

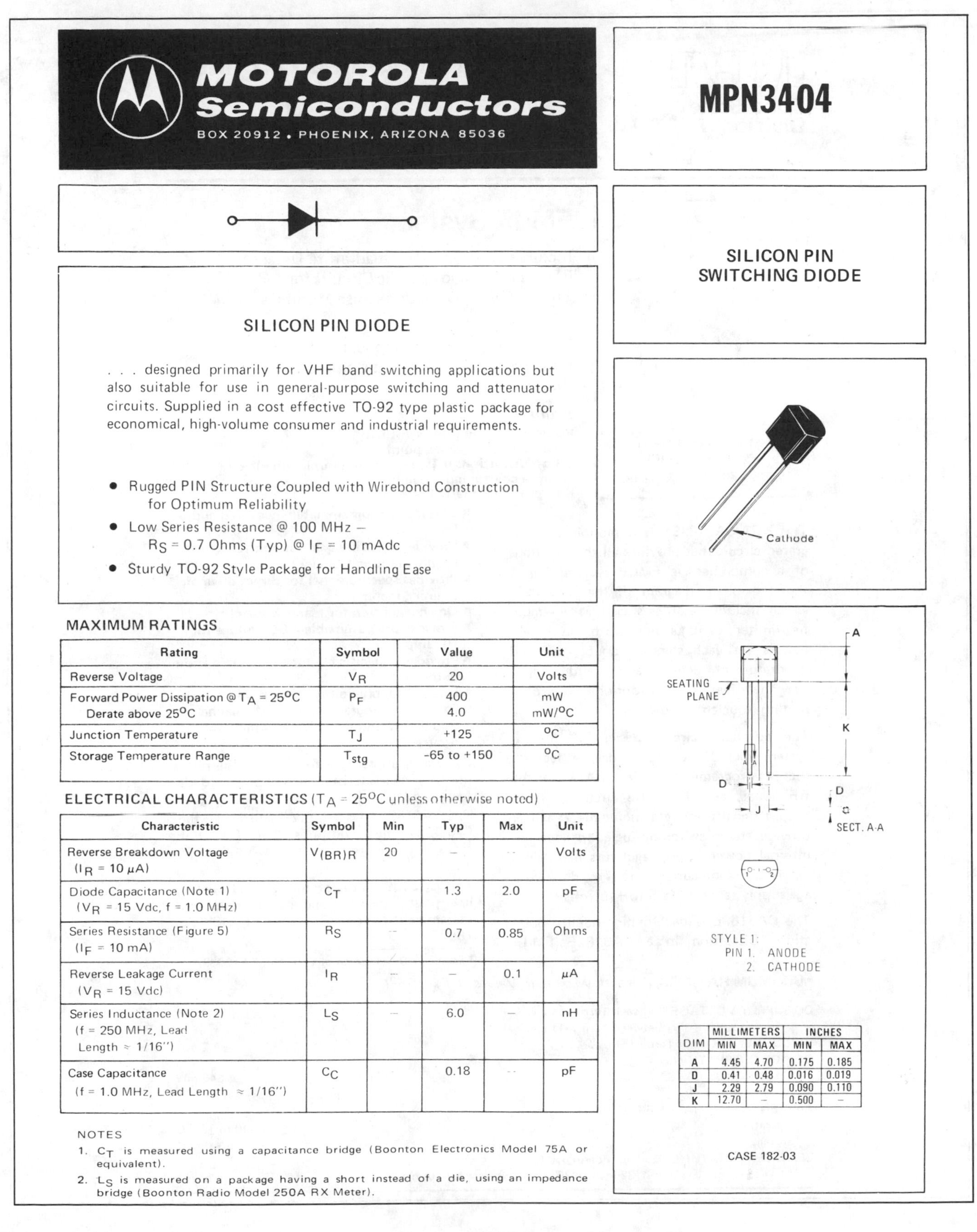

MOTOROLA
Semiconductors
BOX 20912 • PHOENIX, ARIZONA 85036

MPN3404

SILICON PIN SWITCHING DIODE

SILICON PIN DIODE

. . . designed primarily for VHF band switching applications but also suitable for use in general-purpose switching and attenuator circuits. Supplied in a cost effective TO-92 type plastic package for economical, high-volume consumer and industrial requirements.

- Rugged PIN Structure Coupled with Wirebond Construction for Optimum Reliability
- Low Series Resistance @ 100 MHz — R_S = 0.7 Ohms (Typ) @ I_F = 10 mAdc
- Sturdy TO-92 Style Package for Handling Ease

MAXIMUM RATINGS

Rating	Symbol	Value	Unit
Reverse Voltage	V_R	20	Volts
Forward Power Dissipation @ T_A = 25°C Derate above 25°C	P_F	400 4.0	mW mW/°C
Junction Temperature	T_J	+125	°C
Storage Temperature Range	T_{stg}	-65 to +150	°C

ELECTRICAL CHARACTERISTICS (T_A = 25°C unless otherwise noted)

Characteristic	Symbol	Min	Typ	Max	Unit
Reverse Breakdown Voltage (I_R = 10 μA)	$V_{(BR)R}$	20	–	–	Volts
Diode Capacitance (Note 1) (V_R = 15 Vdc, f = 1.0 MHz)	C_T		1.3	2.0	pF
Series Resistance (Figure 5) (I_F = 10 mA)	R_S	–	0.7	0.85	Ohms
Reverse Leakage Current (V_R = 15 Vdc)	I_R	–	–	0.1	μA
Series Inductance (Note 2) (f = 250 MHz, Lead Length ≈ 1/16")	L_S	–	6.0	–	nH
Case Capacitance (f = 1.0 MHz, Lead Length ≈ 1/16")	C_C		0.18	–	pF

NOTES
1. C_T is measured using a capacitance bridge (Boonton Electronics Model 75A or equivalent).
2. L_S is measured on a package having a short instead of a die, using an impedance bridge (Boonton Radio Model 250A RX Meter).

STYLE 1:
PIN 1. ANODE
2. CATHODE

DIM	MILLIMETERS MIN	MILLIMETERS MAX	INCHES MIN	INCHES MAX
A	4.45	4.70	0.175	0.185
D	0.41	0.48	0.016	0.019
J	2.29	2.79	0.090	0.110
K	12.70	–	0.500	–

CASE 182-03

(Courtesy Motorola Inc.)

Solid State Division

Linear Integrated Circuits

Monolithic Silicon

CA3189E

File Number 1046

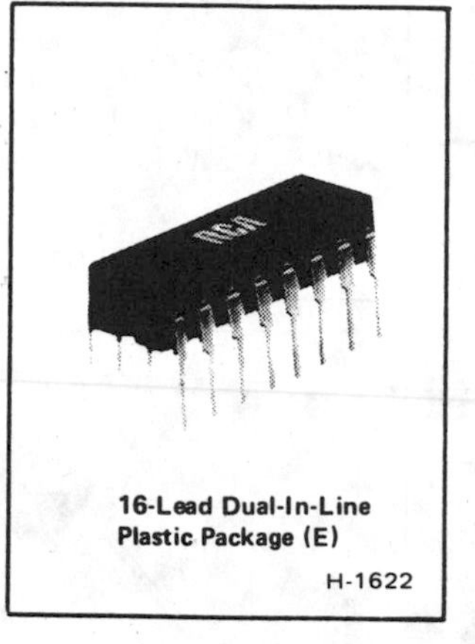

16-Lead Dual-In-Line Plastic Package (E)

H-1622

FM IF System

Includes IF Amplifier, Quadrature Detector, AF Preamplifier, and Specific Circuits for AGC, AFC, Tuning Meter, Deviation-Noise Muting, and ON Channel Detector

For FM IF Amplifier Applications in High-Fidelity, Automotive, and Communications Receivers

Features:

- **Exceptional limiting sensitivity: 12 μV typ. at −3 dB point**
- **Low distortion: 0.1% typ. (with double-tuned coil)**
- **Single-coil tuning capability**
- **Improved S + N/N Ratio**
- **Externally programmable recovered audio level**
- **Provides specific signal for control of inter-channel muting (squelch)**
- **Provides specific signal for direct drive of a tuning meter**
- **On channel step for search control**
- **Provides programmable AGC voltage for RF amplifier**
- **Provides a specific circuit for flexible audio output**
- **Internal supply-voltage regulators**
- **Externally programmable "on" channel step width, and deviation at which muting occurs**

The RCA-CA3189E* is a monolithic integrated circuit that provides all the functions of a comprehensive FM-IF system. Fig. 1 shows a block diagram of the CA3189E, which includes a three-stage FM-IF amplifier/limiter configuration with level detectors for each stage, a doubly-balanced quadrature FM detector and an audio amplifier that features the optional use of a muting (squelch) circuit.

The advanced circuit design of the IF system includes desirable deluxe features such as programmable delayed AGC for the RF tuner, an AFC drive circuit, and an output signal to drive a tuning meter and/or provide stereo switching logic. In addition, internal power-supply regulators maintain a nearly constant current drain over the voltage supply range of +8.5 to +16 volts.

The CA3189E is ideal for high-fidelity operation. Distortion in a CA3189E FM-IF System is primarily a function of the phase linearity characteristic of the outboard detector coil.

The CA3189E has all the features of the CA3089E plus additions. See CA3189E features compared to the CA3089E in Table I.

The CA3189E utilizes the 16-lead dual-in-line plastic package and can operate over the ambient temperature range of −40°C to +85°C.

*Formerly Developmental Type No. TA10038.

MAXIMUM RATINGS, *Absolute-Maximum Values at T_A = 25°C:*

DC SUPPLY VOLTAGE (between Terms. 11 and 4)	16 V
(between Terms. 11 and 14)	16 V
DC CURRENT (Out of Term. 15)	2 mA
DEVICE DISSIPATION:	
Up to T_A = 85°C	640 mW
Above T_A = 85°C	derate linearly at 9.9 mW/°C
AMBIENT-TEMPERATURE RANGE:	
Operating	−40 to +85°C
Storage	−65 to +150°C
LEAD TEMPERATURE (During soldering):	
At distance not less than 1/32 inch (0.79 mm) from case for 10s max.	+265°C

(Courtesy RCA Corporation)

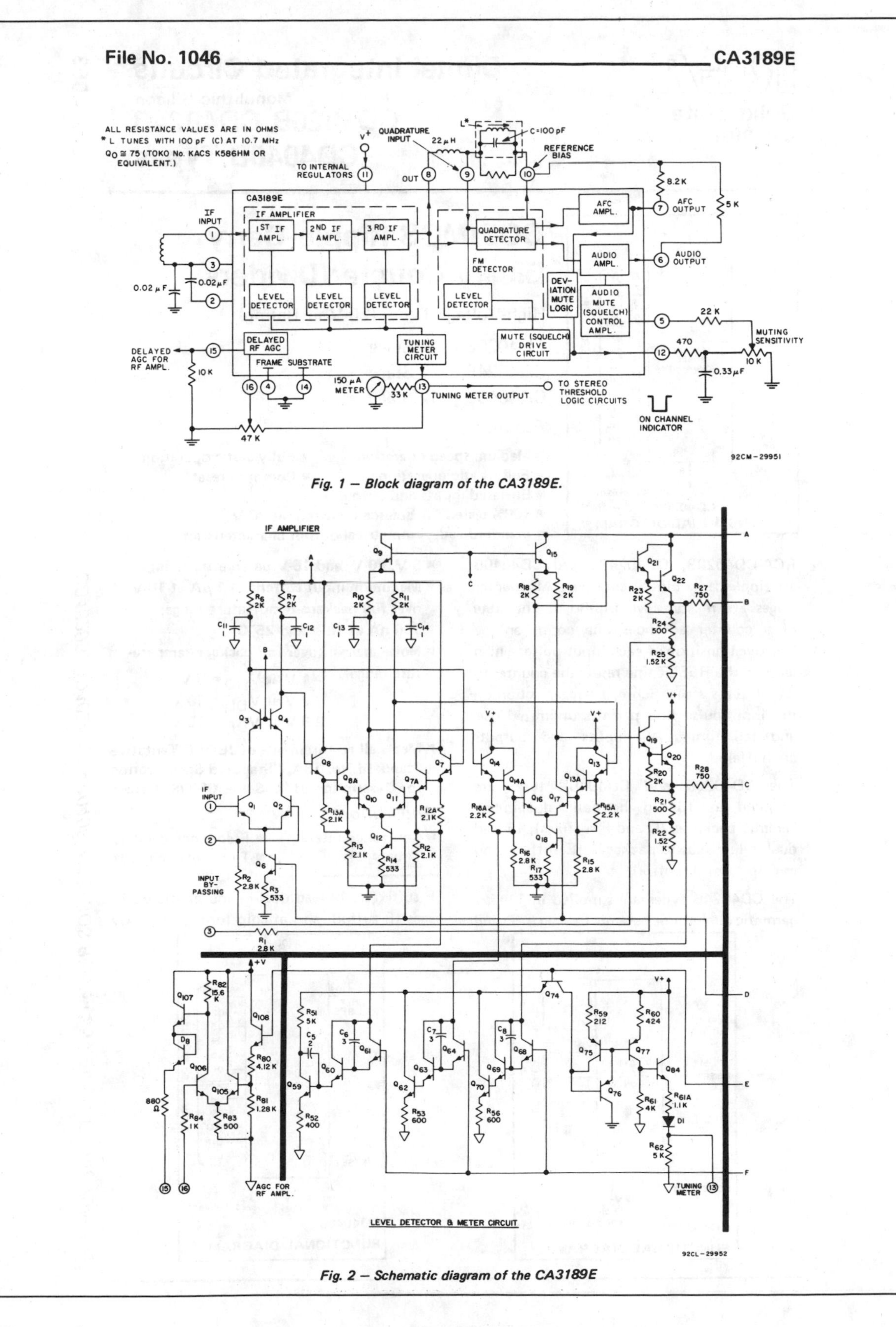

Fig. 1 — Block diagram of the CA3189E.

Fig. 2 — Schematic diagram of the CA3189E

(Courtesy RCA Corporation)

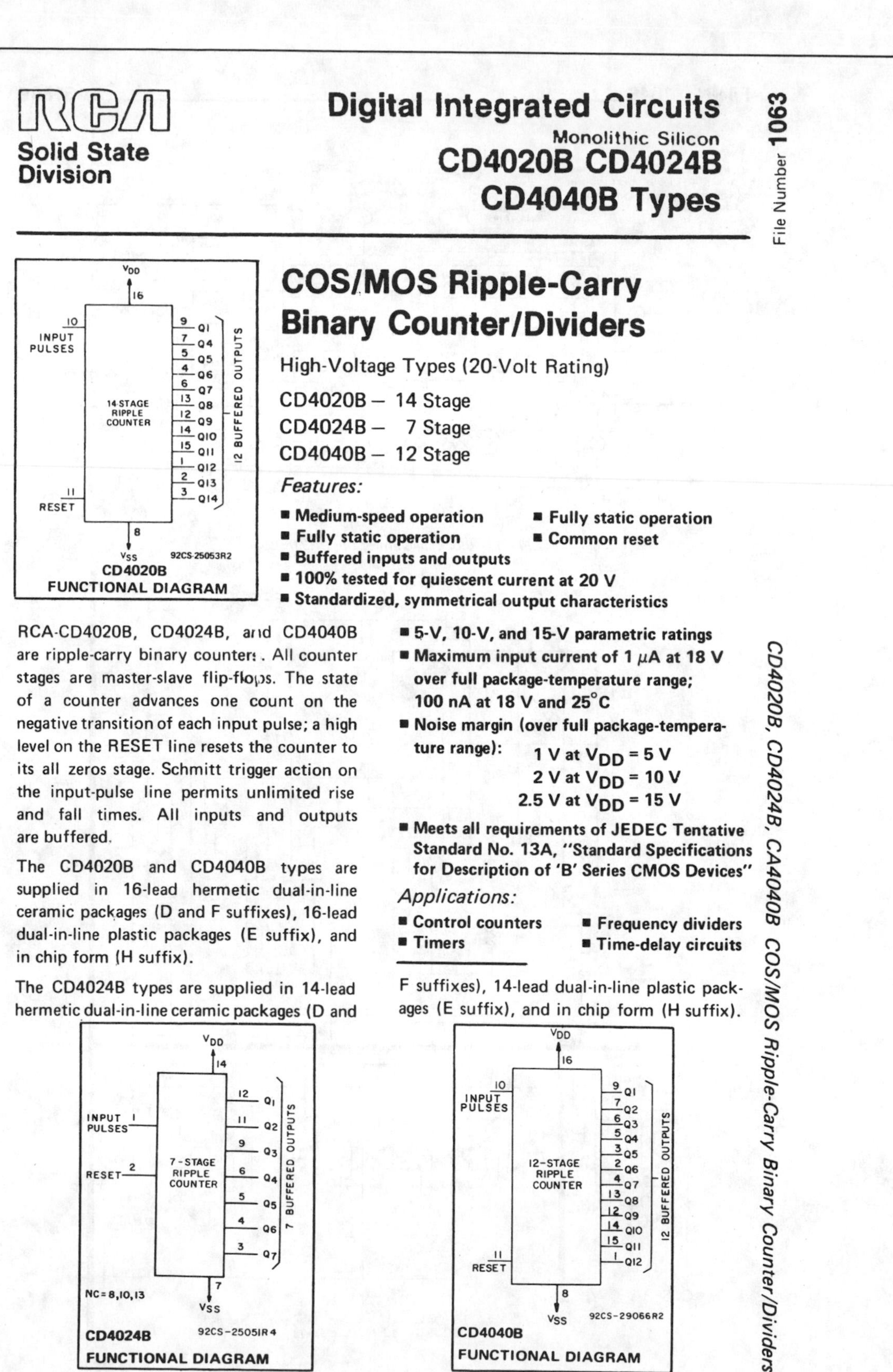

RCA

Solid State Division

Digital Integrated Circuits

Monolithic Silicon

CD4020B CD4024B CD4040B Types

File Number 1063

CD4020B FUNCTIONAL DIAGRAM

COS/MOS Ripple-Carry Binary Counter/Dividers

High-Voltage Types (20-Volt Rating)

CD4020B – 14 Stage
CD4024B – 7 Stage
CD4040B – 12 Stage

Features:

- **Medium-speed operation**
- **Fully static operation**
- **Buffered inputs and outputs**
- **100% tested for quiescent current at 20 V**
- **Standardized, symmetrical output characteristics**
- **Fully static operation**
- **Common reset**
- **5-V, 10-V, and 15-V parametric ratings**
- **Maximum input current of 1 μA at 18 V over full package-temperature range; 100 nA at 18 V and 25°C**
- **Noise margin (over full package-temperature range):** 1 V at V_{DD} = 5 V; 2 V at V_{DD} = 10 V; 2.5 V at V_{DD} = 15 V
- **Meets all requirements of JEDEC Tentative Standard No. 13A, "Standard Specifications for Description of 'B' Series CMOS Devices"**

Applications:

- **Control counters**
- **Timers**
- **Frequency dividers**
- **Time-delay circuits**

RCA-CD4020B, CD4024B, and CD4040B are ripple-carry binary counters. All counter stages are master-slave flip-flops. The state of a counter advances one count on the negative transition of each input pulse; a high level on the RESET line resets the counter to its all zeros stage. Schmitt trigger action on the input-pulse line permits unlimited rise and fall times. All inputs and outputs are buffered.

The CD4020B and CD4040B types are supplied in 16-lead hermetic dual-in-line ceramic packages (D and F suffixes), 16-lead dual-in-line plastic packages (E suffix), and in chip form (H suffix).

The CD4024B types are supplied in 14-lead hermetic dual-in-line ceramic packages (D and F suffixes), 14-lead dual-in-line plastic packages (E suffix), and in chip form (H suffix).

CD4024B FUNCTIONAL DIAGRAM

CD4040B FUNCTIONAL DIAGRAM

CD4020B, CD4024B, CA4040B COS/MOS Ripple-Carry Binary Counter/Dividers

(Courtesy RCA Corporation)

Solid State Division

Digital Integrated Circuits

Monolithic Silicon

CD4051B, CD4052B, CD4053B Types

File Number 902

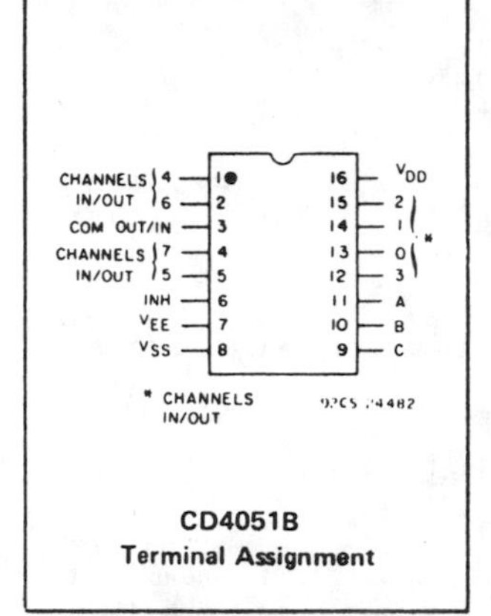

CD4051B
Terminal Assignment

COS/MOS Analog Multiplexers/Demultiplexers*

With Logic-Level Conversion

High-Voltages Types (20-Volt Rating)

CD4051B – Single 8-Channel
CD4052B – Differential 4-Channel
CD4053B – Triple 2-Channel

Applications:

- **Analog and digital multiplexing and demultiplexing**
- **A/D and D/A conversion**
- **Signal gating**

RCA-CD4051B, CD4052B, and CD4053B analog multiplexers/demultiplexers are digitally controlled analog switches having low ON impedance and very low OFF leakage current. Control of analog signals up to 20 V peak-to-peak can be achieved by digital signal amplitudes of 4.5 to 20 V (if V_{DD}-V_{SS} = 3 V, a V_{DD}-V_{EE} of up to 13 V can be controlled; for V_{DD}-V_{EE} level differences above 13 V, a V_{DD}-V_{SS} of at least 4.5 V is required). For example, if V_{DD} = +5 V, V_{SS} = 0, and V_{EE} = −13.5 V, analog signals from −13.5 V to +4.5 V can be controlled by digital inputs of 0 to 5 V. These multiplexer circuits dissipate extremely low quiescent power over the full V_{DD}-V_{SS} and V_{DD}-V_{EE} supply-voltage ranges, independent of the logic state of the control signals. When a logic "1" is present at the inhibit input terminal all channels are off.

The CD4051B is a single 8-channel multiplexer having three binary control inputs, A, B, and C, and an inhibit input. The three binary signals select 1 of 8 channels to be turned on, and connect one of the 8 inputs to the output.

The CD4052B is a differential 4-channel multiplexer having two binary control inputs, A and B, and an inhibit input. The two binary input signals select 1 of 4 pairs of channels to be turned on and connect the analog inputs to the outputs.

* When these devices are used as demultiplexers, the "CHANNEL IN/OUT" terminals are the outputs and the "COMMON OUT/IN" terminals are the inputs.

Features:

- **Wide range of digital and analog signal levels: digital 3 to 20 V, analog to 20 V_{p-p}**
- **Low ON resistance: 125 Ω (typ.) over 15 V_{p-p} signal-input range for V_{DD}-V_{EE} = 15 V**
- **High OFF resistance: channel leakage of ±100 pA (typ.) @ V_{DD}-V_{EE} = 18 V**
- **Logic-level conversion for digital addressing signals of 3 to 20 V (V_{DD}-V_{SS} = 3 to 20 V) to switch analog signals to 20 V p-p (V_{DD}-V_{EE} = 20 V); see introductory text**
- **Matched switch characteristics: R_{ON} = 5 Ω (typ.) for V_{DD}-V_{EE} = 15 V**
- **Very low quiescent power dissipation under under all digital-control input and supply conditions: 0.2 μW (typ.) @ V_{DD}-V_{SS} = V_{DD}-V_{EE} = 10 V**
- **Binary address decoding on chip**
- **5-, 10-, and 15-V parametric ratings**
- **100% tested for quiescent current at 20 V**
- **Maximum input current of 1 μA at 18 V over full package temperature range; 100 nA at 18 V and 25°C**

CD4051B, CD4052B, CD4053B COS/MOS Analog Multiplexers/Demultiplexers

(Courtesy RCA Corporation)

The CD4053B is a triple 2-channel multiplexer having three separate digital control inputs, A, B, and C, and an inhibit input. Each control input selects one of a pair of channels which are connected in a single-pole double-throw configuration.

The CD4051B, CD4052B, and CD4053B are supplied in 16-lead ceramic dual-in-line packages (D and F suffixes), 16-lead plastic dual-in-line packages (E suffix), and in chip form (H suffix).

MAXIMUM RATINGS, *Absolute-Maximum Values:*

DC SUPPLY-VOLTAGE RANGE, (V_{DD})
(Voltages referenced to V_{SS} or V_{EE}, whichever is more negative) . . . . −0.5 to +20 V
INPUT VOLTAGE RANGE, ALL INPUTS . . . . −0.5 to V_{DD} +0.5 V
DC INPUT CURRENT, ANY ONE INPUT . . . . ±10 mA
POWER DISSIPATION PER PACKAGE (P_D):
For T_A = −40 to +60°C (PACKAGE TYPE E) . . . . 500 mW
For T_A = +60 to +85°C (PACKAGE TYPE E) . . . . . Derate Linearly at 12 mW/°C to 200 mW
For T_A = −55 to +100°C (PACKAGE TYPES D,F) . . . . 500 mW
For T_A = +100 to +125°C (PACKAGE TYPES D, F) . . . Derate Linearly at 12 mW/°C to 200 mW
DEVICE DISSIPATION PER OUTPUT TRANSISTOR
FOR T_A = FULL PACKAGE-TEMPERATURE RANGE (All Package Types) . . . . . 100 mW
OPERATING-TEMPERATURE RANGE (T_A):
PACKAGE TYPES D, F, H . . . . −55 to +125°C
PACKAGE TYPE E . . . . −40 to +85°C
STORAGE TEMPERATURE RANGE (T_{stg}) . . . . −65 to +150°C
LEAD TEMPERATURE (DURING SOLDERING):
At distance 1/16 ± 1/32 inch (1.59 ± 0.79 mm) from case for 10 s max. . . . . +265°C

RECOMMENDED OPERATING CONDITIONS AT T_A = 25°C (Unless Otherwise Specified)

For maximum reliability, nominal operating conditions should be selected so that operation is always within the following ranges. Values shown apply to all types except as noted.

CHARACTERISTIC	V_{DD}	Min.	Max.	Units
Supply-Voltage Range (T_A = Full Package-Temp. Range)	—	3	18	V
Multiplexer Switch Input Current Capability*	—	—	25	mA
Output Load Resistance	—	100	—	Ω

* In certain applications, the external load-resistor current may include both V_{DD} and signal-line components. To avoid drawing V_{DD} current when switch current flows into the transmission gate inputs, the voltage drop across the bidirectional switch must not exceed 0.8 volt (calculated from R_{ON} values shown in ELECTRICAL CHARACTERISTICS CHART). No V_{DD} current will flow through R_L if the switch current flows into terminal 3 on the CD4051; terminals 3 and 13 on the CD4052; terminals 4,14, and 15 on the CD4053.

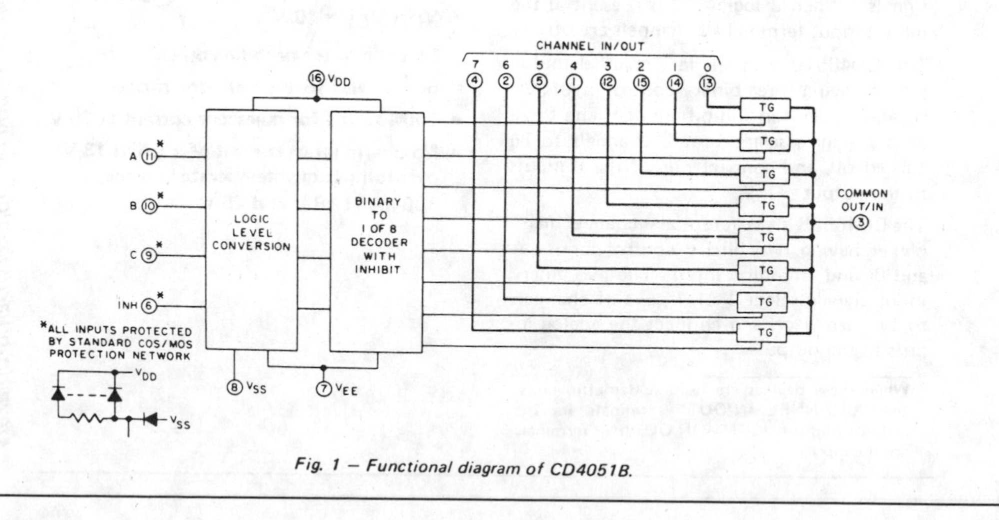

Fig. 1 – Functional diagram of CD4051B.

(Courtesy RCA Corporation)

Solid State
Division

Digital Integrated Circuits

Monolithic Silicon

CD4098B Types

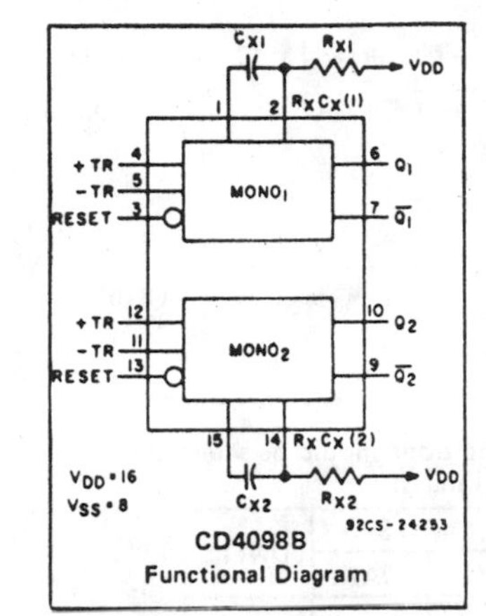

CD4098B
Functional Diagram

COS/MOS Dual Monostable Multivibrator

High-Voltage Types (20-Volt Rating)

Features:

- **Retriggerable/resettable capability**
- **Trigger and reset propagation delays independent of R_X, C_X**
- **Triggering from leading or trailing edge**
- **Q and $\overline{Q}$ buffered outputs available**
- **Separate resets**
- **Wide range of output-pulse widths**
- **100% tested for maximum quiescent current at 20 V**
- **Maximum input current of 1 μA at 18 V over full package-temperature range; 100 nA at 18 V and 25°C**
- **Noise margin (full package-temperature range): 1 V at V_{DD}= 5 V; 2 V at V_{DD}=10 V; 2.5 V at V_{DD}=15 V**
- **5-V, 10-V, and 15-V parametric ratings**
- **Standardized, symmetrical output characteristics**
- **Meets all requirements of JEDEC Tentative Standard No. 13A, "Standard Specifications for Description of 'B' Series CMOS Devices."**

Applications:

- **Pulse delay and timing**
- **Pulse shaping**
- **Astable multivibrator**

The RCA-CD4098B dual monostable multivibrator provides stable retriggerable/resettable one-shot operation for any fixed-voltage timing application.

An external resistor (R_X) and an external capacitor (C_X) control the timing for the circuit. Adjustment of R_X and C_X provides a wide range of output pulse widths from the Q and $\overline{Q}$ terminals. The time delay from trigger input to output transition (trigger propagation delay) and the time delay from reset input to output transition (reset propagation delay) are independent of R_X and C_X.

Leading-edge-triggering (+TR) and trailing-edge-triggering (−TR) inputs are provided for triggering from either edge of an input pulse. An unused +TR input should be tied to V_{SS}. An unused −TR input should be tied to V_{DD}. A RESET (on low level) is provided for immediate termination of the output pulse or to prevent output pulses when power is turned on. An unused RESET input should be tied to V_{DD}. However, if an entire section of the CD4098B is not used, its RESET should be tied to V_{SS}. See Table I.

In normal operation the circuit triggers (extends the output pulse one period) on the application of each new trigger pulse. For operation in the non-retriggerable mode, $\overline{Q}$ is connected to −TR when leading-edge triggering (+TR) is used or Q is connected to +TR when trailing-edge triggering (−TR) is used.

The time period (T) for this multivibrator can be approximated by: $T_X = \frac{1}{2} R_X C_X$ for $C_X \geqslant 0.01$ μF. Time periods as a function of R_X for values of C_X and V_{DD} are given in Fig. 8. Values of T vary from unit to unit and as a function of voltage, temperature, and $R_X C_X$.

The minimum value of external resistance, R_X, is 5 kΩ. The maximum value of external capacitance, C_X, is 100 μF. Fig. 9 shows time periods as a function of C_X for values of R_X and V_{DD}.

The output pulse width has variations of ±2.5% typically, over the temperature range of −55°C to 125°C for C_X=1000 pF and R_X=100 kΩ.

For power supply variations of ±5%, the output pulse width has variations of ±0.5% typically, for V_{DD}=10 V and 15 V and ±1% typically, for V_{DD}=5 V at C_X=1000 pF and R_X=5 kΩ.

The CD4098B types are supplied in 16-lead hermetic dual-in-line ceramic packages (D and F suffixes), 16-lead dual-in-line plastic packages (E suffix) and in chip form (H suffix).

The CD4098B is similar to type MC14528.

(Courtesy RCA Corporation)

MAXIMUM RATINGS, *Absolute-Maximum Values:*

DC SUPPLY-VOLTAGE RANGE, (V_{DD})
(Voltages referenced to V_{SS} Terminal) . . . . . −0.5 to +20 V
INPUT VOLTAGE RANGE, ALL INPUTS . . . . . −0.5 to V_{DD} +0.5 V
DC INPUT CURRENT, ANY ONE INPUT . . . . . ±10 mA
POWER DISSIPATION PER PACKAGE (P_D):
For T_A = −40 to +60°C (PACKAGE TYPE E) . . . . . 500 mW
For T_A = +60 to +85°C (PACKAGE TYPE E) . . . . . Derate Linearly at 12 mW/°C to 200 mW
For T_A = −55 to +100°C (PACKAGE TYPES D, F) . . . . . 500 mW
For T_A = +100 to +125°C (PACKAGE TYPES D, F) . . . . . Derate Linearly at 12 mW/°C to 200 mW
DEVICE DISSIPATION PER OUTPUT TRANSISTOR
FOR T_A = FULL PACKAGE-TEMPERATURE RANGE (All Package Types) . . . . . 100 mW
OPERATING-TEMPERATURE RANGE (T_A):
PACKAGE TYPES D, F, H . . . . . −55 to +125°C
PACKAGE TYPE E . . . . . −40 to +85°C
STORAGE TEMPERATURE RANGE (T_{stg}) . . . . . −65 to +150°C
LEAD TEMPERATURE (DURING SOLDERING):
At distance 1/16 ± 1/32 inch (1.59 ± 0.79 mm) from case for 10 s max. . . . . . +265°C

RECOMMENDED OPERATING CONDITIONS
For maximum reliability, nominal operating conditions should be selected so that operation is always within the following ranges:

CHARACTERISTIC	V_{DD} V	LIMITS MIN.	LIMITS MAX.	UNITS
Supply-Voltage Range (For T_A = Full Package-Temperature Range)	–	3	18	V
Trigger Pulse Width t_W(TR)	5 10 15	140 60 40	– – –	ns
Reset Pulse Width t_W(R) (This is a function of C_X)	–	See Dynamic Char. Chart and Fig. 10		–
Trigger Rise or Fall Time t_r(TR), t_f(TR)	5-15	–	100	μs

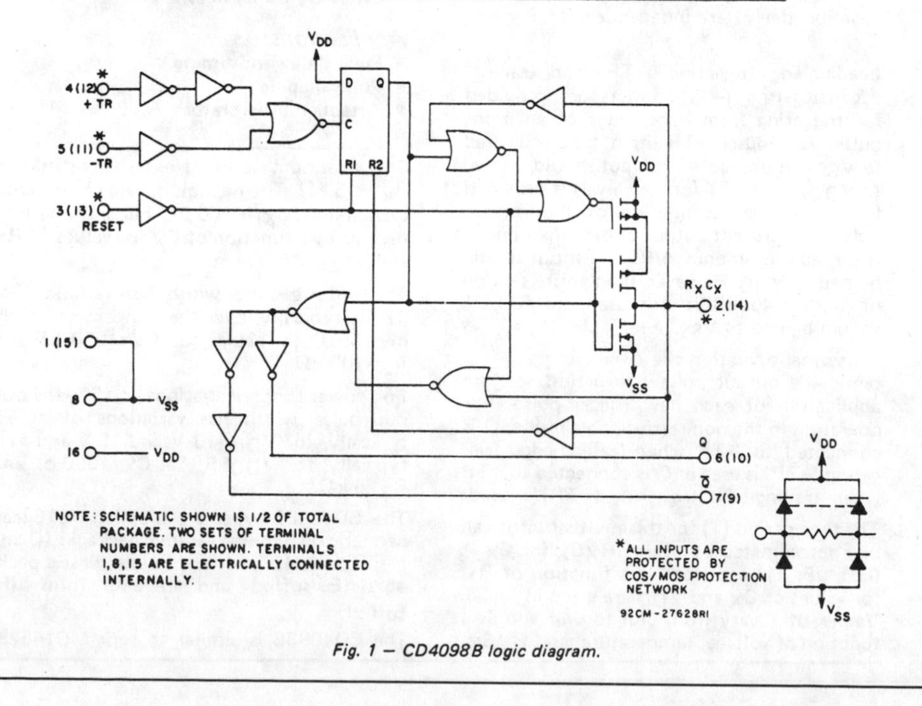

Fig. 1 — CD4098B logic diagram.

(Courtesy RCA Corporation)

TDC1018

Preliminary Information

D/A Converter

8–bit, 125MSPS

The TRW TDC1018 is a 125 MegaSample Per Second (MSPS), 8–bit digital–to–analog converter, capable of directly driving a 75 Ohm load to standard video levels. Most applications require no extra registering, buffering, or deglitching. Four special level controls make the device ideal for video applications. All data and control inputs are ECL compatible.

The TDC1018 is built with TRW's OMICRON–B™ 1–micron bipolar process. On–chip data registers and precise matching of propagation delays make the TDC1018 inherently low–glitching. The TDC1018 offers high performance, low power consumption, and video compatibility in a 24 lead DIP package.

Features

- "Graphics–Ready"
- 125MSPS Conversion Rate
- 8–Bit
- 1/2 LSB Linearity
- Power Supply Noise Rejection $>$ 50dB
- Registered Data And Video Controls
- Differential Current Outputs
- Video Controls: SYNC, BLANK, BRighT, Force High
- Inherently Low Glitch Energy
- ECL Compatible
- Multiplying Mode Capability
- Power Dissipation $<$ 940mW
- Available In 24 Lead DIP Package
- Single –5.2V Power Supply

Applications

- RGB Graphics
- High Resolution Video
- Raster Graphic Displays
- Digital Synthesizers
- Automated Test Equipment
- Digital Transmitters/Modulators

Functional Block Diagram

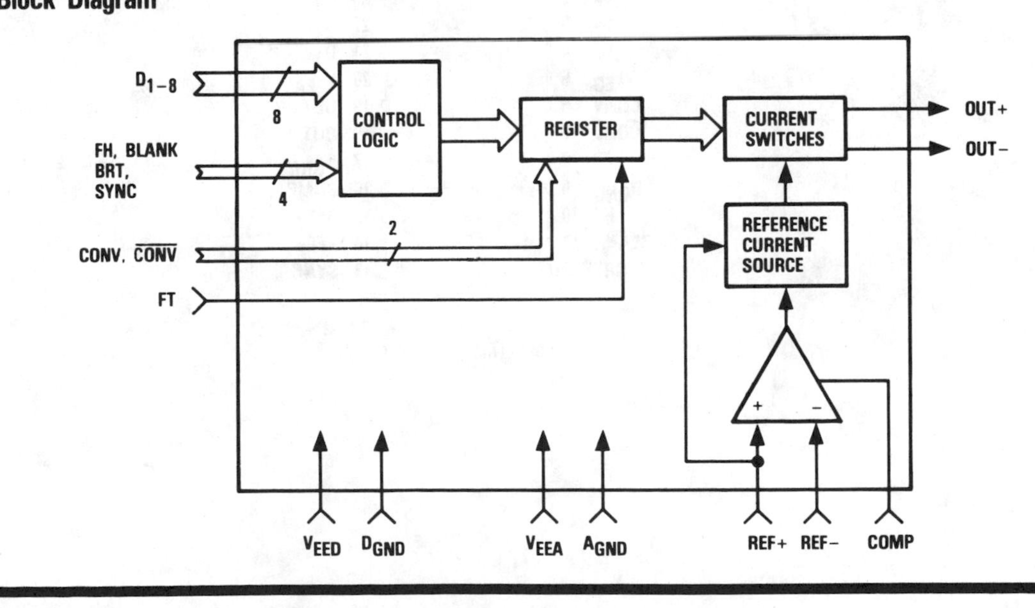

(Reproduced by Permission of TRW LSI Products Division)

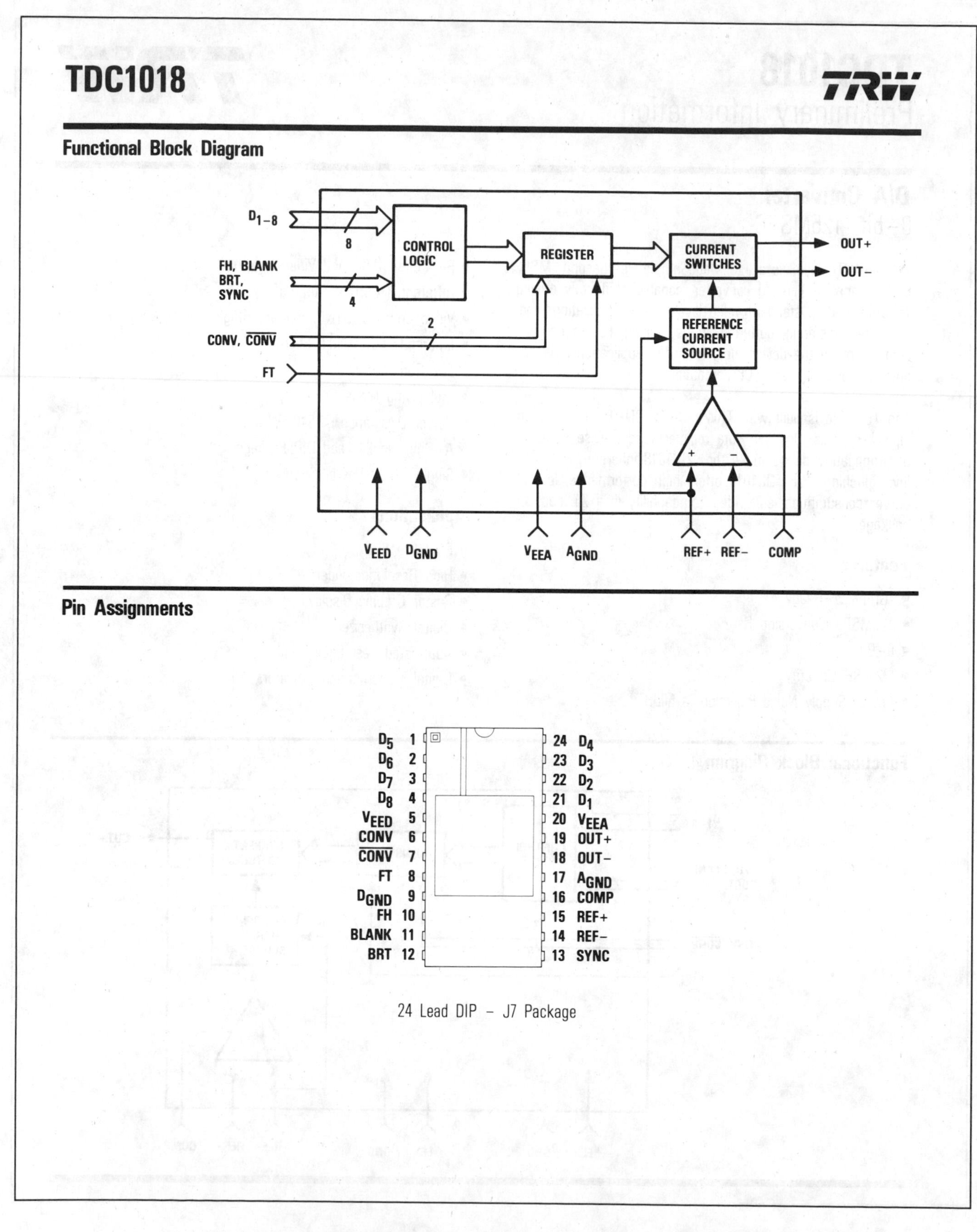

(Reproduced by Permission of TRW LSI Products Division)

Functional Description

General Information

The TDC1018 develops complementary analog output currents proportional to the product of the digital input data and analog reference current. All data and control inputs are compatible with standard ECL logic levels. FeedThrough control (FT) determines whether data and control inputs are synchronous or asynchronous. If FT is LOW, each rising edge of the CONVert clock (CONV) latches decoded data and control values into an internal D-type register. The registered values are then converted into the appropriate analog output by switched current sinks. When FT is HIGH, data and control inputs are not registered, and the analog output asynchronously tracks the input values. FT is the only asynchronous input, and is normally used as a DC control.

The TDC1018 uses a segmented approach in which the four MSBs of the input data are decoded into a parallel "Thermometer" code, which drives fifteen identical current sinks to produce sixteen coarse output levels. The LSBs of the input drive four binary-weighted current switches, with a total contribution of one-sixteenth of full scale. The LSB and MSB currents are summed to provide 256 analog output levels.

Special control inputs, SYNC, BLANK, Force High (FH) and BRighT (BRT), drive appropriately weighted current sinks which add to the output current to produce specific output levels especially useful in video applications.

Power

To provide highest noise immunity, the TDC1018 operates from separate analog and digital power supplies, V_{EEA} and V_{EED}, respectively. Since the required voltage for both V_{EEA} and V_{EED} is −5.2V, these may ultimately be connected to the same power source, but individual high-frequency decoupling for each supply is recommended. A typical decoupling network is shown in Figure 7. The return for I_{EED}, the current drawn from the V_{EED} supply, is D_{GND}. The return for I_{EEA} is A_{GND}. All power and ground pins MUST be connected.

Although the TDC1018 is specified for a nominal supply of −5.2V, operation from a +5.0V supply is possible provided that the relative polarities of all voltages are maintained.

Name	Function	Value	J7 Package
V_{EEA}	Analog Supply Voltage	−5.2V	Pin 20
V_{EED}	Digital Supply Voltage	−5.2V	Pin 5
A_{GND}	Analog Ground	0.0V	Pin 17
D_{GND}	Digital Ground	0.0V	Pin 9

Reference

The TDC1018 has two reference inputs: REF+ and REF−, which are noninverting and inverting inputs of an internal reference buffer amplifier. The output of this operational amplifier serves as a reference for the current sinks. The feedback loop is internally connected around one of the current sinks to achieve high accuracy (see Figure 4).

The analog output currents are proportional to the digital data and reference current, I_{REF}. The full-scale output value may be adjusted over a limited range by varying the reference current. Accordingly, the stability of the analog output depends primarily upon the stability of the reference. A method of achieving a stable reference is shown in Figure 7.

The reference current is fed into the REF+ input, while REF− is typically connected to a negative reference voltage through a resistor chosen to minimize input offset bias current effects.

A COMPensation input (COMP), is provided for external compensation of the TDC1018's reference amplifier. A capacitor (C_C) should be connected between COMP and the V_{EEA} supply, keeping lead lengths as short as possible. The value of the compensation capacitor determines the effective bandwidth of the amplifier. In general, decreasing C_C increases bandwidth and decreases amplifier stability. For applications in which the reference is constant, C_C should be large, while smaller values of C_C may be chosen if dynamic modulation of the reference is required.

Name	Function	Value	J7 Package
REF−	Reference Current − Input	Op-Amp Virtual Ground	Pin 14
REF+	Reference Current + Input	Op-Amp Virtual Ground	Pin 15
COMP	COMPensation Input	C_C	Pin 16

Controls

The TDC1018 has four special video control inputs: SYNC, BLANK, Force High (FH), and BRighT (BRT), in addition to a clock FeedThrough control (FT). All controls are standard ECL level compatible, and include internal pulldown resistors to force unused controls to a logic LOW (inactive) state.

Typically the TDC1018 is operated in the synchronous mode, which assures the highest conversion rate and lowest spurious output noise. By asserting FT, the input registers are disabled, allowing data and control changes to asynchronously feed through to the analog output. Propagation delay from input change (control or data) to analog output is minimized in the asynchronous mode of operation.

In the synchronous mode, the video control inputs are registered by the rising edge of the CONV clock in a manner similar to the data inputs. The controls, like data, must be present at the inputs for a setup time of t_S (ns) before, and a hold time of t_H (ns) after the rising edge of CONV in order to be registered. In the asynchronous mode, the setup and hold times are irrelevant and minimum pulse widths HIGH and LOW become the limiting factor.

Asserting the video controls produces various output levels which are used for frame synchronization, horizontal blanking, etc., as described in video system standards such as RS−170 and RS−343A. The effect of the video controls on the analog outputs is shown in Table 1. Special internal logic governs the interaction of these controls to simplify their use in video applications. BLANK, SYNC, and Force High override the data inputs. SYNC overrides all other inputs, and produces full negative video output. Force High drives the internal digital data to full scale, giving a reference white video level output. The BRT control creates a "whiter than white" level by adding 10% of the full scale value to the present output level, and is especially useful in graphics displays for highlighting cursors, warning messages, or menus. For non-video applications, the special controls can be left unconnected.

Name	Function	Value	J7 Package
FT	Register FeedThrough Control	ECL	Pin 8
FH	Data Force High Control	ECL	Pin 10
BLANK	Video BLANK Input	ECL	Pin 11
BRT	Video BRighT Input	ECL	Pin 12
SYNC	Video SYNC Input	ECL	Pin 13

Data Inputs

Data inputs to the TDC1018 are standard single-ended ECL level compatible. Internal pulldown resistors force unconnected data inputs to logic LOW. Input registers are provided for synchronous data entry and lowest differential data propagation delay (skew), which minimizes glitching.

In the registered mode, valid data must be present at the input a setup time t_S (ns) before, and a hold time t_H (ns) after the rising edge of CONV. When FT is HIGH, data input is asynchronous and the input registers are disabled. In this case the analog output changes asynchronously in direct response to the input data.

Name	Function	Value	J7 Package
D_1	Data Bit 1 (MSB)	ECL	Pin 21
D_2		ECL	Pin 22
D_3		ECL	Pin 23
D_4		ECL	Pin 24
D_5		ECL	Pin 1
D_6		ECL	Pin 2
D_7		ECL	Pin 3
D_8	Data Bit 8 (LSB)	ECL	Pin 4

Convert

CONVert (CONV) is a differential ECL compatible clock input whose rising edge synchronizes data and control entry into the TDC1018. Within the constraints shown in Figure 2, the actual switching threshold of CONV is determined by $\overline{\text{CONV}}$. CONV may be driven single-ended by connecting $\overline{\text{CONV}}$ to a suitable bias voltage (V_{BB}). The bias voltage chosen will determine the switching threshold of CONV. However, for best performance, CONV must be driven differentially. This will minimize clock noise and power supply/output intermodulation. Both clock inputs must normally be connected, with $\overline{\text{CONV}}$ being the complement of CONV.

Name	Function	Value	J7 Package
CONV	CONVert Clock Input	ECL	Pin 6
$\overline{\text{CONV}}$	CONVert Clock Input, Complement	ECL	Pin 7

Analog Outputs

The two analog outputs of the TDC1018 are high-impedance complementary current sinks which vary in proportion to the input data, controls, and reference current values. The outputs are capable of directly driving a dual 75 Ohm load to standard video levels. The output voltage will be the product of the output current and effective load impedance, and will usually be between 0V and −1.07V in the standard configuration (see Figure 5). In this case, the OUT− output gives a DC shifted video output with "sync down." The corresponding output from OUT+ is also DC shifted and inverted, or "sync up."

Name	Function	Value	J7 Package
OUT−	Output Current −	Current Sink	Pin 18
OUT+	Output Current +	Current Sink	Pin 19

(Reproduced by Permission of TRW LSI Products Division)

TDC1048

Preliminary Information

Monolithic Video A/D Converter

8-bit, 20MSPS

The TRW TDC1048 is a 20 MegaSample Per Second (MSPS) full-parallel (flash) analog-to-digital converter, capable of converting an analog signal with full-power frequency components up to 7MHz into 8-bit digital words. A sample-and-hold circuit is not necessary. Low power consumption eases thermal considerations, and board space is minimized with a 28 lead package. All digital inputs and outputs are TTL compatible.

The TDC1048 consists of 255 clocked latching comparators, combining logic, and an output buffer register. A single convert signal controls the conversion operation. The unit can be connected to give either true or inverted outputs in binary or offset two's complement coding.

Features

- 8-Bit Resolution
- 20MSPS Conversion Rate
- Low Power Consumption, 1.6W (Worst Case)
- Sample-And-Hold Circuit Not Required
- Differential Phase 1 Degree
- Differential Gain 2%
- 1/2 LSB Linearity
- TTL Compatible
- Selectable Output Format
- Available In 28 Lead DIP, CERDIP, Or Contact Chip Carrier

Applications

- Low-Cost Video Digitizing
- Radar Data Conversion
- Data Acquisition
- Medical Imaging

Functional Block Diagram

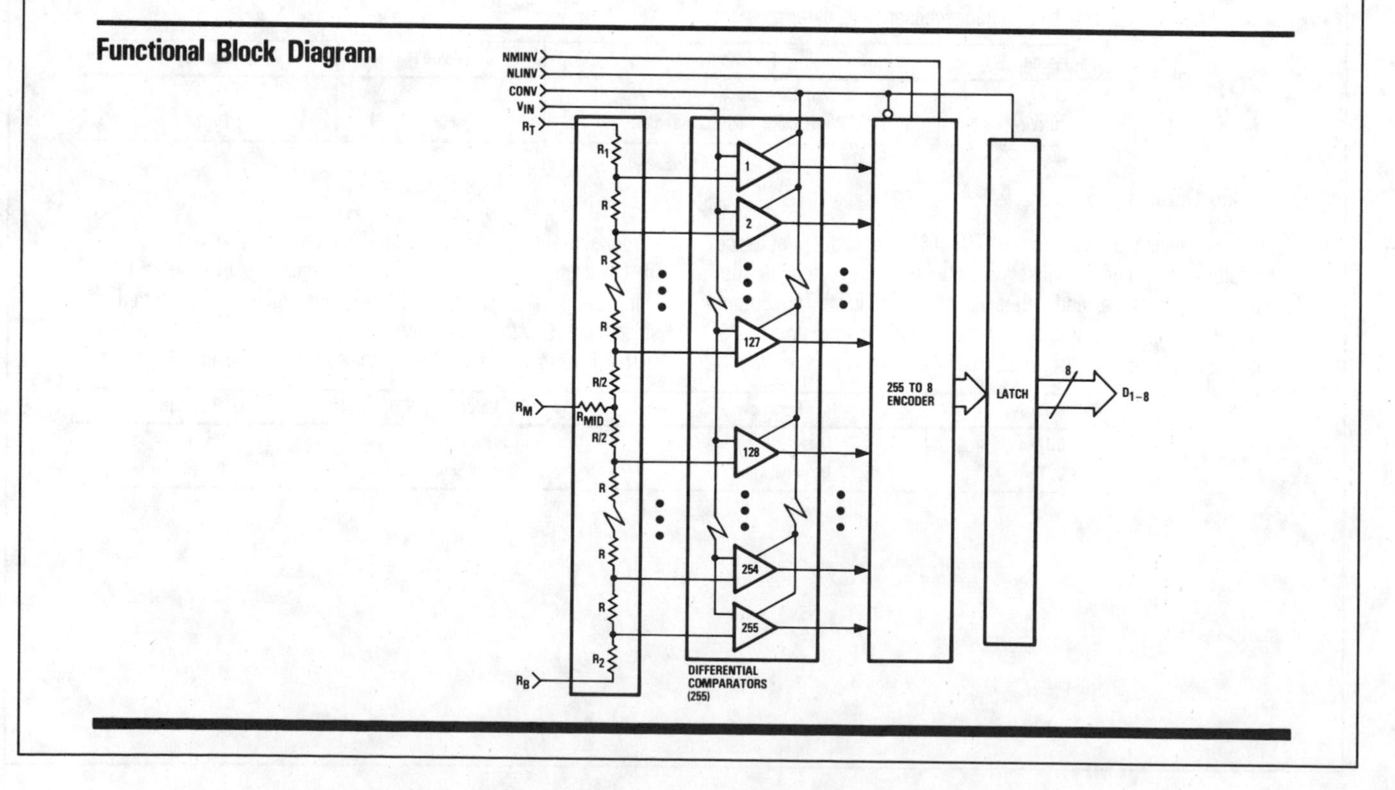

(Reproduced by Permission of TRW LSI Products Division

Functional Block Diagram

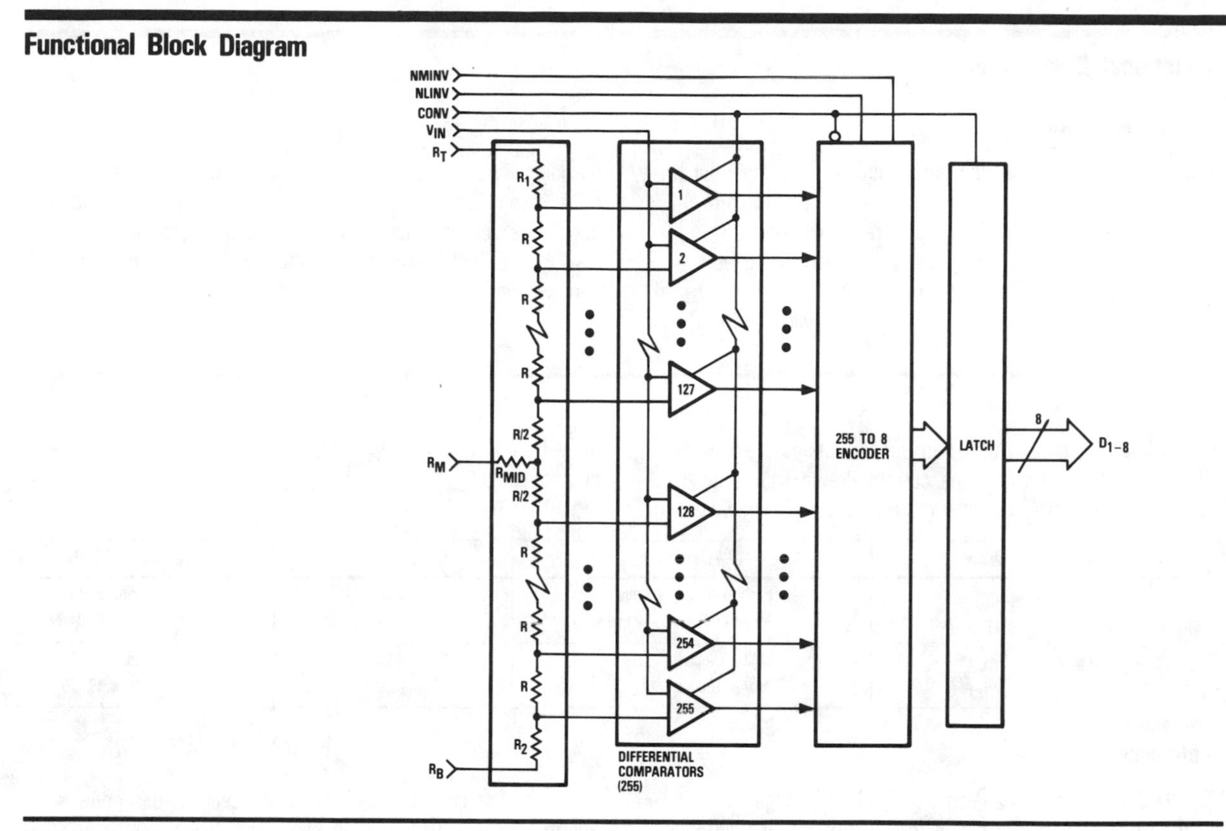

Pin Assignments

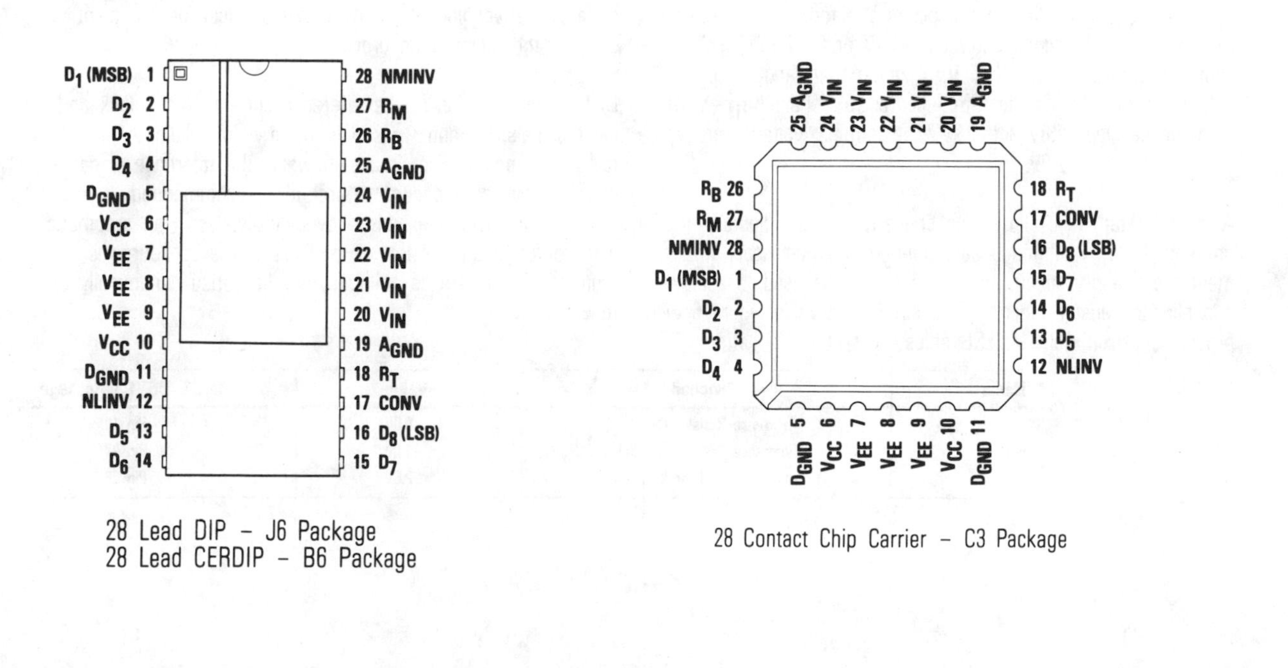

28 Lead DIP – J6 Package
28 Lead CERDIP – B6 Package

28 Contact Chip Carrier – C3 Package

(Reproduced by Permission of TRW LSI Products Division)

Functional Description

General Information

The TDC1048 has three functional sections: a comparator array, encoding logic, and output latches. The comparator array compares the input signal with 255 reference voltages to produce an N-of-255 code (sometimes referred to as a "thermometer" code, as all the comparators below the signal will be on, and all those above the signal will be off). The encoding logic converts the N-of-255 code into binary or offset two's complement coding, and can invert either output code. This coding function is controlled by DC signals on pins NMINV and NLINV. The output latch holds the output constant between updates.

Power

The TDC1048 operates from two supply voltages, +5.0V and −5.2V. The return for I_{CC}, the current drawn from the +5.0V supply, is D_{GND}. The return for I_{EE}, the current drawn from the −5.2V supply, is A_{GND}. All power and ground pins must be connected.

Name	Function	Value	J6, B6, C3 Package
V_{CC}	Positive Supply Voltage	+5.0V	Pins 6, 10
V_{EE}	Negative Supply Voltage	−5.2V	Pins 7, 8, 9
D_{GND}	Digital Ground	0.0V	Pins 5, 11
A_{GND}	Analog Ground	0.0V	Pins 19, 25

Reference

The TDC1048 converts analog signals in the range $V_{RB} \leq V_{IN} \leq V_{RT}$ into digital form. V_{RB} (the voltage applied to the pin at the bottom of the reference resistor chain) and V_{RT} (the voltage applied to the pin at the top of the reference resistor chain) should be between +0.1V and −2.1V. V_{RT} should be more positive than V_{RB} within that range. The voltage applied across the reference resistor chain ($V_{RT}-V_{RB}$) must be between 1.8V and 2.2V. The nominal voltages are V_{RT} = 0.0V, V_{RB} = −2.0V.

A midpoint tap, R_M, allows the converter to be adjusted for optimum linearity, although adjustment is not necessary to meet the linearity specification. It can also be used to achieve a nonlinear transfer function. The circuit shown in Figure 5 will provide approximately 1/2 LSB adjustment of the linearity midpoint. The characteristic impedance seen at this node is approximately 220 Ohms, and should be driven from a low-impedance source. Note that any load applied to this node will affect linearity, and noise introduced at this point will degrade the quantization process.

Due to the variation in the reference currents with clock and input signals, R_T and R_B should be low-impedance-to-ground points. For circuits in which the reference is not varied, a bypass capacitor to ground is recommended. If the reference inputs are exercised dynamically, (as in an automatic gain control circuit), a low-impedance reference source is required. The reference voltages may be varied dynamically up to 5MHz.

Name	Function	Value	J6, B6, C3 Package
R_T	Reference Resistor (Top)	0.0V	Pin 18
R_M	Reference Resistor (Middle)	−1.0V	Pin 27
R_B	Reference Resistor (Bottom)	−2.0V	Pin 26

Control

Two function control pins, NMINV and NLINV are provided. These controls are for DC (i.e. steady state) use. They permit the output coding to be either straight binary or offset two's complement, in either true or inverted sense, according to the Output Coding table on page 121. These pins are active LOW, as signified by the prefix "N" in the signal name. They may be tied to V_{CC} for a logic "1" and D_{GND} for a logic "0."

Name	Function	Value	J6, B6, C3 Package
NMINV	Not Most Significant Bit INVert	TTL	Pin 28
NLINV	Not Least Significant Bit INVert	TTL	Pin 12

Convert

The TDC1048 requires a convert (CONV) signal. A sample is taken (the comparators are latched) within 15ns after a rising edge on the CONV pin. This time is t_{STO}, Sampling Time Offset. This delay varies by a few nanoseconds from part to part and as a function of temperature, but the short-term uncertainty (jitter) in sampling offset time is less than 100 picoseconds. The 255 to 8 encoding is performed on the falling edge of the CONV signal. The coded result is transferred to the output latches on the next rising edge. Data is held valid at the output register for at least t_{HO}, Output Hold Time, after the rising edge of CONV. New data becomes valid after a Digital Output Delay, t_D, time. This permits the previous conversion result to be acquired by external circuitry at that rising edge, i.e. data for sample N is acquired by the external circuitry while the TDC1048 is taking input sample N + 2.

Name	Function	Value	J6, B6, C3 Package
CONV	Convert	TTL	Pin 17

Analog Input

The TDC1048 uses strobed latching comparators which cause the input impedance to vary with the signal level, as comparator input transistors are cut-off or become active. As a result, for optimal performance, the source impedance of the driving device must be less than 25 Ohms. The input signal will not damage the TDC1048 if it remains within the range of V_{EE} to +0.5V. If the input signal is between the V_{RT} and V_{RB} references, the output will be a binary number between 0 and 255 inclusive. A signal outside this range will indicate either full-scale positive or full-scale negative, depending on whether the signal is off-scale in the positive or negative direction. All five analog input pins must be connected together.

Name	Function	Value	J6, B6, C3 Package
V_{IN}	Analog Signal Input	0V to −2V	Pins 20, 21, 22, 23, 24

Outputs

The outputs of the TDC1048 are TTL compatible, capable of driving four low-power Schottky TTL (54/74 LS) unit loads or the equivalent. The outputs hold the previous data a minimum time (t_{HO}) after the rising edge of the CONVert signal.

Name	Function	Value	J6, B6, C3 Package
D_1	MSB Output	TTL	Pin 1
D_2		TTL	Pin 2
D_3		TTL	Pin 3
D_4		TTL	Pin 4
D_5		TTL	Pin 13
D_6		TTL	Pin 14
D_7		TTL	Pin 15
D_8	LSB Output	TTL	Pin 16

14

U.S. Patents and Scrambling Systems

This chapter contains three informative and interesting patents:

United States Patent No. 4,336,553
United States Patent No. 4,405,942
United States Patent No. 4,479,142

These patents include many of the principles discussed in this book. Note that these patents are very general and are written to cover as broad a range of applications as possible. This is done so that the inventor can make the broadest possible claim to various interpretations and versions of his invention.

The patents are interesting to read and contain much information on the subject of scrambling. They refer to related patents, as well. The interested reader can obtain these patents from the appropriate U.S. Governmental Agencies.

United States Patent [19]

den Toonder et al.

[11] **4,336,553**

[45] **Jun. 22, 1982**

[54] **METHOD OF CODING AUDIO AND VIDEO SIGNALS**

[75] Inventors: **Pieter den Toonder,** Dordrecht; **Johannes C. Seltenrijch,** Gorichem, both of Netherlands; **Graham S. Stubbs,** Poway, Calif.; **Richard G. Merrell,** Hebron, Ill.

[73] Assignee: **Oak Industries,** Rancho Bernardo, Calif.

[21] Appl. No.: **149,706**

[22] Filed: **May 14, 1980**

[51] **Int. Cl.**[3] **H04N 7/16**

[52] **U.S. Cl.** **358/120;** 358/118; 358/121; 358/124

[58] **Field of Search** 358/118, 120, 121, 124

[56] **References Cited**

U.S. PATENT DOCUMENTS

2,510,046	5/1950	Ellett et al.	358/120
2,705,740	4/1955	Druz	358/120
3,081,376	3/1963	Loughlin et al.	358/120
3,313,880	4/1967	Bass	358/120
3,485,941	12/1969	Bass	358/120
3,530,232	9/1970	Reiter et al.	358/120
4,081,832	3/1978	Sherman	358/121
4,095,258	6/1978	Sperber	358/118
4,215,366	7/1980	Davidson	358/124

Primary Examiner—S. C. Buczinski

[57] **ABSTRACT**

An audio and video signal coding system in which the normal sync pulses in the horizontal and vertical blanking intervals are suppressed, and clock and control data as well as digital audio data is inserted therein. The data, both audio information and clock data in the horizontal blanking interval is further distorted by the application of a periodic waveform and/or voltage level enhancement of certain portions of the horizontal blanking interval.

11 Claims, 13 Drawing Figures

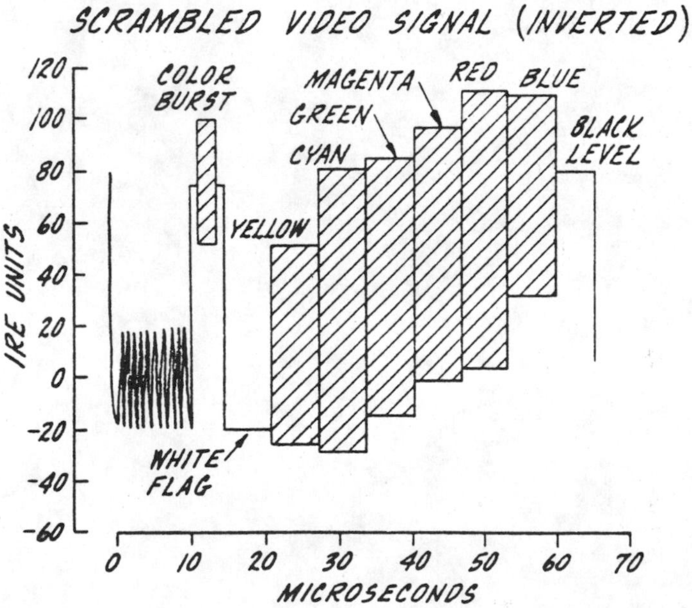

1

METHOD OF CODING AUDIO AND VIDEO SIGNALS

SUMMARY OF THE INVENTION

The present invention relates to audio and video coding systems and in particular to such a system in which the horizontal blanking interval is suppressed, in which certain clock data and digital sound data is inserted therein, and in which there is further distortion by applying a changing voltage level to the horizontal blanking intervals.

Another purpose is a coding system of the type described in which the horizontal blanking interval is used for the transmission of clock data and audio digital data and in which the voltage level of the horizontal blanking interval is changed so as to further confuse receiver sync circuits.

Another purpose is a coding system of the type described in which a very low frequency sine wave is applied to the horizontal blanking interval to provide a changing voltage level.

Another purpose is a coding system of the type described in which the data bytes in the horizontal blanking interval, which include the clock data as well as audio digital data, are selectively randomly enhanced to provide variation in voltage levels during the horizontal blanking intervals.

Other purposes will appear in the ensuing specification, drawings and claims.

BRIEF DESCRIPTION OF THE DRAWINGS

The invention is illustrated diagrammatically in the following drawings wherein:

FIG. **1** is a diagrammatic illustration of the various functional components which together form an entire audio and video television coding system,

FIG. **2** is a block diagram of the input video processor,

FIG. **3** is a block diagram of the input audio processor,

FIG. **4** is a block diagram of the audio and reference data processor,

FIG. **5** is a block diagram of the scene change detector,

FIG. **6** is a block diagram of the scrambling enhancement assembly,

FIG. **7** is a block diagram of the output video processor, and,

FIG. **8** is a block diagram of the decoder.

FIG. **9** is waveform diagram of the standard NTSC video signal illustrating the horizontal sync pulse, the color burst and the color bars.

FIG. **10** is a waveform diagram, similar to FIG. **9**, but illustrating the replacement of horizontal sync with digitized audio, and a bias applied to the color burst,

FIG. **11** is a waveform diagram illustrating the scrambled video signal of FIG. **10** with video polarity reversed, except for the color burst,

FIG. **12** is a waveform diagram illustrating the instantaneous effect of amplitude modulation applied to the three data bytes, and

FIG. **13** is a diagrammatic illustration of the application of a low frequency sine wave to a video signal frame.

2

DESCRIPTION OF THE PREFERRED EMBODIMENT

The present invention relates to subscription television and in particular to a means for scrambling or encoding or distorting both the video and audio portions of a television signal so that the program has no entertainment value unless the subscriber has the proper decoding equipment. The primary means for encoding is the suppression of all synchronizing information in both the vertical and horizontal blanking intervals as described in co-pending application Ser. No. 965,940 assigned to the assignee of the present application. The sound or audio information is placed in digital form and is inserted in the horizontal blanking interval in place of the normal horizontal sync information. The video may be inverted and this inversion may take place on the basis of program scene changes. Additionally, the video may be distorted by both shifting the voltage level of the digital information in the horizontal interval, as well as by varying the voltage level of this portion of a horizontal line by the application of a sine wave phased to vary the amplitude of the horizontal blanking portion of each line. In order to prevent unauthorized reception of the program which might be accomplished by detuning the receiver approximately one MHz in the direction of the chrominance subcarrier, amplitude modulation is applied to the aural carrier in such a way that chrominance subcarrier video information will not provide synchronization.

In the following description certain signals have designated time relationships and frequencies. It should be understood that the invention should not be limited thereto, but such information is only by way of example.

FIG. **1** diagrammatically illustrates the encoding equipment and FIG. **8** diagrammatically illustrates the decoder. In FIG. **1** the input video processor is indicated at **10** and has an input of base band video (FIG. **9**) and outputs of the following signals: a filtered video signal, a 4.0909 MHz clock, a frame reference pulse, a color burst gate signal, and a clamped video output. The use of these various signals will be described in connection with the remaining portions of the circuit.

An input audio processor **12** receives the input audio signal and provides an output of the audio information in digital form. The audio information in digital form is connected to an audio and reference data processor **14** whose output will be data to enable the subscriber decoders as well as the audio information in digital form. The output from processor **14** is directed to output video processor **16** wherein this data is combined with the video signal for subsequent transmission on a suitable carrier. A horizontal timing generator **18** and a vertical timing generator **20** provide various timing signals which coordinate the audio and video processors as well as the operation of a scrambling enhancement assembly **22**. A scene change detector **24** has an input of filtered video and an output designed to control inversion of the video in output video processor **16** in accordance with program scene changes.

Referring to FIG. **2** which details the input video processor, there is an input attenuator **28** which permits manual adjustment of the video gain in order to accommodate operating conditions of different video sources. Attenuator **28** is connected to an amplifier **30** which functions as an isolation stage between the video source and the following video processing circuit, as well as providing a small gain (2X) to allow for low amplitude video signals.

A clamp **32** is connected to amplifier **30** and clamps the video signal at a specific level as is common in television operations. The output from clamp **32** is the video signal clamped at an appropriate level which output is passed directly to the output video processor **16** which will be described in detail hereinafter. A filter **34** is also connected to amplifier **30** and is a low pass filter effectively removing all color signals that may interfere with the various following sync separation circuits. The output from filter **34** is thus a low bandwidth monochrome video signal which will be used in scene change detector **24**. The output from filter **34** is also connected to a second amplifier **36** whose output in turn is connected to a sync separating circuit **38**. One output from sync separating circuit **38** is to a clock circuit **40** which provides a 4.0909 MHz clock signal synchronized with the frequency of the horizontal sync pulses of the incoming video. A pulse processing circuit **42** is also connected to sync separator **38** and provides two outputs. The first, a frame reference signal, is a pulse coincident with the leading edge of the first serration pulse of the vertical interval immediately preceding the odd field. This pulse is required for synchronization of the internal timing signals with the input video. A second output from pulse processor **42** is a color burst gate signal which is coincident with the color burst of each line of the incoming video. The color burst gate will be suppressed during the vertical sync period when no color burst is being received.

Returning to FIG. **1**, the horizontal timing generator **18** will have an input of the clock and frame reference signals from video processor **10**. The timing generator will provide a number of signals all synchronized by its two inputs. Each horizontal line is divided into 260 parts of approximately 250 NS each. The following table indicates the position of the various timing pulses in a horizontal line. In addition to the pulses of the table, a timing generator will provide an approximate 500 KHz and a two MHz signal for operation of certain of the circuits, as described.

	Timing pulse	Start	Stop
SRL	Shift register load	3	4
SS1	First audio sample	14	33
SS2	Second audio sample	144	163
HD	Horizontal drive	9	36
HB	Horizontal blanking	9	59
HW	Horizontal window	60	252

Vertical timing generator **20** will provide four outputs, the first being the field index signal which will be a very short duration pulse at approximately the middle of the fifth line of the vertical interval (FI); a vertical drive signal, a positive pulse beginning at the first line of the vertical interval and extending to the ninth line of that interval (VD); a vertical blanking signal which is a positive pulse beginning with the initiation of the vertical interval and extending until line **21** of the vertical interval (VB); and a vertical window signal which is a positive pulse beginning at line **46** and extending until line **238** (VW).

FIG. **3** illustrates the input audio processor circuit. The audio signal is directed to an attenuator **44** which functions in a manner similar to attenuator **28** and the output of the attenuator is connected to a low pass filter **46** which limits the pass band to approximately 12 KHz, the audible range. Higher frequency signals would cause distortion in the subsequent digitizing process. A sample and hold circuit **48** is connected to filter **46** and is gated by the sound sample gate signals from horizontal timing generator **18**. Circuit **48** will sample the sound during the period that it is gated and will hold the amplitude level of the sound until the next sound sample. As indicated in the previous table, the first audio sample will be made approximately 3.5 microseconds after the start of the horizontal line, with the second audio sample being made approximately 35 microseconds after the beginning of the horizontal line. The sound samples will be converted to digital form by an analog to digital converter **50** which is clocked by a 500 KHz signal from horizontal timing generator **18**. Alternate outputs from ADC **50** are connected, in parallel form, to storage registers **52** and **54**. The data from the storage registers will be transferred to audio and reference data processor circuit **14** in accordance with the operation of a flip-flop **56**. Flip-flop **56** will be gated by the sound sample and horizontal drive (HD) outputs from horizontal timing generator **18**. For example, each of the sound samples may be an eight-bit digital word and the samples may be taken at a rate of approximately 31,500 per second.

The digital audio is transferred in parallel form to audio and reference data processor **14** (FIG. **4**). A storage register **58** has three sections, one for sound byte **1** (the first sound sample), indicated at **60**, a second for sound byte **2** (the second sound sample), indicated at **62**, and a third for a digital receiver clock sync pattern, indicated at **64**. The sync pattern will be hard-wired into the storage register and will in binary form provide the clock signal for the decoder. The parallel information in storage register **58** will be moved, again in a parallel manner, to a shift register **66** upon being gated by the shift register load pulse from horizontal timing generator **18**. A second input for shift register **66** is provided by storage register **68** which has a hard-wired vertical drive reference pattern, which code sequence, again in binary form, is used by the decoder to recognize the existence of an encoded video signal and to reset the decoder time sequence. The field index signal from vertical timing generator **20** is used to move the reference pattern from storage register **68**, once each frame, into shift register **66**. The data in shift register **66** will be gated to the output video processor in accordance with the presence of either field index or shift register load signals at the input of an OR gate **70** which is connected to shift register **66**. The information will be shifted out in accordance with the input four MHz clock signal.

FIG. **10** illustrates the three-byte data insertion into the horizontal blanking interval in place of the horizontal sync pulse illustrated in FIG. **9**.

As indicated previously, in order to enhance distortion or scrambling of the video signal and to insure that unauthorized receivers cannot in some way view subscription programming, the video is inverted or not inverted in accordance with changes of scene of the actual program. The scene change detector (FIG. **5**) has an input of low bandwidth monochrome video from the input video processor and this signal is connected to a voltage comparator **72**. Analog comparator **72** compares the instantaneous brightness of the video signal with the average brightness over a period of time, for example three frames. The output from comparator **72** is sampled at a rate of 2,048 samples per field and these samples are stored in shift register **74**. In fact, the binary video at the output of comparator **72** is sampled at a rate of 32 samples in one out of every three lines over a period of **192** lines in each field.

This sampling process is controlled by the horizontal and vertical timing generators. A divide by three circuits **76** is clocked by the horizontal drive and reset by the vertical window. The vertical window in addition to resetting the divide by three circuit, thus insuring the same starting point in every frame, also prevents counting and blocks the output of this circuit during the vertical interval. Thus, divide by three circuit **76** produces a pulse during every third line except during the vertical interval. A divide by six circuit **78** is driven by the 4 MHz clock and reset by divide by three circuit **76** and the horizontal window. Accordingly, the divide by six circuit **78** produces output pulses only every third line and only during the horizontal window. Since the horizontal window lasts for **192** clock pulses and divide by six circuit **78** produces one output pulse for every six clock pulses, there are 32 sample pulses every third line except during the vertical interval.

A digital comparator **80** is connected to the output of shift register **74** and compares the output binary number from shift register **74** with the output binary number from comparator **72**. Thus, the brightness level of one field is compared with the brightness level of the preceding field at each of the same locations in the field. The output from digital comparator **80** which will be either high or low, depending upon whether the brightness levels are the same or different, is connected to a clocked counter **82**. Counter **82** receives the output from divide circuits **76** and **78** and thus is clocked at the same rate as shift register **74**. Clocked counter **82** will count pulses at the described sample rate when the comparator output from circuit **80** is high indicating dissimilar inputs. Thus, whenever there is a difference in the brightness levels from one field to the next, that indication of a brightness change will be registered by clocked counter **82**. The counter is reset by the vertical drive signal so that a new count begins for each field. Clocked counter **82** is connected to a digital comparator **84** which has a preset number, as provided by a series of manual switches diagrammatically indicated at **86**. Thus, the threshold for recognition of a scene change can be varied. The number from clocked counter **82**, when it exceeds the number provided by preset switches **86** is indicative of a scene change as there have been a sufficient number of changes in the brightness level from one field to the next to indicate a scene change. The output from digital comparator **84** is a pulse indicating that in fact a scene change has taken place and this pulse is connected to a time delay **88**. Time delay **88** may typically have a three second period and thus will not register a new scene change unless three seconds have elapsed. In this way, fast moving objects or the like will not trigger a polarity change. Time delay circuit **88** is connected to a field sync circuit **90** which is gated by the vertical drive signal from vertical timing generator **20**. Thus, a scene change, which will cause inversion or a change of polarity of the video signal as described, will only take place at the end of a field and such inversion will not take place at a greater frequency than every three seconds. The scene change detector output of field sync **90** is connected to output video processor **16**.

FIG. **6** illustrates certain circuits which can be utilized to further enhance the scrambling of the video signal. A data swing oscillator **92** is a free running generator oscillating at a frequency of for example approximately 15 Hz. This variable signal will be applied to the data to vary the level thereof at the output of video processor **16**. The second circuit in scrambling enhancement assembly **22** is an aural amplitude modulating oscillator **94** which provides a frequency of approximately 15.75 KHz, which frequency will be varied approximately 15–30 Hz on either side of the base frequency. Such a swept frequency will be applied to the aural carrier at the transmitter. Such modulation on the aural carrier will cause it to interfere with the reception of the chrominance subcarrier, thus distorting any information on it and preventing an unauthorized subscriber from being able to obtain chrominance information which might in fact provide a usable picture. A third signal in scrambling enhancement assembly **22** is provided by a random data modulator **96**. This circuit has inputs of horizontal drive, vertical drive, and the four MHz clock. Modulator **96** has three outputs, only one of which will be high during each horizontal drive period. The pattern as to which of the three outputs will be high will only be repeated after approximately 65,000 patterns. The horizontal drive pulse gates the circuit into operation and the vertical drive pulse will advance the sequence one step. The sequence is continually changing at the vertical drive rate of 60 Hz.

FIG. **7** illustrates the output video processor. An inverter is indicated at **98** and receives one input from scene change detector **24** and a second input of the standard clamped video from input video processor **10**. Inverter **98** will either reverse the polarity of the video signal or not depending upon the output from scene change detector **24**. The scrambled non-inverted video signal is illustrated in FIG. **10** and the scrambled inverted video signal is illustrated in FIG. **11**. The video signal as applied to inverter **98** is also applied to a switch **100** which will normally block the video signal except during the period of the color burst as controlled by the color burst gate signal applied from input video processor **10**. Thus, the output from switch **100** will be the video color burst. A burst bias circuit **102** has inputs of vertical drive, horizontal drive and the color burst gate. The burst bias circuit, when gated by the color burst gate and not inhibited by either the vertical drive or horizontal drive signals, will provide a DC level or bias voltage for the color burst but will not bias the data. Burst bias circuit **102** is connected to the output of switch **100** so as to provide the bias for the color burst signal. Compare the bias of the color burst signal in the unscrambled waveform of FIG. **9** and the scrambled waveform of FIG. **10**.

The data information from audio and reference data processor **14** provides one input to an amplifier **104** whose gain is controlled by the three outputs from random data modulator **96**. Thus, which of the three data bytes will have an enhanced amplitude is determined by which output is high from modulator **96**. FIG. **12** illustrates one of the three data bytes with an enhanced amplitude. The output from amplifier **104** is connected to a swing circuit **106** which receives the output from data swing oscillator **92**. The three data bytes, in addition to having one of the three enhanced in amplitude, will in total have their bias level varied in accordance with the 15 Hz signal from oscillator **92**. FIG. **13** diagrammatically illustrates the effect of the 15

Hz signal on a single frame of the video signal. The output from swing circuit **106** is connected to switch **108** as are the outputs from switch **100** and burst bias **102**. Switch **108** normally passes the video signal from inverter **98**. However, during the horizontal blanking interval, as determined by the horizontal blanking gate applied to the switch, the switch will pass the inputs from swing circuit **106**, burst bias **102** and switch **100**. Thus, in the horizontal blanking interval, the output from the switch will be the three data bytes enhanced as described and the color burst, all at a predetermined bias level. The output from switch **108** is connected to an amplifier **110**, with the output from the amplifier going to the transmitter.

The output from amplifier **110** is a video signal with all horizontal and vertical sync information removed, which video signal will be polarity inverted or not, depending upon changes in scene of the actual picture. The horizontal blanking interval will be filled with sound data bytes and the conventional color burst as well as the receiver clock sync pattern which is used to control the clock of each decoder. During the vertical blanking interval, the vertical drive reference pattern will be inserted, which enables the decoders to recognize the existence of an encoded video signal. The data in the blanking interval will vary, as described, as effected by the data swing oscillator and the random data modulator. Such variations of signals during the horizontal blanking interval will make it impossible for the receiver to sync onto any repetitive signals in the blanking intervals, thus preventing a usable picture at a non-authorized receiver. Not only is the conventional sync information removed from the video signal, but the information or signals substituted in the horizontal and vertical blanking intervals will prevent the receiver from attaining any synchronization. The polarity reversal caused by scene changes is essentially impossible of detection for anyone not having information as to the switch setting used in digital comparator **84**.

The decoder is illustrated in FIG. **8**. Typically, subscription programs will be carried on either a UHF or VHF station and such programs will only be broadcast during a portion of the station's overall air time. The input for the decoder is a UHF or VHF tuner **120** which provides an output IF signal, for example at frequencies of 41.25 MHz and 45.75 MHz, respectively. Although the program audio is coded, the audio carrier may in fact be used for other purposes, such as additional sound, or as a barker channel. The output from tuner **120** is connected to an IF amplifier **122** whose output is connected to a video detector **124** which provides base band video and a 4.5 MHz audio carrier.

Assuming first that a non-encoded program is being received, the video information will pass through a switch **126** directly to a modulator **128** which will provide an output usable in a TV receiver. The audio signal will pass through a filter **129** and an amplifier **130** whose output is also connected to modulator **128**. In the commercial mode the entire program of both audio and video will pass in the conventional manner. The decoder will have no effect upon either signal.

Assuming now that a program is encoded, the output from video detector **124** is connected to a data separator **132** which provides an output with three different types of information. In effect, the data separator provides a signal which allows the vertical reference pattern detector **134** to recognize the existence of coded video and provides a reset pulse for sync generator **136**. Sync generator **136** will provide the complete series of horizontal and vertical sync pulses necessary to properly control the video information so that it may be recognizably displayed on a TV receiver. There will be a horizontal drive signal, a vertical drive signal, a composite sync signal and a composite blanking signal. Sync generator **136** is controlled by a clock **138** which is synchronized by the sync pattern which has been transmitted as one of the three data bytes in the horizontal blanking interval. This clock signal will properly regulate the operation of the sync generator as gated by the vertical pattern recognition circuit.

The third output from data separator **132** is the audio information in the form of the two data bytes. This information is passed to a first shift register **140** and a second shift register **142** whose outputs are both connected to a digital-to-analog converter **144** whose output is the audio information in analog or conventional audio form. The operation of the shift registers are controlled by clock **138** and by a timer **146** which is gated by the horizontal drive output from sync generator **136**. The timer provides an internally generated clock which consists of two 15.734 KHz signals of opposite phase which alternates operation of the shift registers and is gated or controlled as described by the horizontal drive signal. The data goes into the two shift registers in serial form and comes out in a parallel manner where it is converted by the digital-to-analog converter into conventional audio information.

The output from digital-to-analog converter **144** goes to an FM modulator **145** which will provide the conventional FM signal normally associated with a television program. The output from FM modulator **145** is connected to modulator **128** and to a frequency comparison circuit **147**. The basis for frequency comparison is the horizontal drive signal which will be at a very specific 15.734 KHz. This is compared with the FM carrier of 4.5 MHz divided by **286** and any difference is used to control the FM modulator so that it stays precisely on frequency.

An inversion detector **148** is also connected to the output of video detector **124** and the presence of an inverted video signal may, for example, be determined by the level of line **23** in the vertical blanking interval. The manner in which a video inversion control signal is transmitted to a receiver may vary. Such a signal may occupy a portion of a horizontal line in the vertical interval or it may be transmitted with address information in the manner shown in U.S. Pat. Nos. 4,145,717 and 4,112,464. The output from inversion detector **148** is connected directly to modulator **128** where it is effective to cause inversion of the video signal in accordance with inversions of that signal at the transmitter.

Switch **126** receives all of the necessary sync information from sync generator **136**. This switch will pass the video signal except as it is gated during the horizontal and vertical blanking intervals to pass only the sync information from sync generator **136**. Thus, the output from switch **126** will be the video signal as transmitted with the proper synchronization information inserted therein, which output will subsequently either be inverted or not, depending upon the condition of inversion detector **148**. In the case of a signal inversion, the sync will also have to be inverted, which function is also performed by switch **126**.

As indicated above, the video signal has been reconstituted by the addition of the sync information deleted at the transmitter. The video signal is inverted or not in accordance with the output of the inversion detector. The audio information is detected, converted to an analog form and placed on a controlled FM carrier. The decoder or data separator ignores the varying level of the three data bytes, as brought about by the data swing generator and similarly ignores any enhancement of one of the three data bytes as controlled by the random data modulator. This is brought about by appropriate bias control in the data separator. However, a receiver without an appropriate decoder cannot ignore such variations in signal level during the horizontal blanking intervals and, as described, will be unable to sync on any repetitive signal.

The vertical reference pattern recognition circuit is arranged to recognize the binary reference pattern as provided by storage register **68** in the audio and reference data processor. As indicated above, such recognition effectively permits the decoder to operate in the manner described.

The embodiments of the invention in which an exclusive property or privilege is claimed are defined as follows:

1. Means for coding video and audio signals including means for deriving clock and internal timing signals from the video signal, means controlled by said internal timing signals for providing digital data representation of the audio signal,

   means for suppressing the video signal during horizontal blanking intervals, means for inserting the digital audio data in pulse form into the suppressed horizontal blanking intervals, and means for applying a changing voltage level to the video signal and audio data pulses during the horizontal blanking intervals to enhance video signal coding.

2. The coding means of claim **1** further characterized in that the means for changing the voltage level during horizontal blanking intervals includes means for applying a periodically varying waveform thereto.

3. The coding means of claim **1** or **2** further characterized in that the means for changing the voltage level during horizontal blanking intervals includes means for changing the voltage level during only a portion of the horizontal blanking interval.

4. The coding means of claim **1** further characterized in that the means for changing the voltage level during horizontal blanking intervals includes means for applying a very low frequency sine wave thereto.

5. The coding means of claim **1** further characterized by and including means for providing data representative of a decoder clock signal and means for inserting said decoder clock signal data into the suppressed horizontal blanking intervals.

6. The coding means of claim **5** further characterized in that the horizontal blanking interval includes at least two data insertions, one for the decoder clock signal and another for the digital sound data, the means for changing the voltage level of the video signal during horizontal blanking intervals includes means for changing the level of one of said data insertions.

7. The coding means of claim **6** further characterized in that the means for changing the voltage level during the horizontal blanking interval includes means for applying a low frequency periodic waveform thereto.

8. The coding means of claim **1** further characterized by and including means for suppressing the video signal during the vertical blanking intervals, and means for inserting vertical reference signal data in the suppressed vertical blanking intervals.

9. Means for coding video signals including means for suppressing the video signal during horizontal blanking intervals, means for inserting data pulses into the suppressed horizontal blanking intervals, and means for applying a changing voltage level to the video signal and data pulses during the horizontal blanking intervals to enhance video signal coding.

10. The coding means of claim **9** further characterized in that the means for changing the voltage level during horizontal blanking intervals includes means for applying a periodically varying waveform thereto.

11. The coding means of claims **9** or **10** further characterized in that the means for changing the voltage level during horizontal blanking intervals includes means for changing the voltage level during only a portion of the horizontal blanking interval.

* * * * *

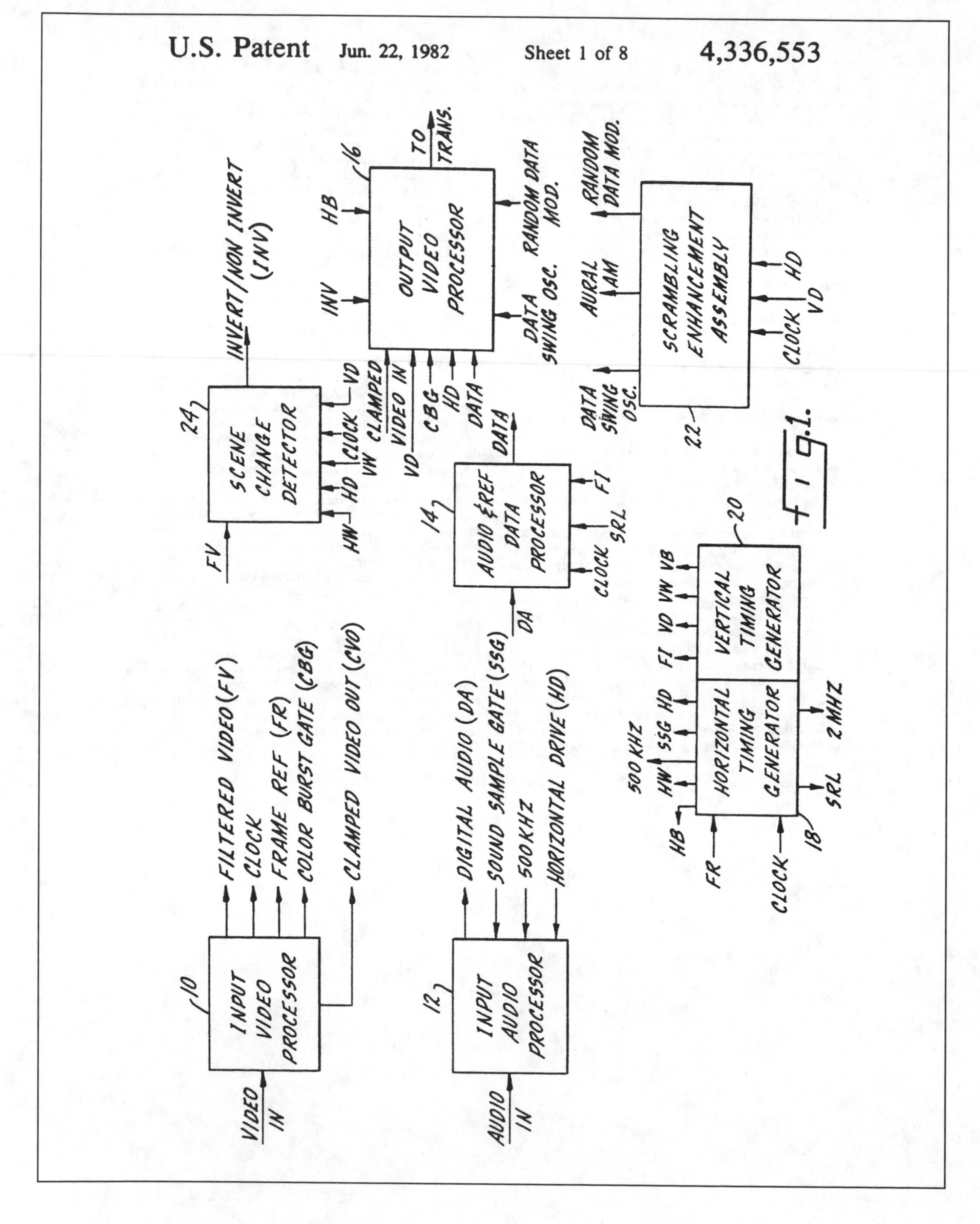

U.S. Patent Jun. 22, 1982 Sheet 1 of 8 4,336,553
INPUT VIDEO PROCESSOR
10
VIDEO IN
FILTERED VIDEO (FV)
CLOCK
FRAME REF (FR)
COLOR BURST GATE (CBG)
CLAMPED VIDEO OUT (CVO)
INPUT AUDIO PROCESSOR
12
AUDIO IN
DIGITAL AUDIO (DA)
SOUND SAMPLE GATE (SSG)
500 KHZ
HORIZONTAL DRIVE (HD)
SCENE CHANGE DETECTOR
24
FV
HW
HD
CLOCK
VW
VD
INVERT/NON INVERT (INV)
AUDIO & REF DATA PROCESSOR
14
DA
CLOCK
SRL
FI
DATA
OUTPUT VIDEO PROCESSOR
16
INV
HB
TO TRANS.
CLAMPED VIDEO IN
VD
CBG
HD
DATA
DATA SWING OSC.
RANDOM DATA MOD.
HORIZONTAL TIMING GENERATOR
VERTICAL TIMING GENERATOR
18
20
HB
FR
CLOCK
HW
500 KHZ
SSG
HD
FI
VD
VW
VB
SRL
2 MHZ
SCRAMBLING ENHANCEMENT ASSEMBLY
22
DATA SWING OSC.
AURAL AM
RANDOM DATA MOD.
CLOCK
VD
HD
Fig. 1.

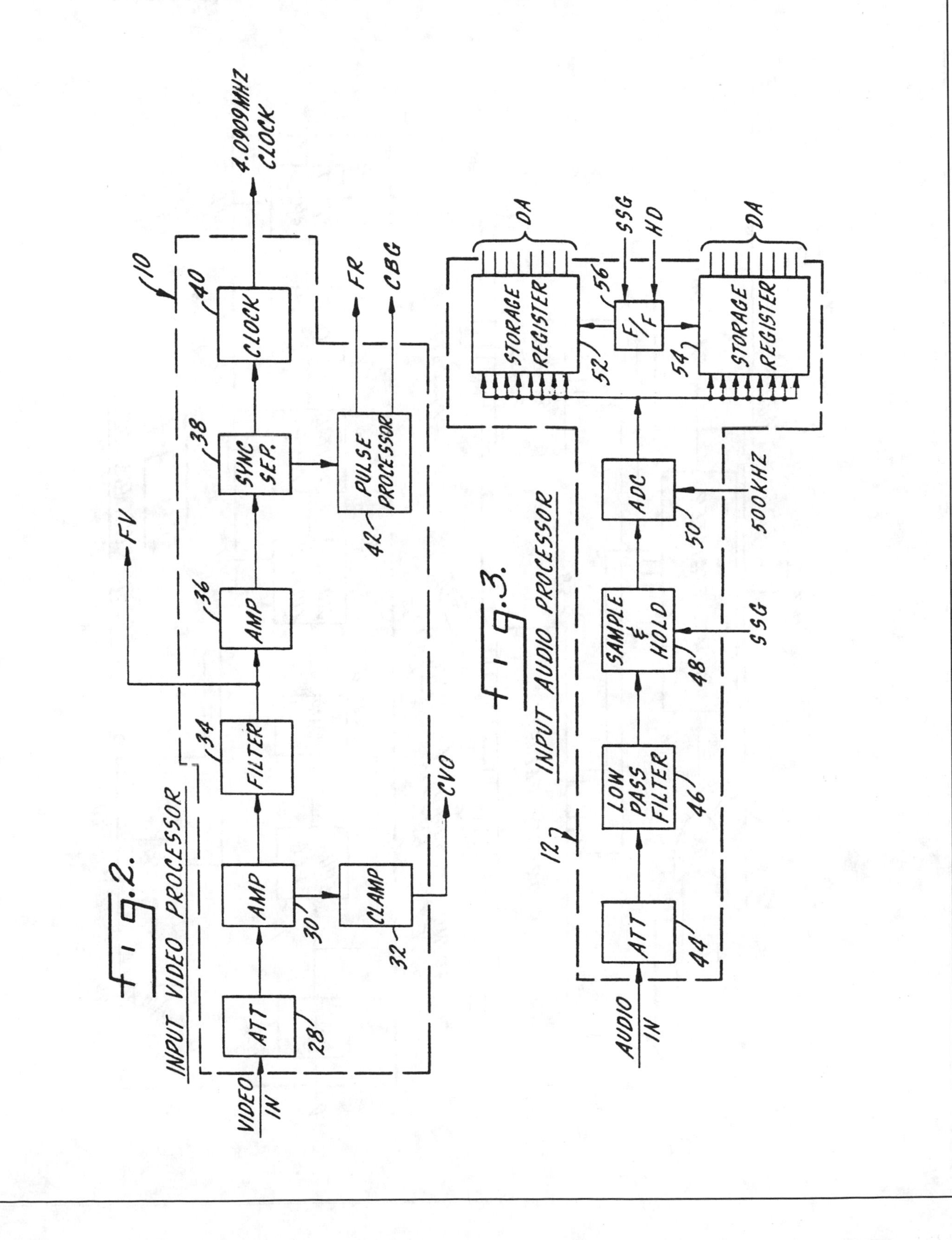
Fig. 2.
INPUT VIDEO PROCESSOR
VIDEO IN
ATT
28
AMP
30
CLAMP
32
CVO
FILTER
34
AMP
36
FV
SYNC SEP.
38
CLOCK
40
4.0909MHZ CLOCK
10
PULSE PROCESSOR
42
FR
CBG
Fig. 3.
INPUT AUDIO PROCESSOR
12
AUDIO IN
ATT
44
LOW PASS FILTER
46
SAMPLE & HOLD
48
SSG
ADC
50
500KHZ
STORAGE REGISTER
52
F/F
56
SSG
HD
STORAGE REGISTER
54
DA
DA

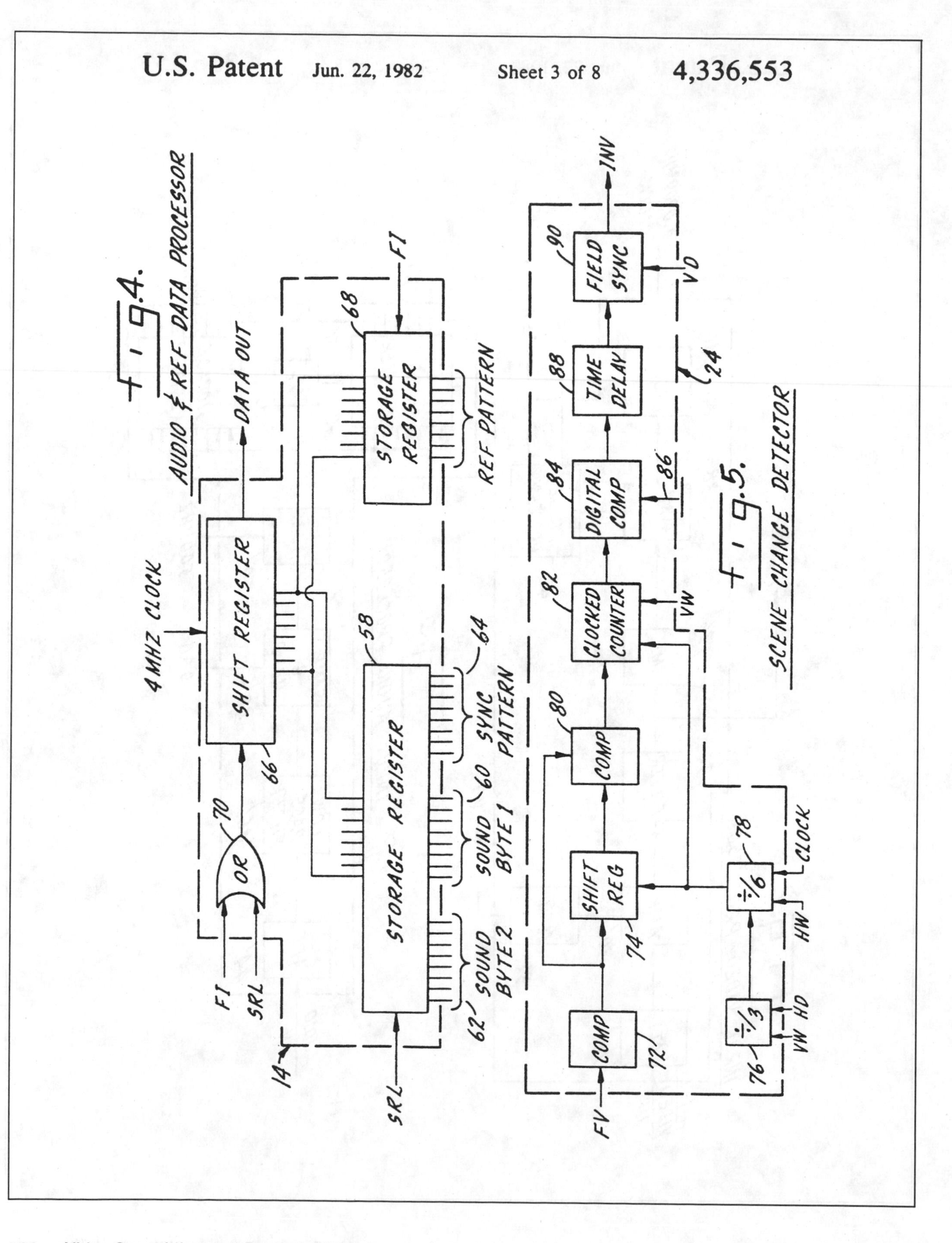
U.S. Patent Jun. 22, 1982 Sheet 3 of 8 4,336,553
Fig. 4.
AUDIO & REF DATA PROCESSOR
DATA OUT
4MHZ CLOCK
SHIFT REGISTER
66
OR
70
FI
SRL
14
STORAGE REGISTER
68
FI
REF PATTERN
STORAGE REGISTER
58
SYNC PATTERN
64
SOUND BYTE 1
60
SOUND BYTE 2
62
SRL
Fig. 5.
SCENE CHANGE DETECTOR
INV
FIELD SYNC
90
VD
TIME DELAY
88
24
DIGITAL COMP
84
86
CLOCKED COUNTER
82
VW
COMP
80
SHIFT REG
74
COMP
72
FV
÷6
78
CLOCK
HW
÷3
76
VW HD

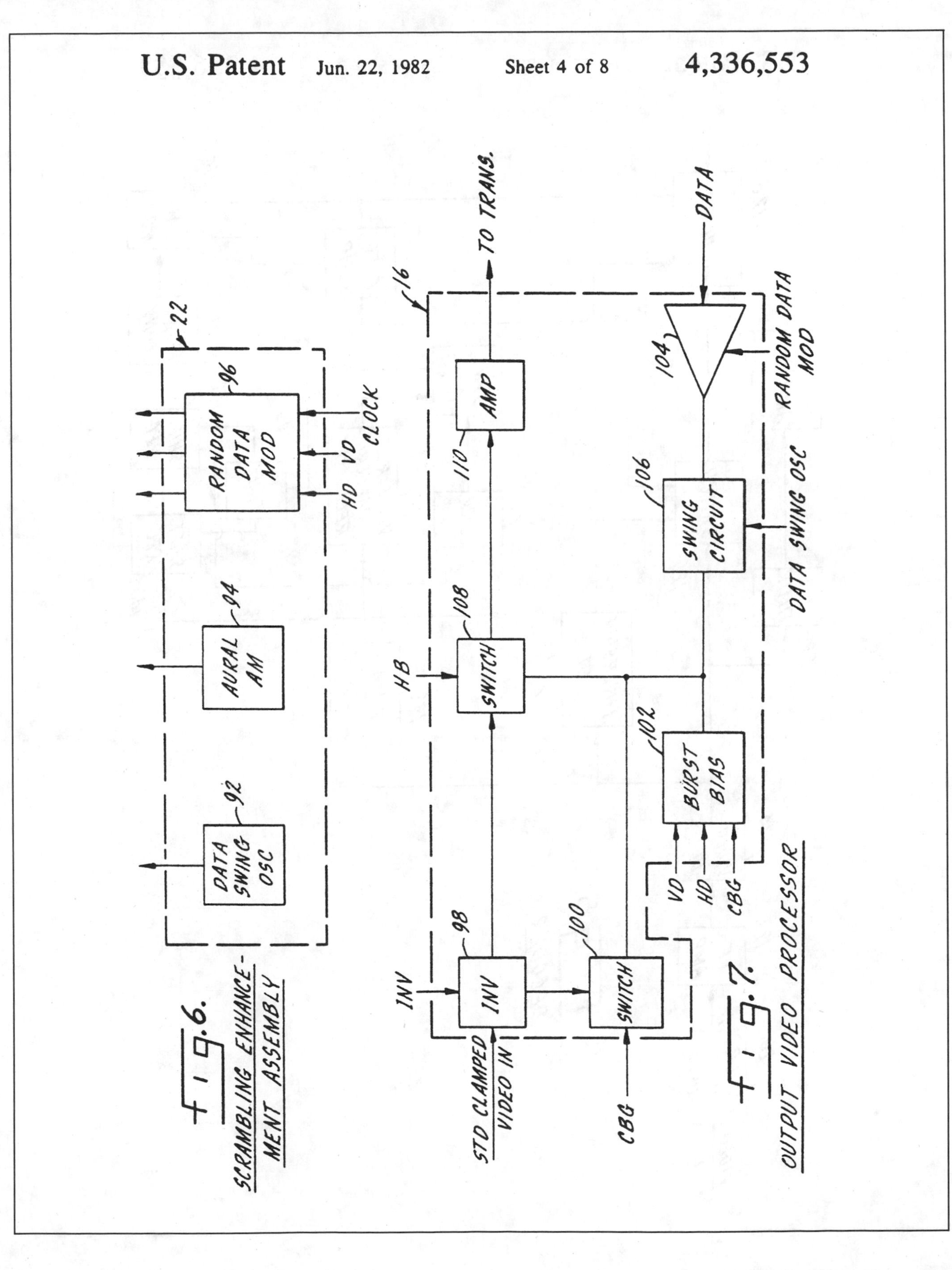
U.S. Patent Jun. 22, 1982 Sheet 4 of 8 4,336,553
Fig. 6.
SCRAMBLING ENHANCEMENT ASSEMBLY
22
DATA SWING OSC
92
AURAL AM
94
RANDOM DATA MOD
96
HD
VD
CLOCK
Fig. 7.
OUTPUT VIDEO PROCESSOR
16
INV
INV
98
STD CLAMPED VIDEO IN
SWITCH
100
CBG
SWITCH
108
HB
AMP
110
TO TRANS.
BURST BIAS
102
VD
HD
CBG
SWING CIRCUIT
106
DATA SWING OSC
104
RANDOM DATA MOD
DATA

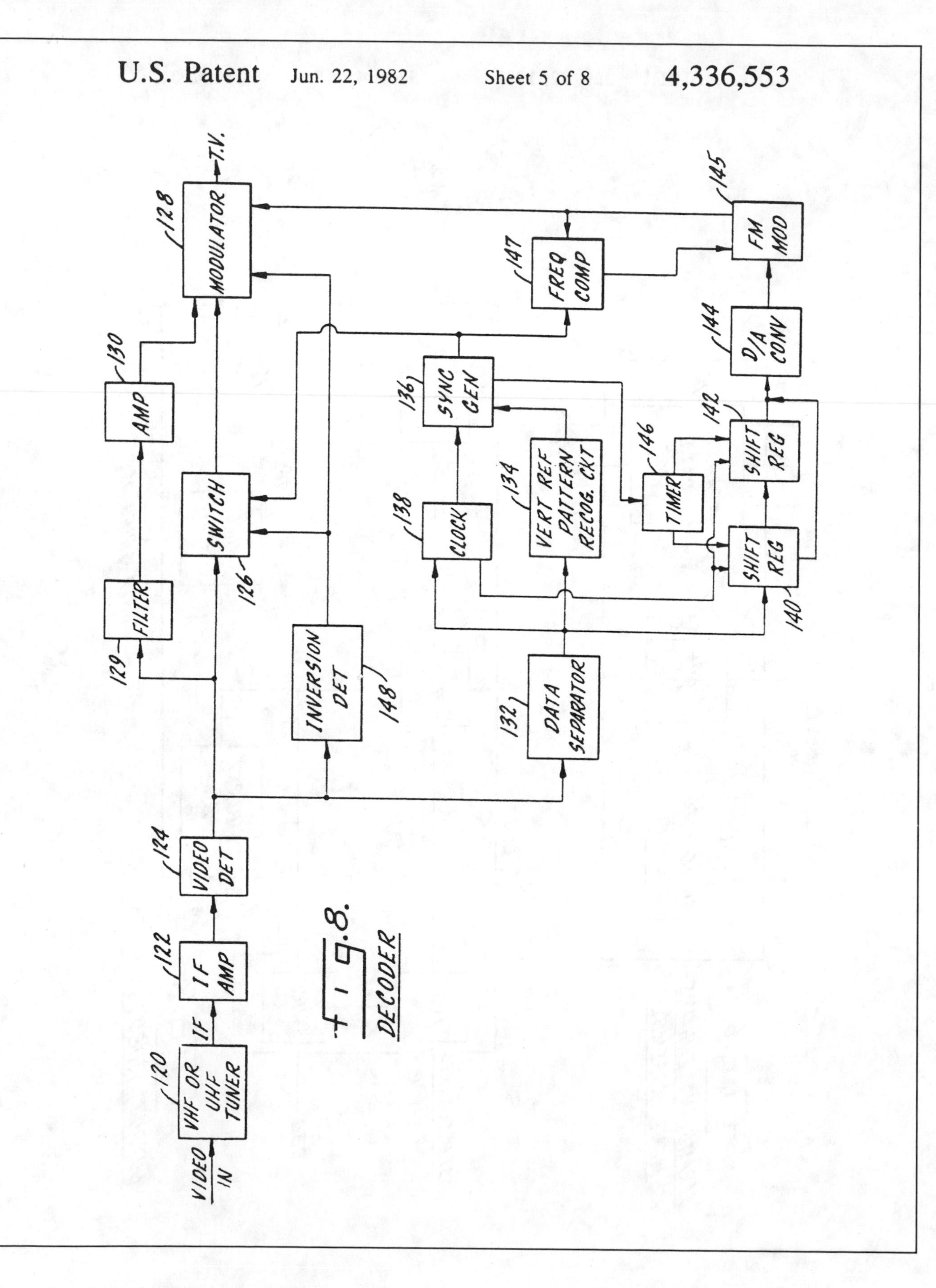
U.S. Patent Jun. 22, 1982 Sheet 5 of 8 4,336,553
VIDEO IN
VHF OR UHF TUNER
120
IF
I.F AMP
122
VIDEO DET
124
FILTER
129
AMP
130
SWITCH
126
MODULATOR
128
T.V.
INVERSION DET
148
DATA SEPARATOR
132
VERT REF PATTERN RECOG. CKT
134
CLOCK
138
SYNC GEN
136
FREQ COMP
147
TIMER
146
SHIFT REG
140
SHIFT REG
142
D/A CONV
144
FM MOD
145
Fig. 8.
DECODER

STANDARD NTSC VIDEO SIGNAL (COLOR BARS)

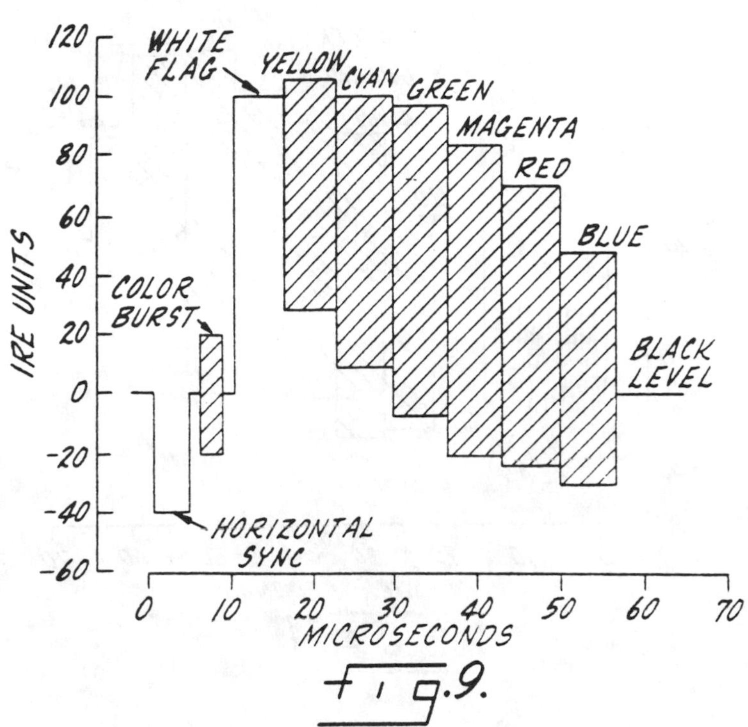

Fig. 9.

SCRAMBLED VIDEO SIGNAL (COLOR BARS, NON-INVERTED)

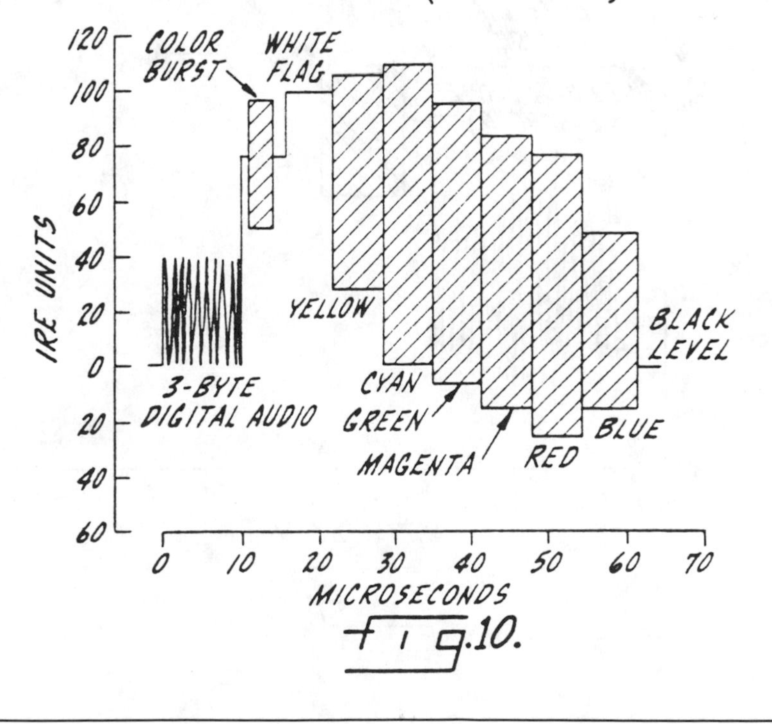

Fig. 10.

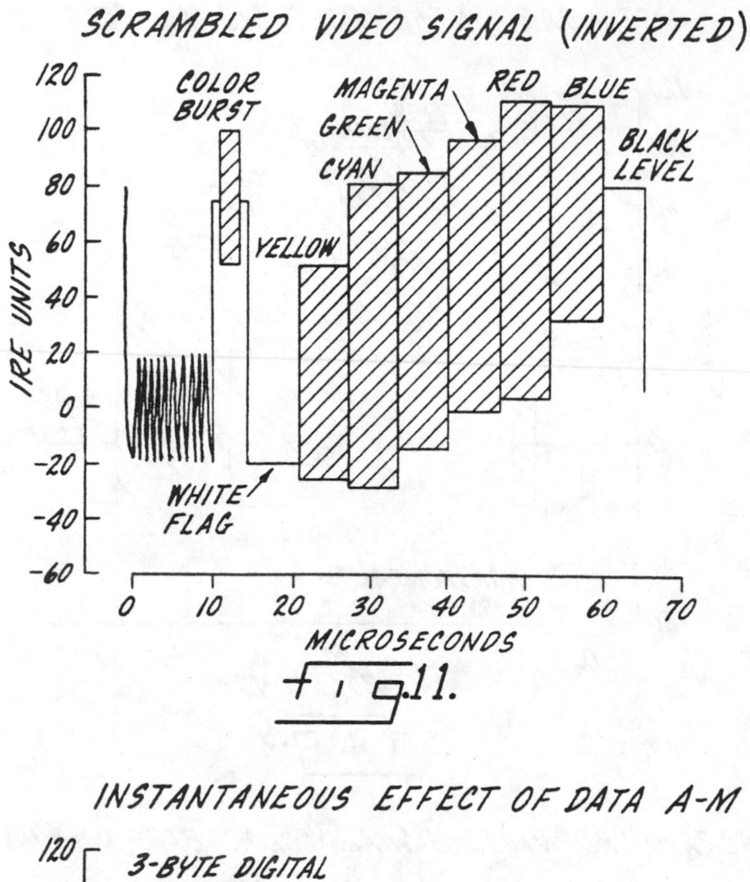
SCRAMBLED VIDEO SIGNAL (INVERTED)
COLOR BURST
MAGENTA
RED
BLUE
GREEN
CYAN
BLACK LEVEL
YELLOW
WHITE FLAG
IRE UNITS
MICROSECONDS

Fig. 11.

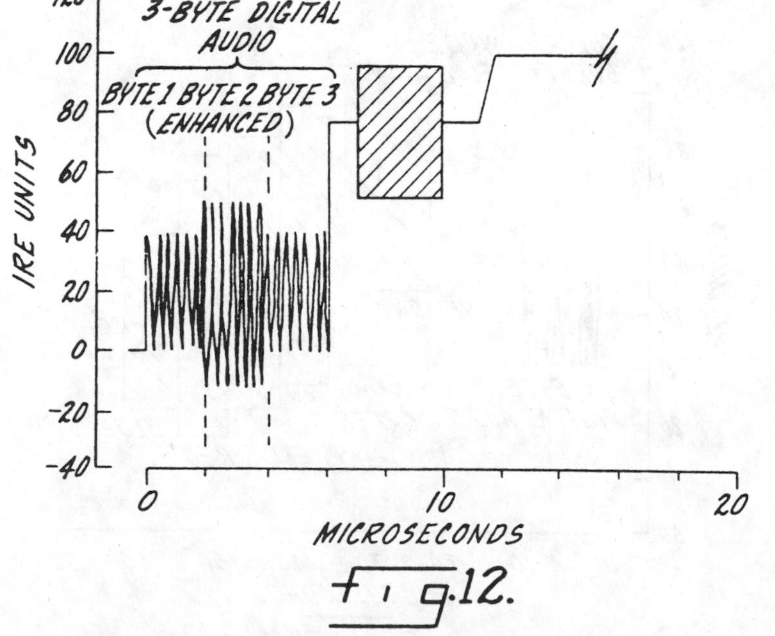
INSTANTANEOUS EFFECT OF DATA A-M
3-BYTE DIGITAL AUDIO
BYTE 1 BYTE 2 BYTE 3
(ENHANCED)
IRE UNITS
MICROSECONDS

Fig. 12.

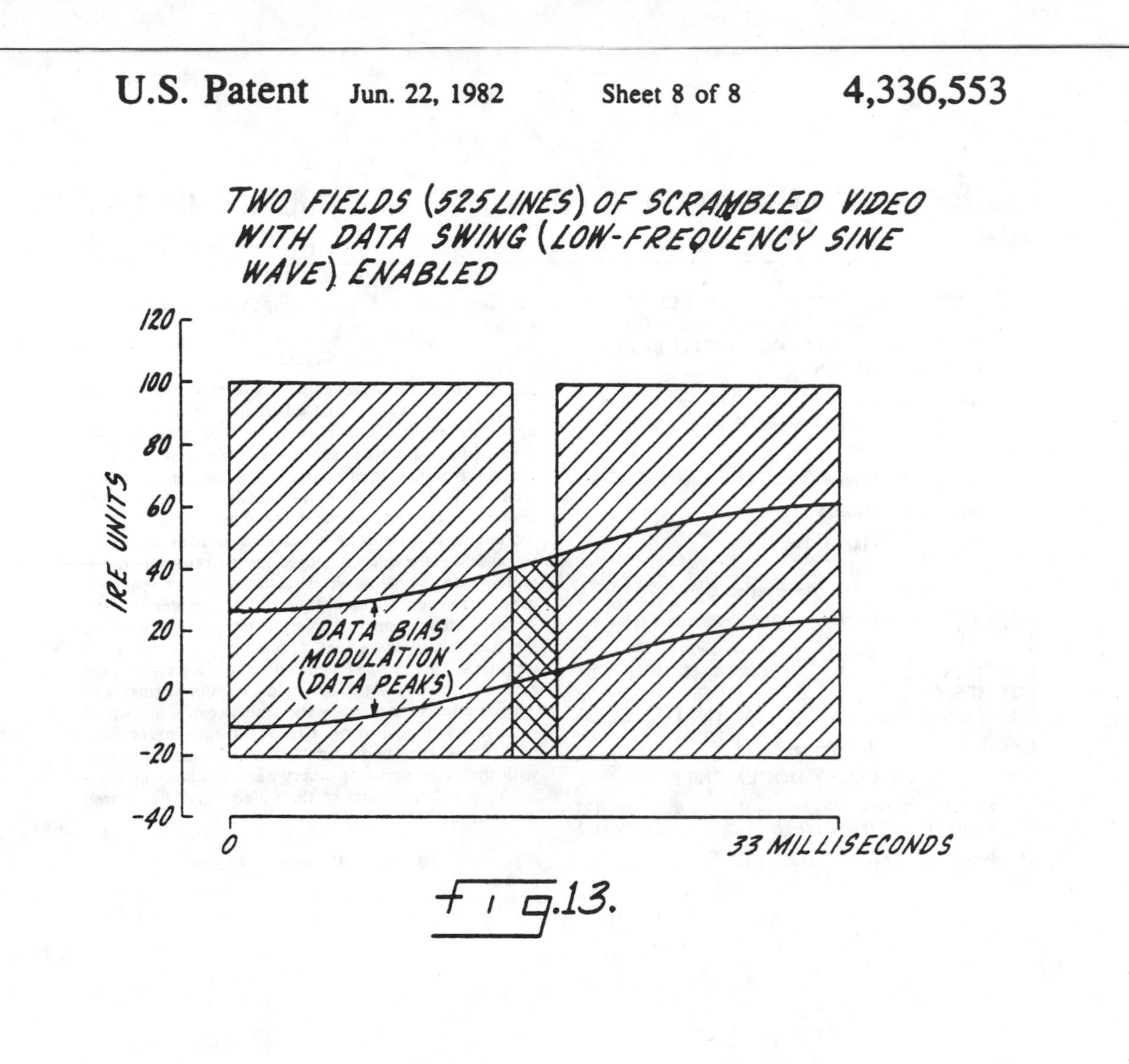
TWO FIELDS (525 LINES) OF SCRAMBLED VIDEO WITH DATA SWING (LOW-FREQUENCY SINE WAVE) ENABLED
120
100
80
60
40
20
0
-20
-40
IRE UNITS
DATA BIAS MODULATION (DATA PEAKS)
0
33 MILLISECONDS

Fig. 13.

United States Patent [19]

Block et al.

[11] **4,405,942**

[45] **Sep. 20, 1983**

[54] **METHOD AND SYSTEM FOR SECURE TRANSMISSION AND RECEPTION OF VIDEO INFORMATION, PARTICULARLY FOR TELEVISION**

[75] Inventors: **Robert S. Block,** Marina Del Ray, Calif.; **John R. Martin,** Milwaukee, Wis.

[73] Assignee: **Telease, Inc.,** Los Angeles, Calif.

[21] Appl. No.: **354,376**

[22] Filed: **Mar. 3, 1982**

Related U.S. Application Data

[63] Continuation of Ser. No. 124,656, Feb. 25, 1981, abandoned.

[51] **Int. Cl.**³ **H04N 7/16; H04K 1/04**

[52] **U.S. Cl.** **358/119;** 358/123

[58] **Field of Search** 358/114, 117, 119, 123

[56] **References Cited**

U.S. PATENT DOCUMENTS

2,972,008 2/1961 Ridenour et al. 358/123

4,070,693 1/1978 Shutterly 358/123

Primary Examiner—S. C. Buczinski

Attorney, Agent, or Firm—Burns, Doane, Swecker & Mathis

[57] **ABSTRACT**

A method and a system for secure transmission and reception of a video signal wherein parts of the video signal arranged in a first predetermined sequence are delayed in relation to each other to form an encoded video signal having its parts rearranged in sequence relative to their positions in the first predetermined sequence. Rearrangement of the video signal parts is accomplished by storing successive parts of the video signal and retrieving the stored parts so that the transmitted video signal has in parts arranged in a sequence other than their normal sequence. A decoder at a remote location is provided for receiving and restoring the transmitted sequence of video signal parts to its original sequence. In a television scrambling system, the parts of the video signal are fields, lines or segments of lines and, rearrangement of the sequence of video signal parts is accomplished on a line segment, line or field basis. Codes may be transmitted to the remote location with the video signal or otherwise to indicate the manner in which the parts of the video signal have been rearranged.

40 Claims, 19 Drawing Figures

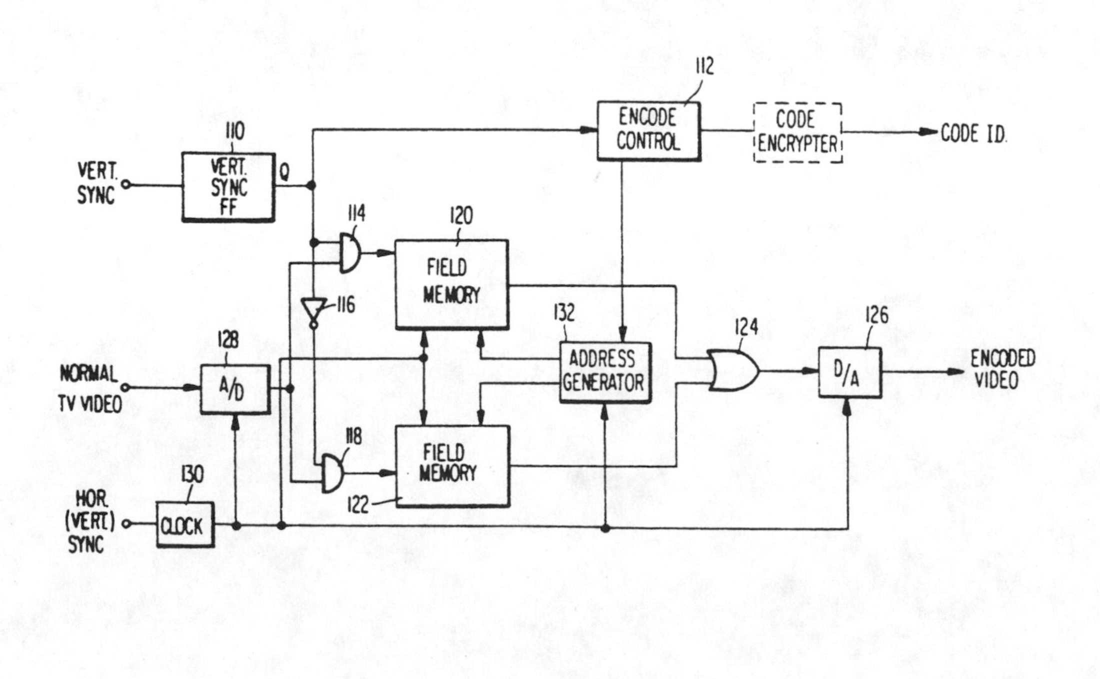

METHOD AND SYSTEM FOR SECURE TRANSMISSION AND RECEPTION OF VIDEO INFORMATION, PARTICULARLY FOR TELEVISION

This is a continuation of application Ser. No. 124,656, filed 2/25/81, now abandoned.

BACKGROUND OF THE INVENTION

The present invention relates to the encoding and decoding of video information and, more particularly, to a method and system for secure transmission of television signals for subscription television or similar video services in which only authorized viewers are permitted to view a video program.

With the increased interest and activity in the field of subscription or pay video transmissin of all types (e.g., broadcast and cable subscription television, long distance satellite transmission, television transmission of textual information, etc.), there has arisen a need for more secure transmission of high quality video information. To this end, numerous video encoding or scrambling techniques have been developed and some are now in use, particularly in broadcast systems where there is no control over who receives the signals and thus the signals must be encoded to prevent unauthorized use of the received signals.

One broadcast subscription television system now in use in Los Angeles transmits a video signal that has been modulated by a sine wave signal such that the blanking and synchronizing levels cannot be recognized by a normal television receiver. The display, without decoding, is thus unintelligible or at least very annoying to a viewer. However, it has been reported that by rather simple means available in most electronics stores, the system may be readily defeated by non-subscribers. As this becomes more widely known to the public, the number of unauthorized viewers grows and the incentive to pay for the services diminishes. This, in turn, detracts from the desirability on the part of video program producers to permit the use of their programming, particularly if they are paid as a function of authorized viewers or revenues collected.

Other approaches to television signal scrambling have proven more secure and may, in fact, make it practically impossible to unscramble the video signals without highly sophisticated and extremely expensive equipment. One such approach is to invert lines or fields of video information on some basis that can be reproduced at the subscriber location to permit viewing of a normal picture. Thus, for example, one known system inverts alternate parts of the video information and a decoder at the subscriber location can reinvert those inverted portions to reconstitute the original video. Another known system inverts fields of video information on a random basis and sends a code with the scrambled video to instruct the decoder as to how the received video has been inverted.

Security tends to be adequate in approaches to video scrambling in which the video information is randomly inverted and a secure code is transmitted with the video so that the decoder can properly reinvert, but difficulties arise with respect to picture quality. For example, inversion and reinversion of video signals may result in a reconstituted video signal that varies in d.c. level from line-to-line or field-to-field. Because of this variation, a

flicker or other annoying effect appears in the television display making it unpleasant for viewing.

Various measures have been employed to eliminate or at least reduce this problem with varying degrees of success. Clamping the video signals to the same d.c. level has been somewhat successful, but some annoying effects may still remain. Also, the additional circuitry required to eliminate or reduce the annoying effects of these types of scrambling add cost and complexity to the decoders. Less expensive and perhaps more effective approaches, such as less frequent inversion, have been suggested, but they seem to have a tendency to reduce the security of transmission and only serve to reduce the annoyance, not eliminate it.

OBJECTS AND SUMMARY OF THE INVENTION

It is accordingly an object of the present invention to provide a novel method and system for scrambling or encoding television signals wherein the difficulties of the prior art systems mentioned above are obviated.

It is a further object of the present invention to provide a novel television video encoding and decoding method and system in which there is an extremely high amount of security, wherein the encoding is sufficient to render an encoded television signal unintelligible or at least extremely annoying to watch, and wherein the decoded video signal is devoid of annoying abnormalities introduced during encoding, yet no special efforts must be made to eliminate encoding abnormalities.

It is another object of the present invention to provide a novel method and system for encoding and decoding television signals wherein the foregoing objects are accomplished with simplicity and at relatively low cost.

It is a more specific object of the present invention to provide a novel television signal scrambling system particularly suitable for broadcast pay television or other television systems meant only for authorized users where the video signal is encoded by delaying parts of the signal relative to other parts in a determinable manner to thereby rearrange the sequence of transmission of the parts in a fashion that can be reproduced at a remote location.

These and other objects and advantages are accomplished in accordance with the invention by a method, and a system that operates in accordance with the method, wherein parts of a video signal arranged in a first predetermined sequence are delayed in relation to each other to form an encoded video signal having its parts rearranged in sequence relative to their positions in the first predetermined sequence. Broadly, this rearrangement is accomplished by storing successive parts of the video signal and retrieving the stored parts in an order or sequence other than their normal sequence.

In accordance with one embodiment of the invention, rearrangement of the sequence of video signal parts is accomplished on a line basis or a field basis. For example, several lines of a video signal are stored in order as they are received. However, the lines are retrieved in a sequence different from that in which they are received and stored. Similarly, several fields of the video signal may be stored and retrieved in a sequence different from that in which they are received. Of course, as in the above-discussed embodiments, the rearrangement can be accomplished by the manner of retrieval or the manner of storage.

In another specific embodiment of the invention, the parts of the video signal chosen for delay are sequential samples or segments of the video signal either in an analog or digital form. The sequence of the samples is rearranged by delaying some longer than others, either individually or in groups, preferably by storing the samples sequentially as they are received and then retrieving the stored samples in a sequence other than that of receipt, or by storing the samples in a sequence other than the one in which they are received and then retrieving the stored samples in the stored sequence.

In one form of apparatus for encoding the video signal, a predetermined portion or part of the video signal (e.g. a horizontal line) is stored in a first storage location and then is transferred to a second storage location for storage therein. Before the portion of the video signal is stored in the second storage location, it is altered (e.g. by a gating arrangement), and the altered video signal is transmitted from the second storage location.

More specifically, in one form of the invention, a first group of samples representing one horizontal line of video (e.g., 256 segments which together constitute one line of video information) is serially stored in sequence in a serial in/parallel out storage device. This storage device may be a binary shift register if the samples are digital representations of the video, or an analog "bucket brigade" if the samples are analog representations of the video. When one full line of samples or video segments is stored, the segments are shifted in parallel into a second parallel in/serial out storage device so that the first storage device can accept the samples of the next line of video while the first stored segments are retrieved from the second storage device for transmission.

Rearrangement of the segments prior to transmission may be accomplished in various ways. The contents of the first storage device may be transferred to the second storage device without rearranging the sequence of samples. The stored samples in the second storage device then may be shifted out for transmission in a sequence rearranged in relation to their original sequence of storage in a determinable manner. For example, the samples may be shifted out so that the last received appears first in the transmitted signal or so that some intermediate sample appears first in the transmitted signal. An encoded video signal comprising a line of video having parts rearranged in sequence relative to corresponding parts of the normal video signal is thus transmitted.

Another way of rearranging segments prior to transmission is to transfer the contents of the first storage device into the second storage device in a sequence rearranged in relation to the original sequence of storage in a determinable manner. The contents of the second storage device are then shifted out in sequence resulting in an encoded video signal having parts of a line of video signal rearranged relative to corresponding parts of the normal video signal. This form of the invention permits any sample along a line to be rearranged to any position along that line or, with additional storage capacity, along any other line.

Since, in each embodiment of an encoder according to the invention, the video parts are altered such as by rearrangement of sequence in a determinable manner (either fixed or variable), a decoder which performs an opposite type of rearrangement will reconstitute the original, viewable form of the video signal.

The manner of rearranging the sequence of parts may be predetermined (i.e., fixed or preset) or may be controlled by a code generator that can vary the sequence from line-to-line or from field-to-field or on any other convenient time basis. The code produced by the code generator can be wholly or partially transmitted with the encoded video signal, e.g., in the horizontal or vertical retrace intervals, so that the exact nature of the rearrangement of parts can be determined by a decoder at a subscriber's location so that the original sequence of parts can be there reconstituted.

In one form of the invention, the code that determines the sequence of delay or rearrangement changes in a random manner and is transmitted in its entirety with the encoded video signal. A decoder at the subscriber location receives the code and reverses the delay procedure at the transmitting station in order to place the parts of the video signal in their normal sequence.

In another form of the invention, the code that determines the sequence of delay or rearrangement is predetermined and is only periodically changed, e.g., on a monthly basis. The new code is provided to the subscriber prior to the start of each month so that the subscriber's decoder can properly rearrange the parts of the encoded video signal and thus display a normal television picture. The code itself may also be ciphered so that a combination of the above techniques can be used. Thus, for example, the code that is needed for rearranging the encoded video signal may be itself changed in a predetermined manner before transmission. The manner in which it is changed may be communicated to a subscriber on a monthly basis so that the subscriber's decoder can only use the transmitted code properly if the encipher code is also known.

BRIEF DESCRIPTION OF THE DRAWINGS

FIGS. **1A** and **1B** are illustrations of standard United States color television signals;

FIGS. **1C** and **1D** are pictorial representations of the scanning of two successive fields of video information on a television picture tube in response to the signals of FIGS. **1A** and **1B**;

FIGS. **2–5**, **6A**, **6B**, and **6C** are pictorial representations of a standard television picture tube with lines scanned with signals encoded in accordance with various embodiments of the present inventions;

FIG. **7** is a functional block diagram of one general embodiment of the present invention;

FIG. **8** is a functional block diagram of a more specific embodiment of an encoder according to the present invention wherein fields of video information are rearranged;

FIGS. **9** and **10** are functional block diagrams of particular embodiments of encoders according to the present invention wherein lines of video information are rearranged;

FIG. **11** is a more detailed functional block diagram of an encode control circuit which may be employed in conjunction with the present invention;

FIG. **12** is a function block diagram illustrating one form of a gate circuit which may be used in connection with the present invention;

FIG. **13** is a functional block diagram illustrating an embodiment of an encoder according to the present invention wherein lines of the video signal are encoded upon initial storage; and,

FIG. 14 is a functional block diagram illustrating an encoding station and a decoding station in accordance with the present invention.

DETAILED DESCRIPTION

FIGS. 1A and 1B illustrate a standard or normal color television signal before it is superimposed on a carrier for transmission (or after demodulation).

This illustrated signal is the standard for the United States but it should be noted that the "standard or normal" television signal for other countries may differ somewhat. Accordingly, FIGS. 1A and 1B are illustrative of only one of a possible number of standard signals. As is illustrated in FIG. 1A, the standard color television signal is composed of consecutive fields each including synchronizing signals and video signals which convey color and "contrast" information. Each field commences with a vertical blanking interval comprised of equalizing pulses, vertical synchronization pulses and horizontal synchronization pulses. At the TV receiver, this vertical blanking interval (i.e. the blanking pulse) blanks the cathode ray tube beam as it is returned from the bottom to the top of the display (TV screen), and the horizontal scanning circuits are stabilized in synchronization with the horizontal sync pulses prior to the arrival of the first horizontal line of video information.

At the end of the vertical blanking interval, a horizontal sync pulse signifies the start of the scanning of a line of video information from one edge of the television display or picture tube to the other edge, normally from left to right when viewed from the front of the picture tube. The video signal information modulates the electron beam (or electron beams in a color system) as the beam is scanned. A horizontal blanking pulse blanks the retrace of the electron beam from right to left and a horizontal sync pulse then signifies the start of the next line of video information. Two consecutive lines of video information (actually 1½ consecutive lines of a second field of video information) are shown in FIG. 1B.

The result of scanning two consecutive fields of video information as seen by a viewer on the face of the picture tube is illustrated in FIGS. 1C and 1D. As can be seen in FIG. 1C, the face of the cathode ray picture tube is scanned with video information starting at the top left and ending at the bottom right of the display. Since the first 21 horizontal sync pulses are used to stabilize the horizontal sync circuits during vertical blanking, the first visible scan line of the first field is line 22(L22). Moreover, in the first field, for purposes of interlacing with the second field, the last line scanned is a half line. Thus, the first field can be viewed as containing lines 22 through 262½ to facilitate the description. The second field is similar except that for the purpose of interlacing the very first line scanned is a half line and the last line is a while line. Thus, the second field can be viewed as containing lines 22½ though 263. In FIGS. 1C and 1D, the beginning and end of each horizontal scan line is designated B and E, respectively, again to facilitate an understanding of the present invention.

In accordance with the invention, the normal TV video signal is encoded or scrambled such that if the encoded signal were displayed by a standard television receiver without decoding, the resultant picture presented to the viewer would have parts rearranged in sequence relative to their normal positions in the scanning sequence. The parts that are rearranged may be individual scan lines (e.g., the sequence of scan lines in a field) the fields, or other parts of the video signal such as individual segments of a scan line. As will be appreciated this rearrangement of sequence of video signal parts will result in a picture that is unintelligible or at least very annoying to watch.

To facilitate an understanding of the invention, the manner in which video information, encoded in accordance with the present invention would be presented on a TV display without proper decoding is illustrated in FIGS. 2–6.

FIG. 2 generally illustrates one display resulting from the encoding technique of the present invention in which the sequence of individual horizontal lines of video information is rearranged in a determinable manner relative to the normal sequence of lines in a field of video information. Normally, the first line appearing at the top of the video display would be line 22 (or 22 ½) and the last line would be 262 ½ (or 263). In one embodiment of the invention described hereinafter in greater detail, the individual horizontal lines are rearranged so that if they were displayed without decoding, they would reappear in a sequence other than the normal sequence. Thus, for example, line 201 (L201) would be scanned first at the top of the TV display. This would be followed in sequence by line 202, line 28, line 22, etc., as illustrated in FIG. 2. The resultant display would thus be unintelligible unless the lines were rearranged to their normal sequence prior to display.

FIG. 3 illustrates the display resulting from encoding in accordance with another form of the invention. In FIG. 3, the lines are scanned in their normal sequence, i.e., each horizontal line of video is transmitted and received in its normal order, but the video signal or information content of selected horizontal lines is reversed in a determinable manner so that the video information normally appearing at the end of the line is first and thus appears at the beginning of the line. The selected lines therefore appear on the display as being reversed and when intermixed with normally scanned lines produces a scrambled, unintelligible display unless selective reversal is effected prior to display.

FIG. 5 illustrates a further form of the invention wherein the sequence of scan lines of video information is normal but the order of the fields is rearranged. More specifically, an entire field of information is transmitted out of order relative to its normal sequence so that, for example, the third field is received and displayed prior to the first field and the second field. Rearrangment of fields in this manner may encompass a number of fields in any desired rearrangement as will subsequently be described in greater detail.

It will be appreciated from the foregoing and subsequent description that other rearrangments of parts of the video signal may be employed to encode the television signal. For example, combinations of line reversal and field reversal may be employed either together or on an alternating basis. Moreover, smaller parts of the video signal such as portions of horizontal lines may be rearranged to encode the TV video.

FIG. 6A, for example, illustrates a normal television display with the individual horizontal scan lines divided into a plurality of segments, i.e., 256 segments in the illustrated example. The individual horizontal lines are not normally segmented in this fashion can be readily segmented by suitable analog or digital sampling techniques as will be described hereinafter in greater detail. The sampling that is accomplished will provide the segments in a predetermined order and if this order is maintained throughout transmission and reception, the display will be that of an ordinary television signal. However, in accordance with one embodiment of the present invention, the segments are rearranged relative to their normal order prior to broadcast or other transmission as is illustrated in FIGS. **6B** and **6C**.

One manner or rearranging the segments shown in FIG. **6B** provides for the rearrangement of segments in a different fashion for each scan line. As is shown in FIG. **6C**, however, the rearrangement of the segments of the horizontal line can be maintained, once established for the first scan line, throughout any number of scan lines including all the lines that make up one or more fields.

All of the foregoing encoding techniques illustrated in FIGS. **2–6** involve the delay of some part of a video signal relative to other similar parts so that the various parts are rearranged in sequence upon transmission and produce an abnormal display if not properly decoded. Of course, the manner of rearranging the parts is determinable either in the sense that it is fixed (predeterminedly) and can be reversed by a similar fixed decoding technique or in the sense that it varies in response to an identifiable code that is provided to the decoder. According, in either case a decoder operating in a reverse manner can reconstitute the original signal from the rearranged parts.

Moreover, it should be understood that while the normal or standard TV video signal is often thought of as a composite signal comprising both sync signals and video information, the term video signal as used herein refers to the video information that determines picture content and thus is peculiar to a particular line or field, unless it is stated otherwise. Therefore, when rearrangement of a part of the video signal is discussed, it is contemplated that the sync pulses and other television signal components that are not peculiar to any particular line or field are not encompassed by this term. It will, however, be appreciated that these other signal components are also be rearranged for convenience or for other purposes in addition to or in conjunction with the video signal rearrangement contemplated herein without departing from the essential characteristics of this invention.

Various embodiments of the invention for delaying and rearranging parts of the video signal are described hereinafter in greater detail in connection with FIGS. **7–11**. In general, the manner of carrying out the present invention involves rearranging the sequence of parts of the normal video signal through the use of a suitable mechanism that will delay the video signal parts in a controllable manner. A storage medium such as a digital memory device, a sample and hold device or like digital or analog storage devices may be arranged to provide such controlled delay.

Referring to FIG. **7**, for example, the normal TV video signal may be supplied to a suitable storage device **100** and there stored in predetermined storage locations. These locations may be, for example, sequential locations in the sense that the addresses of stored sequential parts of the video signal are sequential. An address generator **102** controlled by clock signals from a timing control circuit **104** reads the stored parts of the TV video signal from the storage device **100** and applies the stored signals to the input terminals of a conventional multiplexer **106** in synchronism with normal video timing (synchronization) signals to reconstitute a television signal that appears normal in the sense that the overall timing of lines and fields is normal. However, the addresses produced by the address generator **102** are such that the stored signals are read in a sequence other than the sequence in which the signals were stored. For example, the address generator **102** may generate a sequence of addresses different from the sequence of memory addresses in which the normal TV video signal was stored initially thus reading the video information from memory in a sequence differing from its normal sequence. Thus, the video content of the television signal at the output terminal of the multiplexer, while conveying the same information, is in a sequence other than its normal sequence.

The address generator **102** may be any suitable conventional circuit that produces a series of digital words or other suitable addresses of a reprocucible sequence (e.g., a psuedo-random sequence) in which case the decoder need only reproduce this same, known sequence to rearrange the video signal parts in their proper sequence. For additional security or as an alternative for ease of decoding, a code from an encode control circuit **108** may control the sequence of addresses generated by the address generator **102** and a code ID signal identifying the mode of operation of the address generator may also be supplied to the multiplexer **106** for transmission with the signals read from the storage device **100**. In either case, the signals from the multiplexer **106** form an encoded or scrambled video signal that has parts rearranged from their normal sequence in a determinable manner relative to other parts.

For example, several entire fields of video may be stored in their normal sequence in the storage device **100** and then addressed for output to the multiplexer **106** in a sequence other than the sequence of storage. While the fields of video information are read from memory, the normal TV video arriving at the storage device is continuously stored so that none of the information is lost. The encode control may alter the code ID to change the sequence of addresses from the address generator **102** at the end of some predetermined number of fields or on some other basis synchronized with the video sync information so that the sequence of fields is changed periodically. Similarly, lines of video information or other parts of the video signal may be stored in the storage device **100** and clocked or read out in a sequence other than the sequence in which they are stored.

It will be appreciated that the normal TV video signal may be stored in the storage device **100** in a sequence of locations other than consecutive locations and then the storage locations of the storage device **100** may be read consecutively. This also results in an output having a sequence other than the input sequence. In this latter arrangement, the encoding of addresses is accomplished on the input to the storage device as opposed to the output thereof.

FIG. **8** illustrates one specific form of the invention wherein the normal TV video is converted to digital form for storage and rearrangement and is then reconverted to analog form for transmission. The embodiment of FIG. **8** is specifically arranged to provide the line rearrangement encoding of the FIGS. **2–4**, but it should be noted that by adding additional field memories, the field rearrangement embodiment of the invention illustrated in FIG. **5** can also be implemented.

Referring now to FIG. **8**, the vertical sync signal is applied to a toggle input terminal of a conventional flip-flop **110** (designated vertical sync flip-flop) and the binary "1" or Q output signal from the flip-flop **110** is applied to an encode control circuit **112**, to an input terminal of an AND-gate **114** and through an inverter **116** to one input terminal of AND-gate **118**. The output signal from the AND-gate **114** is supplied to an input terminal of a memory **120** of sufficient capacity to store an entire field of video information. The output signal from the AND-gate **118** is applied to a data input terminal of a similar field memory **122**. The output terminals of the field memories **120** and **122** are connected to respective input terminals of an OR-gate **124**, the output signal from which is applied to an input terminal of a digital-to-analog converter **126**.

The normal TV video signal is converted to digital form in an analog to digital converter **128** and the digitized TV video signal is supplied to the other input terminals of the AND-gates **114** and **118**. A TV video signal sync signal (either the horizontal or vertical sync) is applied to a synchronized clock **130** which generates the clock signals are supplied from the clock **130** to the analog to digital converter **128**, the field memories **120** and **122**, the address generator **132** and a digital to analog converter **126**.

In operation, the normal TV video signal is digitized in a conventional manner by the analog to digital converter **128**. For example, 16 distinct levels of video signal between black and white may be detected and encoded as four bit digital words by the converter **128**. Thus, for example, the white information may be represented by an all zero's code (0000). The black information may be represented by an all one's code (1111). Intermediate levels in 14 intermediate steps may be represented by other combinations of zero's and one's. It is contemplated that about 256 samples of the video information levels will be taken and represented by a digital word for each horizontal line of video, since 256 is a convenient number in digital work and seems to be a sufficient number of samples to provide adequate resolution. However, it will be appreciated that 32 or some other convenient number of levels of video may be selected for the digital conversion, and a greater number of sampler along each line may be taken (e.g., 512).

During a first field of incoming TV video information, the flip-flop is in one binary state (either set or reset) and the first field is loaded into one of the field memories **120** and **122**. Immediately before the next field of video information arrives, the flip-flop **110** changes state and the next field of video information is loaded into the other of the memories **120** and **122**. Simultaneously, the address generator **132** clocks information out of the first loaded field memory (i.e. reads the contents of the first field memory to the OR-gate **124**) in a sequence determined by the encode control **118**. As this information is clocked out of the field memory, it is supplied through the OR-gate **124** to the digital to analog converter **126** for reconversion to analog form and transmission as the encoded video signal.

When all of the information for a field has been clocked out of the first loaded field memory, the second field memory will then contain a second field of video information. The sync flip-flop **110** changes state causing the next incoming field to be loaded into the memory that was just read while the previously loaded memory is read.

It will be appreciated that the address generator **132** may be controlled to clock the stored information out of the field memories in any number of ways. For example, the address generator may clock the horizontal lines out of memory on a line-by-line basis in a pseudo-random fashion wherein the starting point of the pseudo-random address sequence is controlled by the encode control **118**. The resultant encoded video information would thus be arranged as illustrated in FIG. **2**. Alternately, the address generator may be controlled to clock an entire field of information out of the field memories line-by-line in reverse of the order of storage (i.e. first line in, last line out) resulting in a display such as that illustrated in FIG. **4**. Further encoding may be achieved by reversing the beginning and end of each line (e.g. by clocking the last stored segment out first and proceeding in this reverse sequence) so that without proper decoding the field is displayed upsidedown and backwards. By controlling this reversal of fields so that some are reversed and some are not, normal reception is impossible without knowledge of the existence and manner of the reversals, which information is provided by the code ID or by some predetermined, fixed order.

It will also be appreciated that by clocking the lines of video information out of the field memories in the normal line sequence but with some determinable reversal of line beginning and end (e.g. by reversing the segment order or otherwise mixing the segment order in a determinable fashion), another form of encoding such as that illustrated in FIG. **3**, may be achieved with the FIG. **8** circuit. Moreover, since the video signal will be stored as a number of samples of each horizontal line, these samples or segments may be rearranged on an individual basis to produce encoding like that shown in FIG. **6**. Of course, as will be seen hereinafter, entire fields of video information need not be stored to achieve line reversal or line segment encoding.

If the manner of rearranging the video signal sequence is fixed and predetermined, the decoder at a subscriber station need not receive any code information to properly restore the video to its normal sequence. On the other hand, if the video signal sequence is rearranged in response to a code that varies at the transmitter location, this code must be conveyed to the subscriber location. In this regard, the code may be transmitted with the television signal or by way of a separate communication link (e.g., by telephone) or by mail or by a combination of such means. Suitable techniques for supplying code information for unscrambling television signals are described, by way of example, in U.S. Pat. Nos. 4,025,948 and 4,068,264.

FIGS. **9** and **10** illustrate specific implementations of the invention in which the parts of the video signal that are rearranged are segments of horizontal scan lines. In the circuit of FIG. **9**, the horizontal line segments are rearranged so that the lines are selectively reversed as shown in FIG. **3**. In the FIG. **10** circuit, the segments of a horizontal line of video information are rearranged so that the individual segments within a horizontal line are in other than their normal sequence as shown in FIGS. **6B** or **6C**.

Referring to FIG. **9**, the normal TV video signal is supplied to the data input terminal of a suitable conventional serial is supplied storage device **140**. Segments of a horizontal line of video information (e.g., 256 segments or level samples) are stored by serially clocking the segments into the storage device **140** under the control of a clock signal from a conventional synchronized clock **142**.

When the segments of one full line of video information are stored in the device **140**, the horizontal sync pulse SYNC enables the parallel enable or read terminal PE of a suitable conventional parallel in/serial out storage device **144**. The SYNC signal is also delayed by a conventional delay circuit **146** and resets the storage device **140** so that it is prepared to receive the next horizontal line of video.

As the next horizontal line of video is received and clocked into the storage device **140**, the video information stored in the storage device **144** is clocked out. An encode control circuit **148** provides a forward/reverse signal to the forward/reverse shift input terminal of the storage device **144** and depending upon the value of this signal, the stored video signal is shifted out of the storage device either in a forward direction from the final stage **Q256** or in a reverse direction from the first state **Q1**. An OR gate circuit **150** is connected to receive the signals from the respective first and final stages **Q1** and **Q256** of the storage device **144** to provide one or the other as the encoded video signal.

If the encode control circuit **148** applies the forward control signal to the storage device terminal FWD/RUS, the first segment stored (**S1**) will be clocked out first, followed in sequence by the second through the 256th segments. This, the line of video information segments will be identical in sequence to their normal sequence (i.e., their original sequence of storage). If, however, the reverse control signal is applied to the FWD/RVS terminal, the last stored segment of that line of video will be clocked out first, followed by the next to last through the first in sequence. Thus, the rerverse control signal will result in transmission of the line of video information in its reverse order (e.g., such as line **L23** in FIG. **3**).

By changing the forward/reverse signal from one state to the other, the normal and reversed lines can be interspersed in a manner which is determinable by reference to the CODE I.D. signal. This CODE I.D. signal can determine the reversals on a line-by-line basis in some desired fashion (e.g., random or predetermined) or can determine the reversals on a less frequent basis such as on a field-to-field basis by specifying a particular pattern of reversals for an entire field. In either event, transmission of the CODE I.D. signal with the television signal or over some other convenient communication channel will allow decoding of the television video at a subscriber station provided the decoder at that station is capable of responding properly to the received codes.

In FIG. **10**, wherein like numerical designations are used to denote like components, the segments that comprise a line of the video signal are stored by the storage device **140** and transferred to the storage device **144** prior to the arrival of the next line. However, the segments are transferred through a gating circuit **152** controlled by a control word CTW from an encode control circuit **154**. The control word controls the gate circuit **152** so that the segments **51** through **S256** are stored in the storage device **144** in a sequence which is determined by the control word.

Thus, the segments may be stored in the storage device **144** in any sequence, including their original deuqence. The stored segments will be clocked out of the storage device **144** by shifting them out in one predetermined direction (or if desired for further encoding, in a selected one of two directions) and the encoded video output signal will thus be a sequence of segments arranged in a determinable pattern other than their original, normal sequence. Again, the CODE I.D. will permit the original sequence to be reconstituted at the subscriber location.

It should be understood that the above techniques for rearranging segmental parts of the video signal may be implemented on an analog or digital basis. For analog operation the storage devices **140** and **144** and the gate circuits **150** and **152** may be any suitable analog devices such as "bucket brigades", gated sample and hold circuits and analog gates of conventional design. Suitable conventional digital AND and OR gates as well as conventional digital shift registers may be used for a digital implementation. Such shift registers and gate components are well known and are readily adapted to the uses described.

In an analog implementation, the video signal may be supplied in its normal analog form and sampled at a desired clock rate to produce the desired number of samples or segments per horizontal line. Reconversion to analog form before transmission will not be necessary, of course.

In a digital implementation, e.g., using binary shift registers, the video signal must first be converted to digital form before storing and rearranging the signal parts. Because of present bandwidth constraints on transmission, however, reconversion to analog form for transmission is desirable. Also, reconversion to analog form for transmission prevents the loss of a digital bit of information that might grossly affect the value of a particular segment upon decoding and display.

It will be understood by those skilled in the art that for digital video manipulation the individual segments of a line must be represented by several digital bits and not just one. For example, each segment may be represented by a four-bit binary word. The registers must thus handle several bits (a multiple bit word) for each segment. Since there is a reconversion to analog form before transmission, these digital words may be handled by conventional parallel data storage techniques. This facilitates keeping track of all the bits for a particular segment and facilitates the conversions from analog to digital and back to analog.

The gate circuit **152** may be any suitable conventional gating or switching arrangement that responds to the control word CTW to route the segments **S1–S256** in variable patterns to the terminals **D1–D256** of the storage device **144**. One manner in which this may be accomplished is to provide an eight bit control word which will provide up to 256 variations so that each individual segment can be stored in any one of the 256 locations **D1–D256** (with provision, of course, that there are no duplications for a particular code). Similarly, a four bit word may be used if, for example, only 16 different storage patterns are desired.

Table I below illustrates one possible coding format where an eight bit code word is used to determine the storage location of the individual segments. The gating circuit in the Table I illustration applies the segments **S1–S256** to the respective **D256–D1** locations i.e. without rearrangement) for an all zero's code (00000000) so that the first segment **S1** is read out first followed by the second segment **S2** and so on, in order, when the storage device **144** is clocked serially. For all one's code (11111111) or the binary equivalent of decimal 255, the gating circuit applies the last segment **5156** to the last storage location **D256** so that it is read out first when storage device **144** is serially read. For codes between all zero's and all one's (the decimal number 1-254), the segments will be shifted in location in direct proportion to the code value.

TABLE I

CODE (DECIMAL)	STORAGE LOCATION D1	D2	D3	D4	D5	. . . D256
0	S256	S255	S254	S253	S252	. . . S1
1	S1	S256	S255	S254	S253	. . . S2
2	S2	S1	S256	S255	S254	. . . S3
3	S3	S2	S1	S256	S255	. . . S4
.	.	.	.	.	.	.
.	.	.	.	.	.	.
.	.	.	.	.	.	.
255	S255	S254	S253	S252	S251	. . . S256

Table II below illustrates another possible coding format in which a four bit code word is used to determine the storage locations of the individual segments within one of a possible sixteen patterns (i.e. a four bit word can assume 16 different values). The all zero's code word controls the gate circuit **152** to arrange the segments in the storage device **144** in their original, unaltered sequence as shown in the first line of Table II. The all one's code word (1111), the decimal number 15, controls the gate circuit **152** to arrange the segments in reverse order in storage device **144**. Various other different patterns of storage rearrangement are provided by the code words corresponding to the decimal number 1-14.

TABLE II

CODE (DECIMAL)	STORAGE LOCATION D1	D2	D3	D4	D5	. . . D256
0	S256	S255	S254	S253	S252	. . . S1
1	S10	S105	S203	S2	S8	. . . S100
2	S26	S123	S126	S98	S215	. . . S226
.	.	.	.	.	.	.
.	.	.	.	.	.	.
.	.	.	.	.	.	.
16	S1	S2	S3	S4	S5	. . . S256

The encode control circuit **154** and the gate circuit **152** may be implemented to providing segment rearrangement using known digital techniques. For example, as is shown in FIG. **11**, the encode control circuit **154** may be any suitable random or pseudo-random code generator that produces a multiple bit code word during an appropriate time period (e.g. at the end of each horizontal line) to gate the stored video information from one storage device to the other in accordance with a pattern determined by the code.

In this regard, the encode control may comprise a monostable multivibrator **200** which triggered by an appropriate one of the sync signals, depending upon how often information is transferred. When used in conjunction with the FIG. **10** embodiment, the horizontal sync pulse may be used to trigger the multivibrator.

The binary zero output signal Q from the multivibrator controls the gating of pulses from a free running clock **202** through an AND gate **204** to the clock input terminal of a binary counter **206**. The multivibrator output signal is also inverted by an inverter **208** and applied to a gate **210**. An output signal from each stage of the counter **206** together form the control word CTW and are gated through the gate **210** to form the CODE I.D. signal.

In operation, the multivibrator output signal is normally high or binary ONE and allows the oscillator signal to pass through the AND gate **204**. Thus, before a horizontal sync pulse arrives, the counter is freely counting and the control signal CTW is changing.

When the horizontal sync pulse arrives, the multivibrator signal assumes a low signal level for a period determined by the multivibrator time constant. The binary counter **206** no longer receives clock signals so it stops counting and its output stays the same for the duration of the low level or zero multivibrator output signal.

The CTW signal is applied to the gate circuit **152** of FIG. **10** and, by choosing appropriate delays, the video information is tranferred through the gate circuit while the CTW signal is fixed in value. Also, the CODE I.S. signal is either sampled directly or through the gate **210** for transmission withthe encoded video signal.

At the end of the time period of the multivibrator, the Q output signal assumes a high signal level and the counter **206** once again counts pulses from the oscillator. Since no information is being transferred through the gate circuit **152** during this interval, the changes in the CTW signal do not have any adverse affect.

Any suitable gating arrangement that will transfer signals differently between its input and output terminals in response to a control word may be used for the gate circuit **152**. One possible arrangement parially shown in FIG. **12** permits the input segments **S1-S256** to be output on any of the 256 data lines for storage in any of the 256 storage locations **D1-D256**. It can be seen, for example, that with a conventional binary to decimal decoder **212**, up to 256 distinct control signals are provided from an eight bit control word CTW. Using an AND and OR gate arrangement (an expansion of that shown), any input segment can be applied to any data input terminal **D1-D256** in accordance with the pattern determined by the gates.

Thus, for example, the segment **S256** is applied to terminal **D1** when the decoder output signal is decimal "0". Simultaneously, the segments **S255-S1** are applied to data terminals **D2-D256**, respectively for a decimal "0" output signal from the decoder **212**. In the illustrated arrangement, a decimal 1 shifts the segments one data terminal to the right, but any other pattern of response to the control word CTW may be readily implemented using such a technique.

Since there is no gating circuit in the embodiment of FIG. **9** and only two states of encoding are necessary, it will be appreciated that a very simple circuit such as a controlled bistable flip flop may be used to reverse line retrieval. Moreover, it will be appreciated that with a fixed segment rearrangement pattern, there is no need for an encode control circuit or a gate circuit in the FIG. **10** embodiment. Thus, the gating and encoding may be as simple or complex as desired with the limits of the number of segments available for rearrangement.

FIG. **13** illustrates an embodiment of the invention in which the selected parts (e.g., segments of lines) of the normal TV video signal are stored in the rearranged order desired for transmission. The normal TV video is applied to a plurality of AND gates **220**, one for each input terminal D of a suitable conventional parallel in/parallel out storage device or register **222**. The output terminals Q of the register **222** are connected to corresponding input terminals D of a conventional parallel in/serial out storage device or register **224** whose serial output is the encoded video signal.

The parts of the normal TV video signal will be stored in the register **222** at storage locations determined by the sequence of gating of the AND gates **220**.

One circuit for varying the sequence of storage locations is illustrated in FIG. 13. In this regard, the sync signal (e.g., horizontal sync) is applied to a conventional synchronized clock 226, to an encode control 228, to the preset enable input terminal PRE of a conventional presettable binary counter 230 and through a delay circuit 232 to the parallel read or enable input terminal PE of the register 224.

The output signal from the clock 226 is supplied to the clock input terminal CL of the counter 230 and through a delay circuit 234 to one input terminal of each of the AND gates 220 and to the clock input terminal of the register 224. The output signals from the counter 230 are applied in parallel to a conventional binary to decimal converter 236 and the 0–225 output signals from the converter 236 are applied to the respective input terminals of the AND gates 220.

In operation, the encode control 228 generates binary code word (e.g., as previously discribed in connection with FIG. 11) and this binary word is applied to the counter 230 as a preset signal. In the illustrated embodiment the counter 230 is an eight bit counter so the binary word from the encode control is an eight bit word.

The counter 230 is preset by the code word from the encode control 228 at the start of a horizontal line. The counter then counts 256 bits of the clock signal (if 256 video signal parts are to be stored), starting its count at the preset number. During one horizontal line, the output signal from the binary to decimal converter thus steps from the preset number until it arrives at the number before or after the preset number, depending on whether the counter 230 counts up or down.

. As the counter 236 is stepped from the preset number through all its other outputs, the AND gates 220 are enabled in a corresponding sequence, one at a time. As an AND gate is enabled, the video signal preset at that time is stored in the location in the register 222 connected to that AND gate. At the end of a horizontal line, the stored line of video is shifted into the register 224 and is clocked out of this register for transmission as the encoded video signal while the next line of video is stored in the register 222. The code I.D. signal can also be transmitted with this encoded video signal as often as is necessary (depending on how often it is changed).

FIG. 14 illustrates an overall pay television system which may be constituted in accordance with the present invention. Of course, it will be appreciated that other secure video systems (e.g., "teletext" CATV or satellite) may also be operated in accordance with the techniques, and with the apparatus, disclosed and claimed herein without departing from the essential characteristics of the invention.

Referring to FIG. 14, the synchronizing and blanking signals as well as the normal TV video signal are provided to a video encoder 250 such as one of the previously disclosed encoders. The sync and blanking signals are also provided to a suitable conventional multiplexer 252 together with the encoded video signal and code I.D. from the video encoder 250.

The multiplexed signal from the multiplex 252 is supplied to a suitable conventional transmitter 254 which broadcasts the multiplexed signal at a suitable carrier frequency. This transmitted signal is received by the subscriber decoder and is demodulated by a conventional receiver/demodulator 256. The output from the receiver 256 is thus the encoded video signal with the code I.D. and the sync and blanking signals at appropriate locations in the demodulated signal.

A synch and code I.D. detector 258 detects the synch and blanking signals as well as the code I.D. signal in the demodulated video stream. Conventional television sync and blanking circuits may be utilized for detection of the sync and blanking signals, a since the code I.D. signals are in the received signal at known locations relative to the synch and blanking signals, the code I.D. may be detected in any conventional manner such as in conventional television scrambling and unscrambling systems.

The demodulated video stream from the receiver 256 is also supplied to a video decoder 260 together with the detected sync anc code I.D. signals. The video decoder reconstitutes the normal video TV signal in response to the sync and code I.D. signals reversing the encoding process that was performed at the television transmitter station. In this regard, if the encoder of the FIG. 10 embodiment is used to encode the video signal at the transmitter station, then an identical decoder may be used at the subscriber station.

For example, the detected sync signal would be applied to the input terminal labeled "SYNC" in FIG. 10 and the demodulated (encoded) video from the receiver 256 would be applied to the terminal labeled "NORMAL TV VIDEO" in FIG. 10. The detected code I.D. would apply to the gate circuit 152 as the control word CTW or to a suitable decoder, if necessary, to form a suitable control word that will reverse the previous encoding process.

It will be appreciated that, in the FIG. 14 embodiment, the sync and blanking signals are not encoded but, rather, are multiplexed with the encoded video signal from transmission in their normal form. Depending upon the manner in which the video signal parts are rearranged, this form of sync and blanking singal transmission may or may not be desirable.

Moreover, it should be noted that the Code I.D. may be supplied to the subscriber decoder in some manner other than in real time with the transmitted signal. For example, the code I.D. may be fixed for an entire program or for a series of programs or for a predetermined time period. This code I.D. may be set by the subscriber (e.g., upon receipt each month of a decoding schedule from the subscription TV operator) or may be transmitted to the decoder as a block of information covering a predetermined time interval, with the transmission being effected either over the air or by telephone.

The principles, preferred embodiments and modes of operation of the present invention have been described in the foregoing specification. The invention which is intended to be protected herein, however, is not to be construed as limited to the particular forms disclosed, since these are to be regarded as illustrative rather than restrictive. Variations and changes may be made by those skilled in the art without departing from the spirit of the invention.

We claim:

1. A method for transmitting a television video signal in a nonstandard form comprising the steps of:

(a) sequentially storing signals representing discrete samples of the video signal for one horizontal line in a first storage device;
(b) transferring the stored signals for the one horizontal line from the first storage device into a second storage device while simultaneously modifying the stored signals to form an encoded video signal;
(c) retrieving the stored signals from the second storage device; and

(d) transmitting the encoded video signal.

2. The method of claim 1 including the steps of:

(a) generating a code signal identifying the manner in which the parts of said encoded video signal are modified in relation to said first predetermined sequence; and

(b) transmitting at least a portion of the code signal with the encoded video signal.

3. The method of claim 1 further including the step of storing signals representing samples of the video signal for a second horizontal line in the first storage device during the step of retrieving the stored signals from the second storage device.

4. The method of claim 1 wherein the stored signals are retrieved from the second storage device in a sequence other than the sequence in which they were stored in the first storage device.

5. The method of claim 1 wherein the step of modifying includes rearranging the sequence of the stored signals as they are transferred from the first storage device to the second storage device.

6. A method for decoding a television video signals that is transmitted in a nonstandard form wherein parts of the video signal are delayed in relation to other parts of the video signal and thereby form an encoded video signal having parts rearranged in sequence relative to a normal sequence in a determinable manner, the method comprising the steps of:

(a) receiving the encoded video signal;

(b) sequentially storing signals representing received samples of the encoded video signal for one part of the signal in a first storage device;

(c) transferring the stored signals for the one part of the signal in parallel from the first storage device into a second storage device;

(d) storing signals representing the received samples of the encoded video signal for a second part of the signal in the first storage device while retrieving the stored signals from the second storage device in a selectable sequence opposite that of sequence of rearrangement of said encoded sequence to form a decoded video signal; and

(e) applying said decoded video signal to a television receiver.

7. The method of claim 6 wherein at least a portion of a code signal identifying the manner in which the parts of said encoded video signal are rearranged in relation to said normal sequence is transmitted with the encoded video signal, and wherein the step of decoding includes detecting said at least a portion of the code signal transmitted with the encoded video signal and retrieving said stored signals in a sequence determined according to the detected code signal.

8. A method for transmitting a video signal in an encoded form comprising the steps of:

storing a predetermined portion of the video in a first storage location;

transferring the predetermined portion of the video signal from the first storage location to a second storage location for storage therein and altering the transferred portion in a determinable manner prior to storage in the second storage location; and,

transmitting the predetermined portion of the video signal from the second storage location in its altered form.

9. A system for transmitting a television video signal in a nonstandard form comprising:

means for generating a video signal having a plurality of parts arranged in a first predetermined sequence;

first and second storage means each for sequentially storing signals representing the parts of the video signal;

means for transferring the stored signals from the first storage means into the second storage means and for retrieving the stored signals in parallel from the second storage means in a selectable sequence other than said first predetermined sequence to form an encoded video signal; and

means for transmitting the encoded video signal.

10. The system of claim 9 including:

means for generating a code signal indentifying the manner in which the parts of said encoded video signal are rearranged in relation to said first predetermined sequence; and

means for transmitting at least a portion of the code signal with the encoded video signal.

11. The system of claim 9 including:

means for generating a code signal identifying the parts of said encoded video signal; and

means for transmitting at least portion of said code signal with the encoded video signal.

12. An encoder for transmitting an encoded video signal comprising:

storage means for storing a predetermined portion of the video signal in a first storage location, said storage means having a second storage location;

means for transferring said predetermined portion of the video signal from said first storage location to said second storage location and altering said transferred portion of the video signal in a determinable manner prior to storage in said second storage location; and,

means for transmitting said predetermined portion of the video signal from said second storage location in its altered form.

13. A method for transmitting a television video signal in a nonstandard form comprising the steps of:

(a) storing signals representing discrete samples of the video signals for one horizontal line in a first storage device;

(b) transferring the stored signals for the one horizontal line from the first storage device into a second storage device with the sequence of stored samples in the second storage device differing in a selectable manner from the sequence of samples in the first storage device;

(c) retrieving the stored samples from the second storage device in sequence to form an encoded video signal; and

(d) transmitting the encoded video signal.

14. A method for decoding a television video signal that is transmitted in a nonstandard form wherein parts of the video signal are delayed in relation to other parts of the video signal and thereby form an encoded video signal having parts rearranged in sequence relative to a normal sequence in a determinable manner, the method comprising the steps of:

(a) receiving the encoded video signal;

(b) storing signals representing received parts of the encoded video signal in a first storage device;

(c) transferring the stored signals from the first storage device into a second storage device with the stored sequence of parts in the second storage device differing from the sequence of parts in the first storage device in a manner opposite said determinable manner;

(d) retrieving the stored parts from the second storage device in sequence to form a decoded video signal; and

(e) applying said decoded video signal to a television receiver.

15. The method of claim **6** or **14** wherein said samples of said encoded video signal are discrete sequential samples of said video signal along horizontal lines of a television raster.

16. A system for transmitting a television video signal in a nonstandard form comprising:

means for generating a video signal having a plurality of parts arranged in a first predetermined sequence;

first and second storage means each for storing signals representing parts of the video signal;

means for transferring the stored signals from the first storage means into a second storage means with the stored sequence of parts in the second storage means differing in a selectable manner from the sequence of parts in the first storage means;

means for retrieving the stored parts from the second storage device in sequence to form an encoded video signal; and

means for transmitting the encoded video signal.

17. The apparatus of claim **15** or **16** wherein said parts of said video signal comprise discrete samples of said video signal along horizontal lines of a television raster.

18. A method for transmitting a television signal in a nonstandard form to prevent unauthorized reception of the signal comprising the steps of:

sequentially storing in a first storage means a plurality of successive parts of the television signal;

transferring the stored parts of the television signal from the first storage means into a second storage means wile simultaneously modifying the stored parts in a reversible manner so that a modified version of the first stored parts is stored in the second storage means;

retrieving the stored parts from the second storage means; and

transmitting the retrieved stored parts as a television signal is a nonstandard form.

19. The method of claim **18** wherein the parts of the television signal are successive samples of the video signal comprising one horizontal line of video.

20. The method of claim **18** including the further step of sequentially storing a successive plurality of similar parts of the television signal in the first storage means while retrieving the stored, modified verison of the parts from the second storage means.

21. The method of claim **18** wherein the television signal is modified by modifying the sequence of succession of the stored parts.

22. The method of claim **18** including the steps of:

(a) generating a code signal identifying the parts of said encoded video signal; and

(b) transmitting at least a portion of said code signal with the encoded video signal.

23. A method for decoding a television signal that is transmitted in a nonstandard form wherein parts of the video signal are modified in a determinable manner to form an encoded video signal, comprising the steps of:

receiving the encoded video signal;

sequentially storing a plurality of successive parts of the received video signal in a first storage device;

transferring the stored parts from the first storage means into a second storage means while simultaneously modifying the stored parts in a manner opposite said determinable manner so that a decoded version of the received video signal is stored in the second storage means;

retrieving the stored parts from the second storage means as a decoded video signal; and

applying the decoded video signal to a television receiver.

24. The method of claim **23** wherein the parts of the television signal are successive samples of the video signal comprising one horizontal line of video.

25. The method of claim **23** including the further step of sequentially storing a successive plurality of similar parts of the received video signal in the first storage means while retrieving the decoded video signal from the second storage means.

26. The method of claim **23** wherein the stored parts are modified by rearranging their sequence during the transfer from the first storage device to the second storage device.

27. Apparatus for transmitting a television signal in a nonstandard form to prevent unauthorized reception of the signal, comprising:

means for generating a video signal having a plurality of parts;

first and second storage means each for storing the parts of the video signal;

means for transferring stored parts from the first storage means to the second storage means while simultaneously modifying the parts in a reversible manner so that a modified version of the parts is stored in the second storage means;

means for retrieving the modified version of the stored parts from the second storage device as an encoded video signal; and

means for transmitting the encoded video signal.

28. The apparatus of claim **27** wherein the parts of the television signal are successive samples of the video signal comprising one horizontal line of video.

29. Apparatus for decoding a television signal that is transmitted in a nonstandard form wherein parts of the video signal are modified in a determinable manner to form an encoded video signal, comprising:

means for receiving the encoded video signal;

first and second storage means each for storing the parts of the video signal;

means for transferring stored parts from the first storage means to the second storage means while simultaneously modifying the parts in a manner opposite said determinable manner so that an unmodified version of the parts is stored in the second storage means; and

means for retrieving the unmodified version of the parts from the second storage means as a decoded video signal.

30. The apparatus of claim **29** wherein the parts of the television signal are successive samples of the video signal comprising one horizontal line of video.

31. A method for transmitting a television signal in a nonstandard form to prevent unauthorized viewing of a television program, comprising the steps of:

generating a television signal comprising both timing and video information;

dividing the video information portion of the television signal into successive parts;

storing a first plurality of the parts in a first storage device;

transferring the stored first plurality of parts from the first storage device to the second storage device;

retrieving the stored first plurality of parts from the second storage device in a modified form to thereby provide an encoded video signal while simultaneously storing a second plurality of the parts in the first storage device;

combining the modified parts retrieved from the second storage device with unmodified timing information to form a composite television signal; and

transmitting the composite television signal.

32. A method for transmitting a television signal in a nonstandard form to prevent unauthorized viewing of a television program, comprising the steps of:

generating a television signal comprising both timing and video information;

dividing the video information portion of the television signal into successive parts;

storing a first plurality of the parts in a first storage device;

transferring the stored first plurality of parts from the first storage device to the second storage device;

modifying the plurality of parts as they are transferred from the first storage device to the second storage device so that they are stored in the second storage device in a modified form;

retrieving the stored first plurality of parts from the second storage device in the modified form to thereby provide an encoded video signal while simultaneously storing a second plurality of the parts in the first storage device;

combining the modified parts retrieved from the second storage device with unmodified timing information to form a composite television signal; and

transmitting the composite television signal.

33. The method of claim **31** wherein the plurality of parts are modified upon being stored in the first storage device so that they are stored in both the first and second storage devices in, the modified form.

34. The method of claim **32** or **33** wherein the modification includes rearranging the sequence of the parts.

35. The method of claim **31** wherein the stored parts are successively retrieved from the second storage device in a selectable sequence to provide the modified form.

36. A method for transmitting a television signal in a nonstandard form to prevent unauthorized viewing of a television program, comprising the steps of:

(a) sequentially storing in a first storage means signals representing a first plurality of successive parts of a television video signal while simultaneously modifying the representative signals so as to form an encoded video signal in the first storage means;

(b) transferring the encoded video signal from the first storage means to a second storage means; and

(c) repeating step (a) for a second plurality of parts of the television video signal while simultaneously retrieving the encoded video signal from the second storage means.

37. The method of claim **36** wherein the step of modifying includes gating signals representing successive parts of the video signal into non-sequential memory locations in the first storage means so that the signals are rearranged in sequence relative to the unmodified video signal.

38. Apparatus for transmitting a television signal in nonstandard form, comprising:

means for generating a video signal comprising successive parts;

first and second storage means each for storing a plurality of the successive parts of the video signal;

means for transferring the parts stored in the first storage means to the second storage means in parallel;

means for gating successive parts of the video signal into non-sequential memory locations of one of the first and second storage means so that the parts of the video signal stored in said one storage means are rearranged in sequence relative to the sequence in which they were generated; and

means for sequentially retrieving the plurality of parts stored in the second storage means to form an encoded video signal and transmitting said encoded video signal.

39. The apparatus of claim **38** wherein said gating means gates the parts of the video signal into said first storage means.

40. The apparatus of claim **38** wherein said gating means gates the parts of the video signal from said first storage means into said second storage means.

* * * * *

U.S. Patent Sep. 20, 1983 Sheet 1 of 9 4,405,942

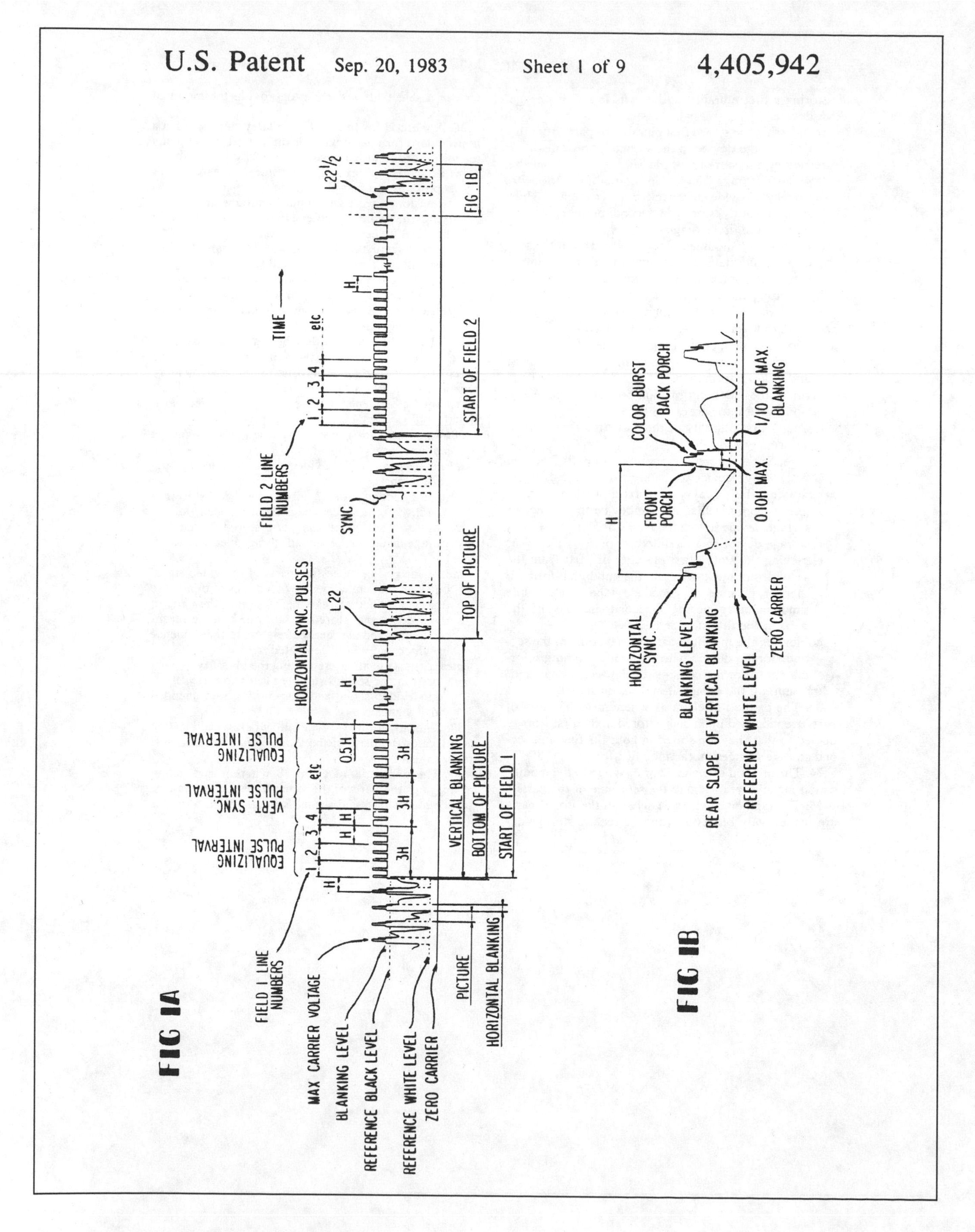

FIG IC

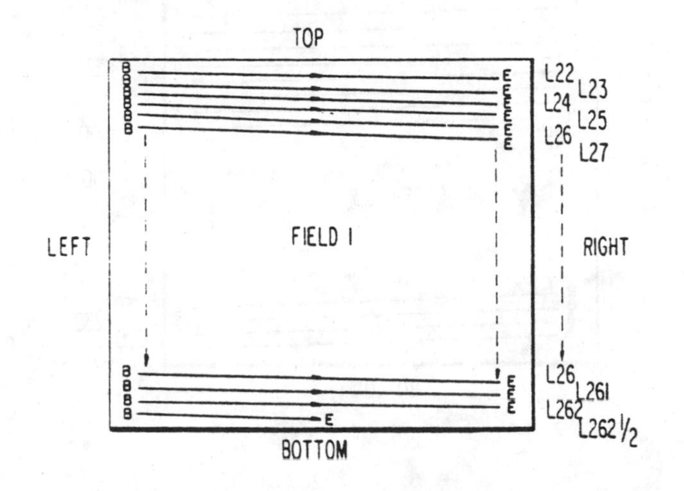
TOP
L22
L23
L24
L25
L26
L27
LEFT
FIELD I
RIGHT
L26
L261
L262
L262 1/2
BOTTOM

FIG ID

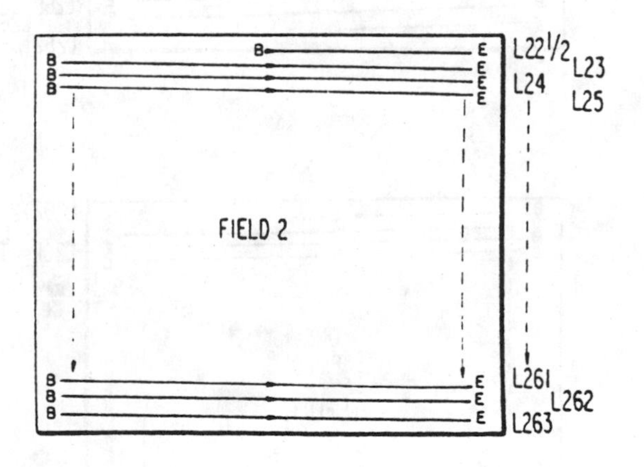
L22 1/2
L23
L24
L25
FIELD 2
L261
L262
L263

U.S. Patent Sep. 20, 1983 Sheet 3 of 9 4,405,942

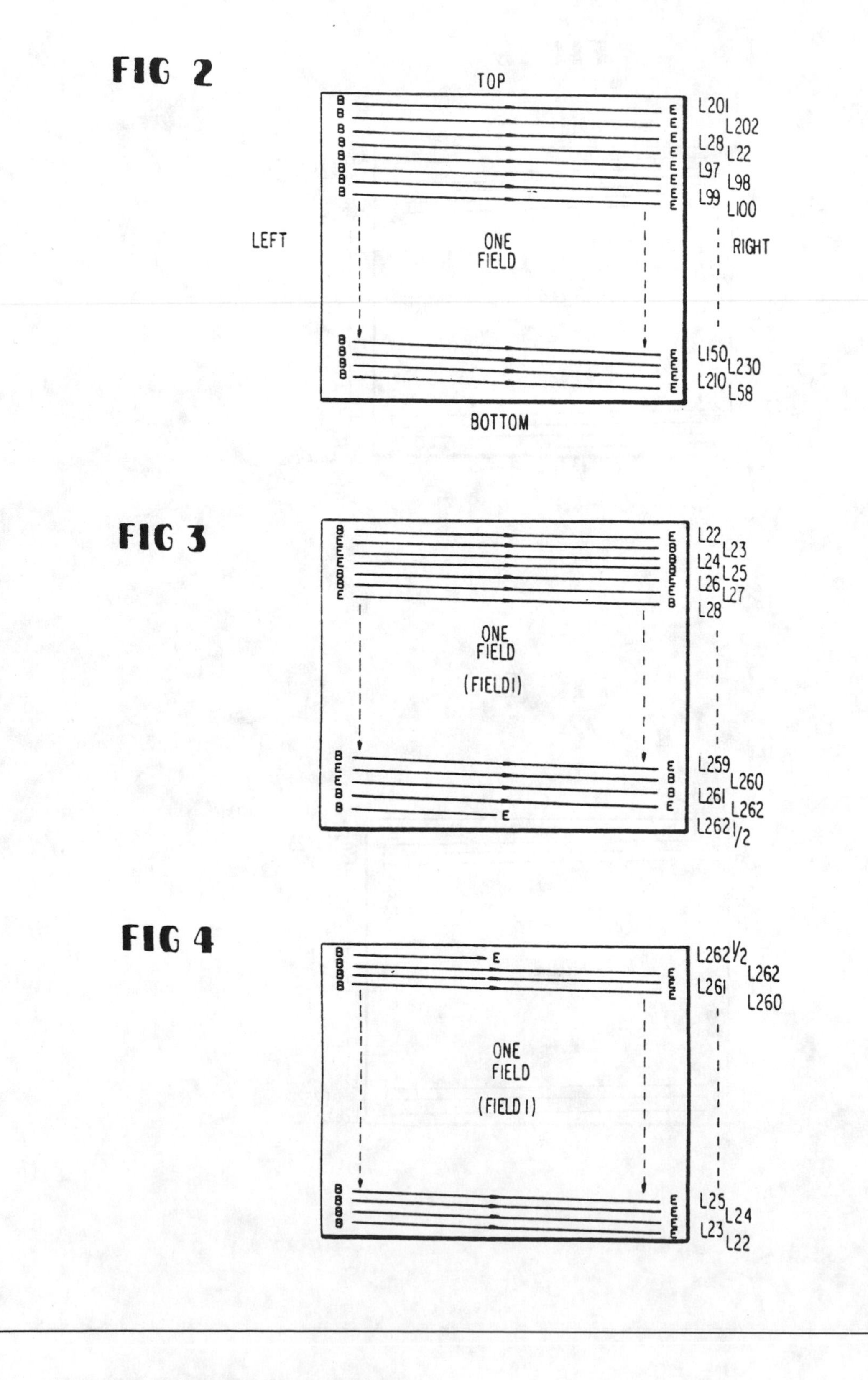

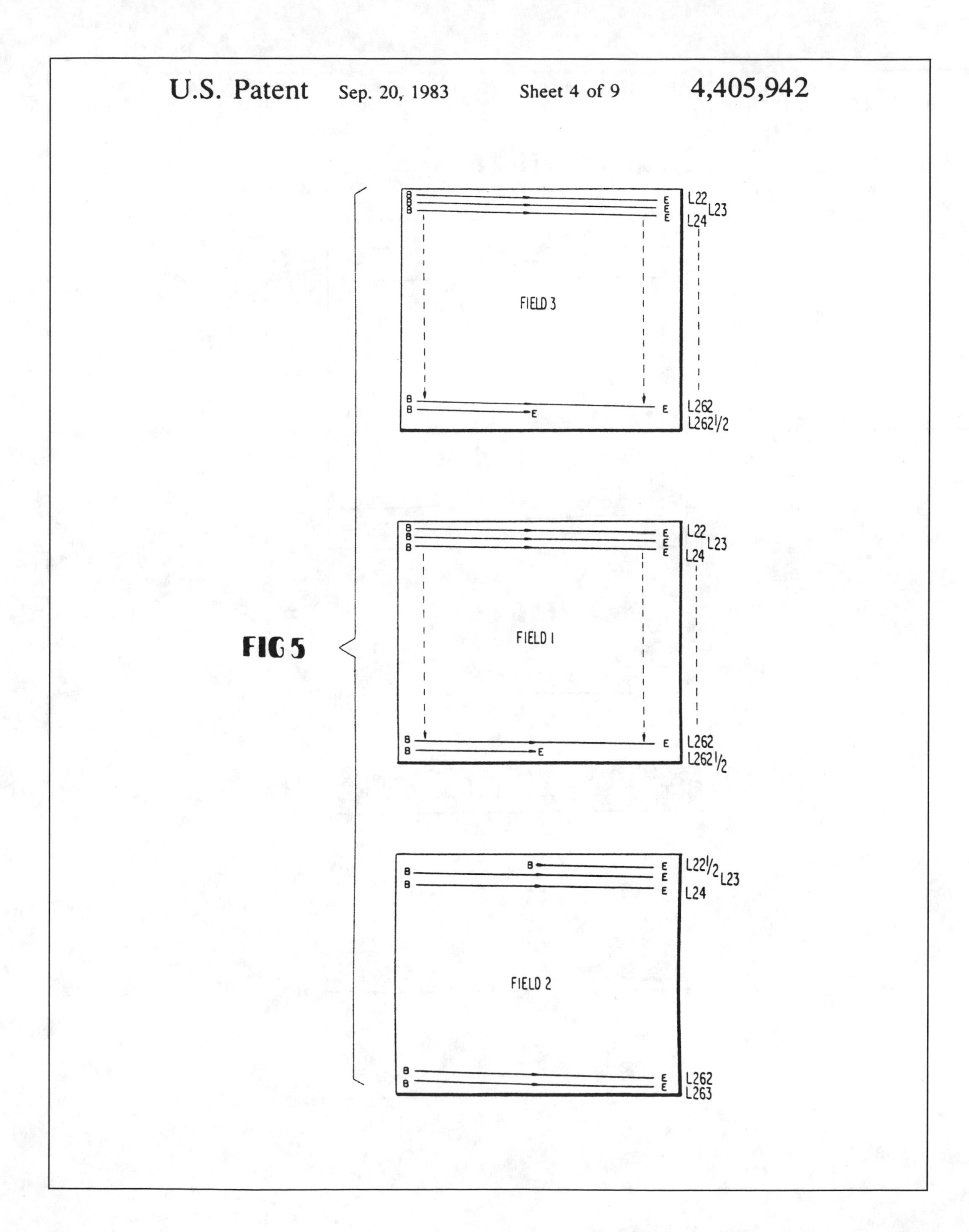
U.S. Patent
Sep. 20, 1983
Sheet 4 of 9
4,405,942
FIG 5
FIELD 3
L22
L23
L24
L262
L262 1/2
FIELD 1
L22
L23
L24
L262
L262 1/2
FIELD 2
L22 1/2
L23
L24
L262
L263

FIG 6A

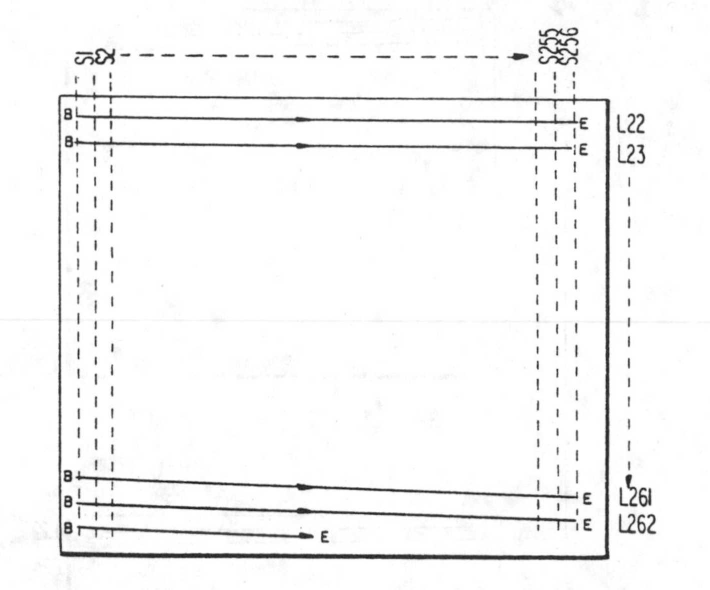

FIG 6B

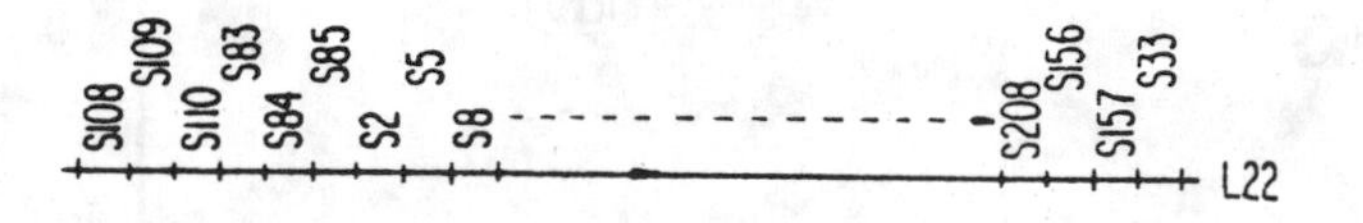

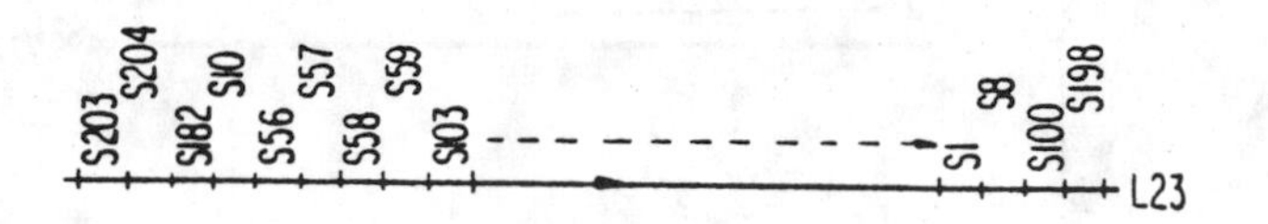

FIG 6C

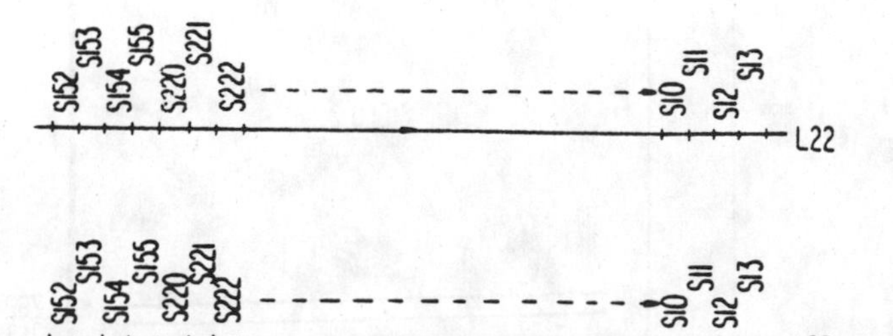

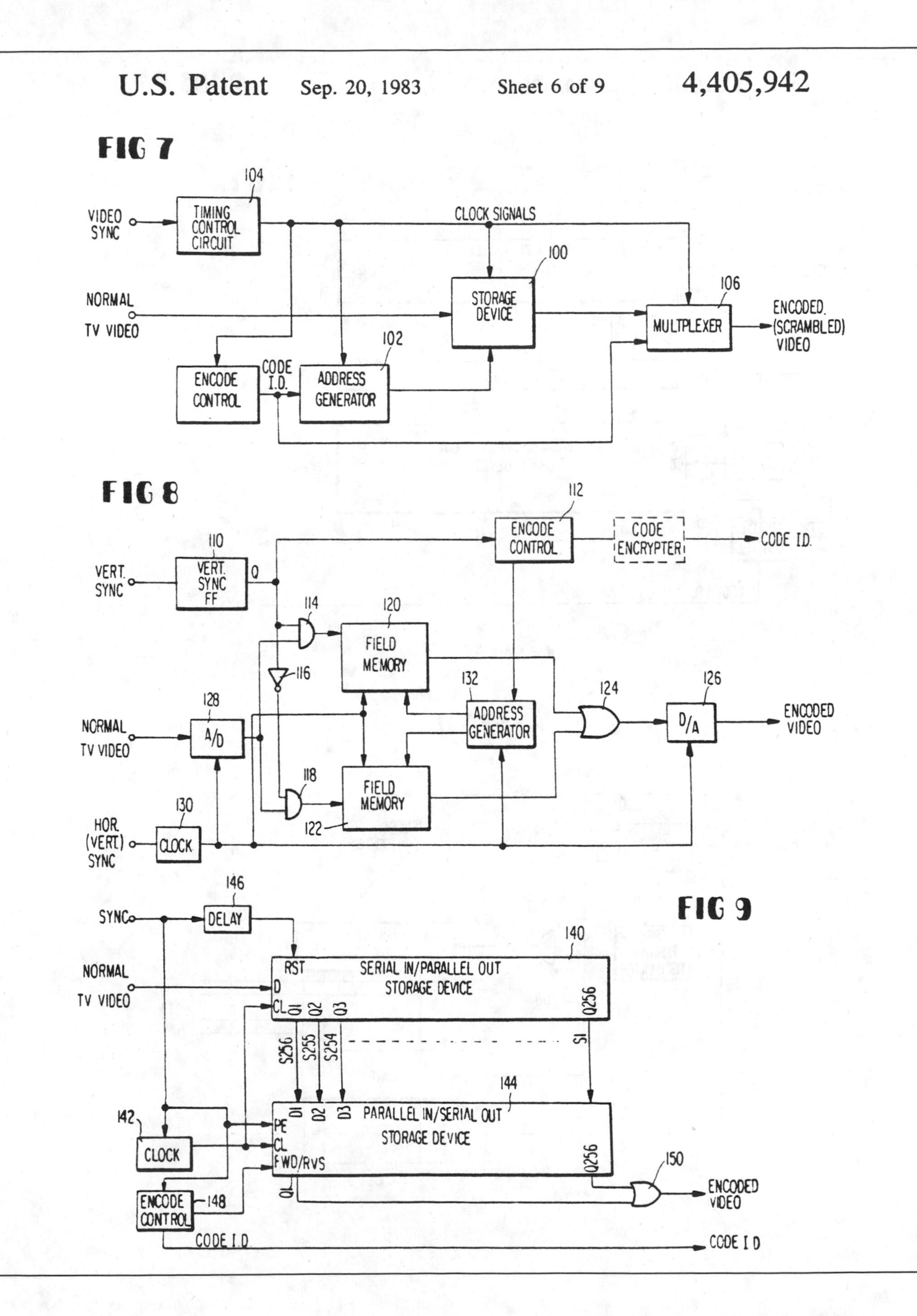
U.S. Patent Sep. 20, 1983 Sheet 6 of 9 4,405,942
FIG 7
104
VIDEO SYNC
TIMING CONTROL CIRCUIT
CLOCK SIGNALS
100
106
NORMAL TV VIDEO
STORAGE DEVICE
MULTPLEXER
ENCODED (SCRAMBLED) VIDEO
102
ENCODE CONTROL
CODE I.D.
ADDRESS GENERATOR
FIG 8
112
110
ENCODE CONTROL
CODE ENCRYPTER
CODE I.D.
VERT. SYNC
VERT. SYNC FF
Q
114
120
FIELD MEMORY
116
132
124
126
128
NORMAL TV VIDEO
A/D
ADDRESS GENERATOR
D/A
ENCODED VIDEO
118
FIELD MEMORY
130
122
HOR. (VERT.) SYNC
CLOCK
146
FIG 9
SYNC
DELAY
140
NORMAL TV VIDEO
RST
SERIAL IN/PARALLEL OUT STORAGE DEVICE
D
CL
Q1 Q2 Q3
Q256
S256 S255 S254
S1
144
142
PE
CL
FWD/RVS
D1 D2 D3
PARALLEL IN/SERIAL OUT STORAGE DEVICE
Q256
CLOCK
150
ENCODE CONTROL
148
QL
ENCODED VIDEO
CODE I.D.
CODE I D

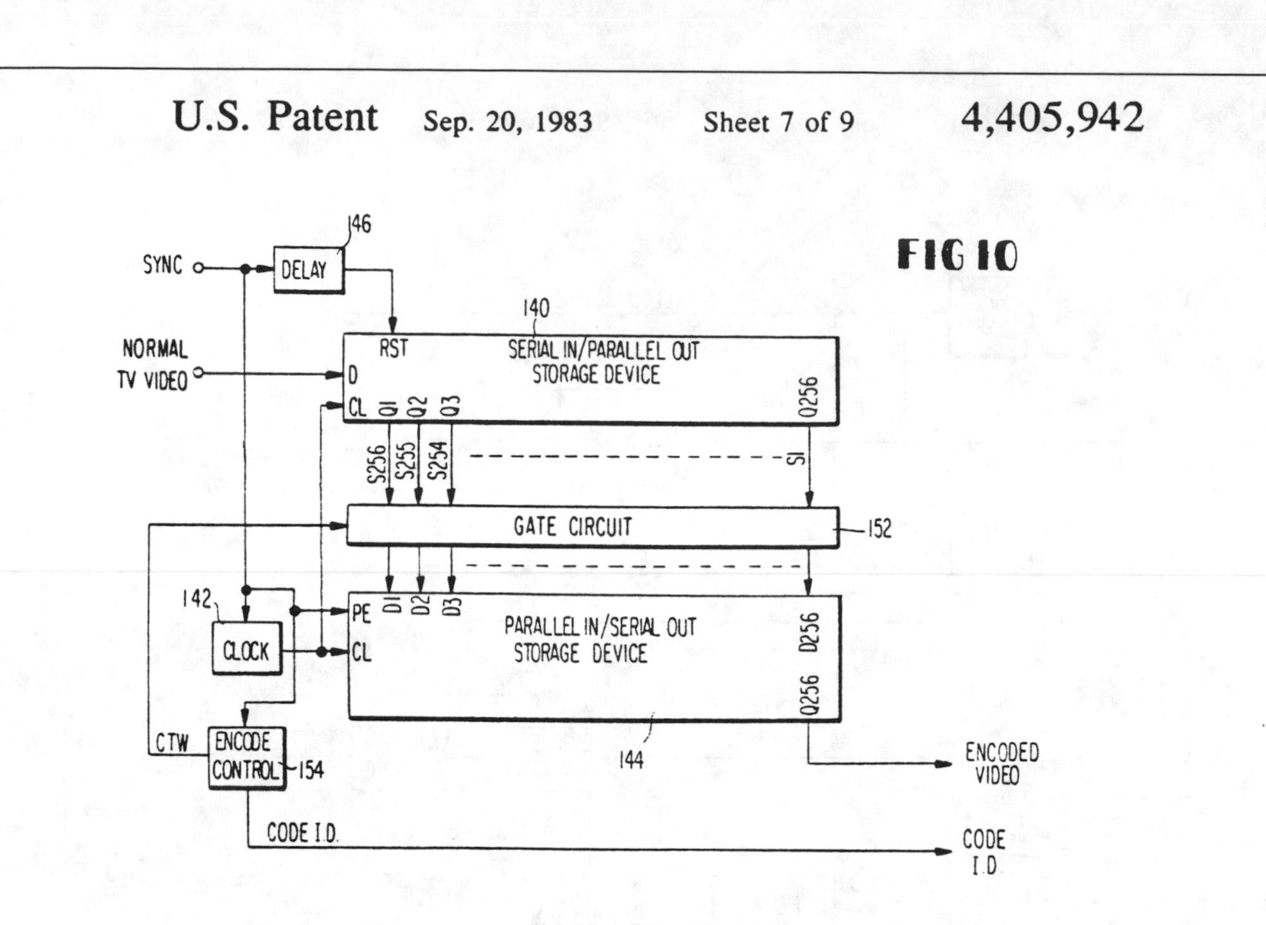

U.S. Patent Sep. 20, 1983 Sheet 7 of 9 4,405,942
FIG 10
SYNC
DELAY
146
NORMAL TV VIDEO
140
RST
SERIAL IN/PARALLEL OUT STORAGE DEVICE
D
CL
Q1
Q2
Q3
Q256
S256
S255
S254
S1
GATE CIRCUIT
152
142
CLOCK
PE
CL
D1
D2
D3
PARALLEL IN/SERIAL OUT STORAGE DEVICE
D256
Q256
CTW
ENCODE CONTROL
154
144
ENCODED VIDEO
CODE I.D.
CODE I.D.

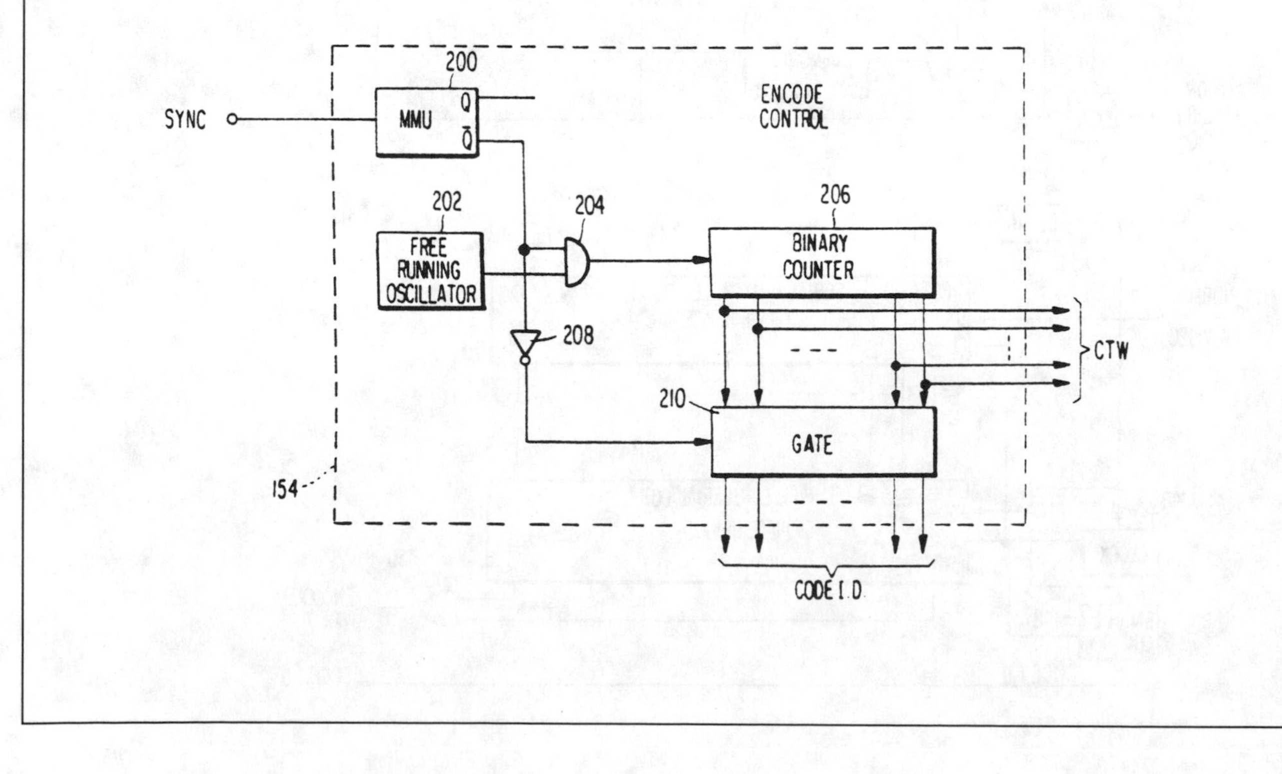

FIG 11
SYNC
200
MMU
Q
Q̄
ENCODE CONTROL
202
FREE RUNNING OSCILLATOR
204
206
BINARY COUNTER
208
CTW
210
GATE
154
CODE I.D.

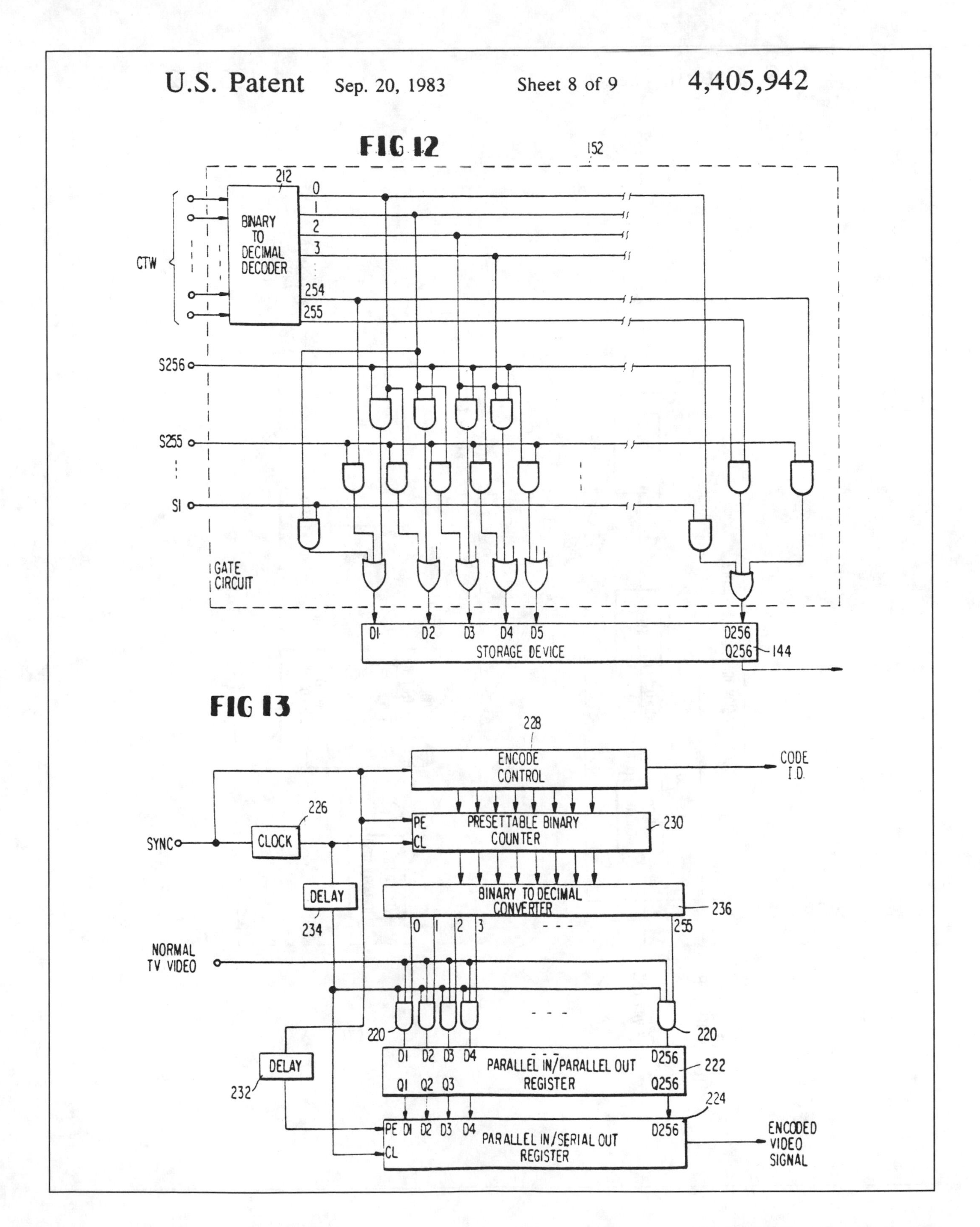
U.S. Patent Sep. 20, 1983 Sheet 8 of 9 4,405,942
FIG 12
152
212
BINARY TO DECIMAL DECODER
CTW
0
1
2
3
254
255
S256
S255
SI
GATE CIRCUIT
D1 D2 D3 D4 D5 D256
STORAGE DEVICE
Q256
144
FIG 13
228
ENCODE CONTROL
CODE I.D.
226
SYNC
CLOCK
PE
CL
PRESETTABLE BINARY COUNTER
230
DELAY
234
BINARY TO DECIMAL CONVERTER
236
0 1 2 3
255
NORMAL TV VIDEO
220
220
DELAY
232
D1 D2 D3 D4
PARALLEL IN/PARALLEL OUT REGISTER
D256
222
Q1 Q2 Q3
Q256
224
PE D1 D2 D3 D4
CL
PARALLEL IN/SERIAL OUT REGISTER
D256
ENCODED VIDEO SIGNAL

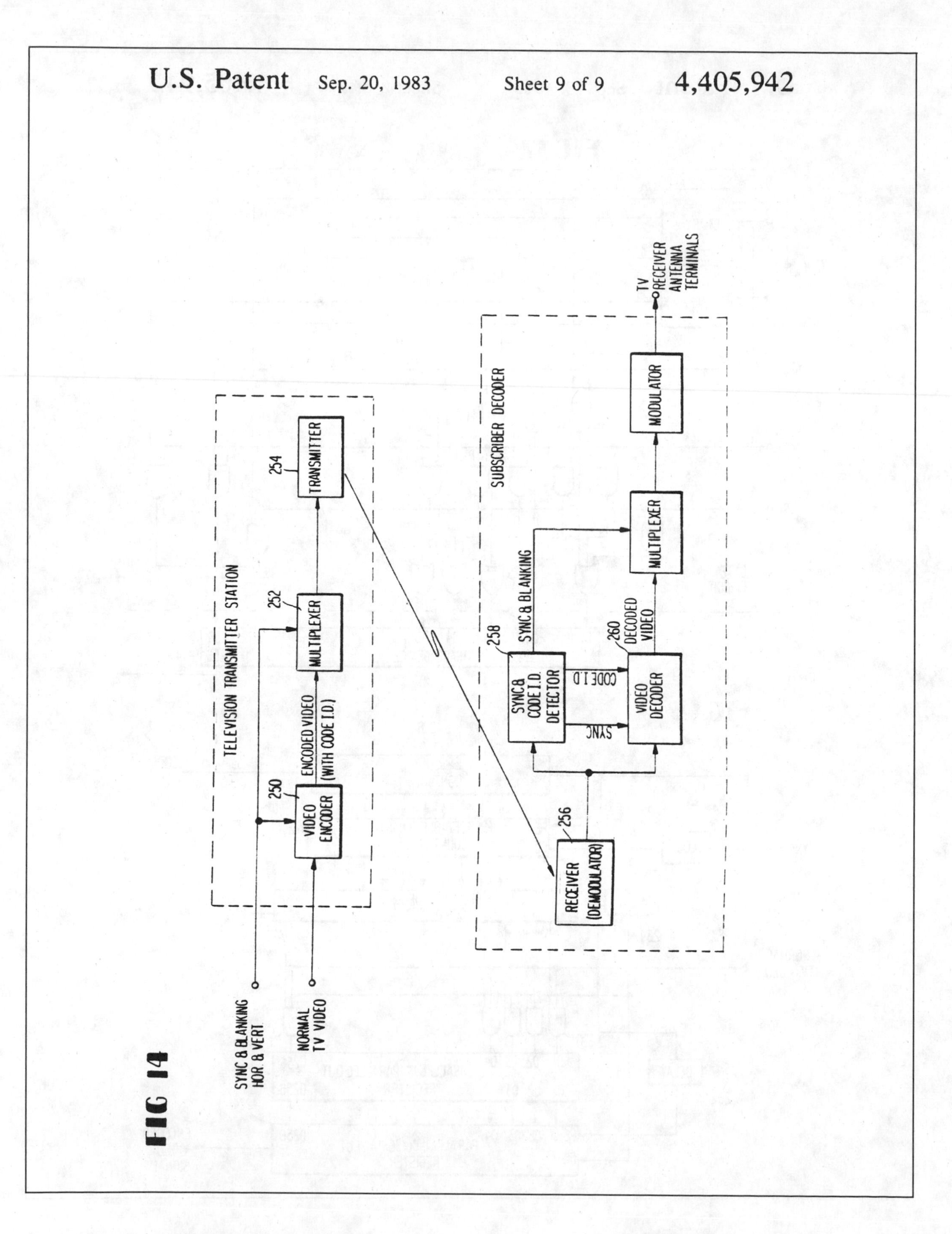
U.S. Patent
Sep. 20, 1983
Sheet 9 of 9
4,405,942
FIG 14
SYNC & BLANKING HOR & VERT
NORMAL TV VIDEO
TELEVISION TRANSMITTER STATION
250
VIDEO ENCODER
ENCODED VIDEO (WITH CODE I.D.)
252
MULTIPLEXER
254
TRANSMITTER
SUBSCRIBER DECODER
256
RECEIVER (DEMODULATOR)
258
SYNC & CODE I.D. DETECTOR
SYNC
CODE I.D
260
VIDEO DECODER
DECODED VIDEO
SYNC & BLANKING
MULTIPLEXER
MODULATOR
TV RECEIVER ANTENNA TERMINALS

United States Patent [19]

Buschman et al.

[11] **Patent Number: 4,479,142**

[45] **Date of Patent: Oct. 23, 1984**

[54] **INTERFACE APPARATUS AND METHOD FOR ASYNCHRONOUS ENCODING OF DIGITAL TELEVISION**

[75] Inventors: **Bob D. Buschman**, Gaithersburg; **Glenn D. Muth**, Frederick; **Ronald T. O'Connell**, Germantown, all of Md.

[73] **Assignee: M/A-COM DCC, Inc.**, Germantown, Md.

[21] Appl. No.: **379,382**

[22] Filed: **May 17, 1982**

[51] **Int. Cl.**[3] **H04N 9/32; H04N 9/42; H04N 7/08**

[52] **U.S. Cl.** **358/13; 358/11; 358/142**

[58] **Field of Search** **358/320, 310, 141, 140, 358/13, 11, 143, 146**

[56] **References Cited**

U.S. PATENT DOCUMENTS

4,122,477 10/1978 Gallo 358/13

Primary Examiner—John C. Martin
Assistant Examiner—Luan Nguyen
Attorney, Agent, or Firm—Pollock, Vande Sande & Priddy

[57] **ABSTRACT**

Digitally encoded (DPCM) NTSC video is transmitted in a standard T3 data format and rate via a transmit interface module. The module is subjected to inputs comprising DPCM encoded NTSC video, horizontal sync and a video sampling clock at $3f_{sc}$, where f_{sc} is the color subcarrier frequency. The sampling clock is used to write the DPCM encoded video and horizontal sync into a buffer. The buffer is read out via a read clock which is a submultiple of the 44736 MHz T3 rate. To assist in receiver decoding, the video data is supplemented with horizontal sync indicating code words, which designate the start of each horizontal scan line. On detection of a horizontal sync word read out of the buffer memory, the read clock is inhibited for the insertion of the horizontal sync indicating code words. The T3 frame is also supplemented with two digitized audio channels, stuff indicators and a stuff opportunity slot as well as parity, frame and multiframe indicators and an alarm channel. The T3 frame also carries data indicating a relationship between the T3 clock and f_{sc}. This is useful at the receiver for ensuring that f_{sc} at the receiver will track f_{sc} at the transmitter. In accordance with standard T3 operation, the frame comprises 56 85-bit words, 28 odd and 28 even. The first bit of each 85-bit word is a control bit. Substantially all of the odd words include a 4-bit digitized audio nibble. Exceptions are two unused 4-bit nibbles, a single 4-bit stuff opportunity nibble and a single 4-bit video sample. Each of the even words includes 21 4-bit video samples which can be either video samples or sync indicating code words.

12 Claims, 10 Drawing Figures

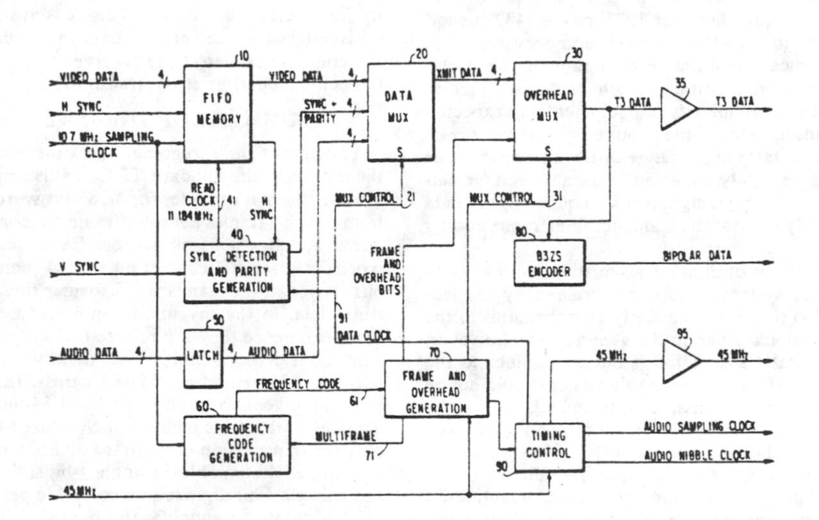

INTERFACE APPARATUS AND METHOD FOR ASYNCHRONOUS ENCODING OF DIGITAL TELEVISION

TECHNICAL FIELD

The present invention relates to the digital transmission of color television signals, especially transmission of digitized NTSC at the standard T3 rate and format.

BACKGROUND ART

The field of digital television transmission, especially that transmission employing bandwidth compression techniques is summarized in "Digital Television Transmission Using Bandwidth Compression Techniques" by Kneko et al. in the *IEEE Communications Magazine* for July, 1980 at pages 14–22; and in "Digital Encoding of Color Video Signals—A Review" by Limb et al. appearing in the *IEEE Transactions on Communications,* Vol. COM-25, No. 11, November 1977 at pages 1349–1385. Unencoded digitized NTSC signals produce an approximately 86 Mbps bit stream. It has long been recognized that bandwidth compression techniques (for example DPCM) can be used to reduce that relatively high rate. For example, Ishiguro, in U.S. Pat. No. 3,843,940 discloses a "Differential Pulse Code Modulation Transmission System" for NTSC, SECAM or PAL video signals. Other examples of bandwidth compression techniques for color television are found in "Interframe Coding for 4 MHz Color Television Signals" by Iinuma et al. appearing in the *IEEE Transactions on Communications,* Vol. COM-23, No. 12, December 1975 at pages 1461 et seq; "1.544 Megabits per Second Transmission of TV Signals by Interframe Coding System" by Yasuda et al, appearing in the *IEEE Transactions on Communications,* Vol. COM-24, October 1976 at pages 1175 et seq; "A 32 Megabit per Second Component Separation DPCM Coding System for NTSC Color TV" by Swada et al. appearing in the *IEEE Transactions on Communications,* Vol. COM-26, No. 4 for April 1978 at pages 458 et seq and "32 Megabit per Second Transmission of NTSC Color TV Signals by Composite DPCM Coding" by Sawada et al. appearing in the *IEEE Transactions on Communications,* Vol. COM-26, No. 10 for October 1978, pages 1432 et seq.

The extent to which bandwidth compression is applied and hence the ultimate data rate of the coded video depends on constraints such as desired picture quality, cost and complexity of equipment and expected mode of transmission. While captive systems have been proposed at bit rates in the range of 1.5–32 megabits per second, it is generally expected that cable and/or network quality television signals will require higher data rates especially if intraframe, and not interframe, coding is used.

For applications of digital transmission techniques to existing video systems, one is constrained by the fact that the video signal is necessarily asynchronous to the transmission clock. Even in the general case, however, since the digital transmission is but an adjunct to the generation and use of the video signals, it should not be expected that the transmission system clock rate can be imposed on the video signal generation and thus, even in general, one should expect that the video signal will be asynchronous to the transmission rate clock.

There are, at the present time, substantial quantities of digital transmission equipment which adhere to common carrier standards. There is for example a T1 standard at 1.544 megabits per second, a T2 standard at 6.13 megabits per second and a T3 standard at 44.736 megabits per second.

Prior art suggestions for transmitting digitized video tie the transmission rate and format to the video line rate, see in this regard the Sawada et al. publication cited above, particularly Vol. COM-26, No. 4, page 462, wherein for transmission purposes, two different frames are defined, a first frame type is exactly equal to two adjacent horizontal line signals and a second frame type is exactly equal to a single line signal. See also Vol. COM-26, No. 10, page 1435.

This technique, while having an appearance of simplifying equipment, actually imposes unnecessary burdens on the transmission equipment. If, for example, standard data rate and format were used to encode the digital television, existing transmission monitoring equipment could be used to monitor the error rate of the data using parity bits provided for in the standard frame format. This practical consideration saves the cost of additional overhead in the transmitted data to perform the monitoring function, thus reducing the overall data rate required.

It is therefore an object of the present invention to provide a method and apparatus for transmitting digitized NTSC signals via standard transmission equipment using standard frame format and rate.

Since the video is DPCM encoded, the receiver requires a decoder to recreate the original video. This decoder can be simplified if it can recognize each video line start. Since the video timing is asynchronous to the T3 clock, we cannot use the transmission clock to identify video lines, as could be the case with the Swada proposals referenced above. Accordingly, we insert into the video stream digital codes uniquely identifying each video line start. Accordingly, it is another object of the invention to provide a method and apparatus for transmitting digitized video which has the capability of inserting code words uniquely identifying each video line start.

To further assist in properly receiving the encoded digitized video, the transmission includes information relating the color subcarrier (or video sampling clock) to the T3 clock. Since the T3 clock is available at both transmitter and receiver, this data can be used to meter the color subcarrier at the receiver to ensure it tracks the color subcarrier at the transmitter.

SUMMARY OF THE INVENTION

To achieve the foregoing, and other advantages of the invention, the standard T3 format is employed. This standard format provides for 56 85-bit words per multiframe, with 7 frames per multiframe. Accordingly, each frame includes 8 85-bit words. Each of these 85-bit words consists of 1 control bit and 84 (nominally) data bits. While in the standard T3 format these 84 bits are truly data, in the invention some are used for other purposes, hence they are referred to as only nominally data. Considering a multiframe, then, we can locate 28 odd and 28 even 84-bit "data" words. In accordance with the invention, each of the odd 84-bit data words includes a first 4-bit nibble which is used substantially only for transmission of digitized audio. Exceptions are two unused 4-bit nibbles, a single 4-bit stuff opportunity and a single 4-bit digitized video nibble per multiframe.

In addition, to simplify the decoding process at a receiver, the method and apparatus of the invention provides for the transmission of supplemental data in the form of a plurality of 4-bit nibbles. This supplemental data actually consists of 3 4-bit nibbles (or a 12-bit code word) signalling the start of each video horizontal scan line. Because the frame duration is fixed (by the T3 standard) that duration is not rationally related to the line rate or scan line duration. However, with the format described above, at least one of every pair of frames includes the code word mentioned above. For those frames which do not include this code word, encoded video is transmitted in its place.

To provide for the generation of the standard T3 rate and format, the apparatus of the invention provides a buffer which operates as a first in, first out memory. The buffer is subjected to four significant input signals, DPCM encoded video samples (since the digitized video is differential encoded, the samples are sometimes referred to as differences or video differences), horizontal sync and a sampling clock, the latter at an integer multiple of the color subcarrier, e.g. 10.7 MHz. This clock is used to write the buffer, and the buffer is read by a read clock which is nominally an integer (4) submultiple of the T3 44.736 MHz rate or 11.184 MHz. Although the read clock is exactly at this rate, the read operation is only nominally performed at this rate because, associated with the buffer is a sync detector which produces a sync detection signal, on reading a horizontal sync signal from the buffer which has the effect of inhibiting the read clock. This allows a supplemental data generator to generate and output the horizontal sync indicating code for insertion in the data stream during the time that the read clock is inhibited. After termination of the read clock inhibition, reading of the buffer continues. The read clock is generated by a timing circuit control which is subjected to a T3 rate clock input. The DPCM encoded video is output from the buffer to a data multiplexer, another input to the data multiplexer is the supplemental data signal from the supplemental data generator.

Accordingly, in one aspect, the invention provides for an interface for a video encoder arranged to output asynchronously encoded video and supplemental data in a T3 data format and rate for interfacing with encoded video, sampled at an integer multiple of f_{sc}, where f_{sc} is the color subcarrier frequency comprising:

a buffer with DPCM encoded video and video sync inputs, and an output, means to write video and video sync into said buffer in response to a sampling clock,

timing circuit control means responsive to a T3 rate clock for producing a data read clock at a submultiple of said T3 rate clock,

coupling means for coupling said data read clock to said buffer for reading out said buffer at said data read clock rate to produce a data stream,

sync detection means responsive to a sync signal read from said buffer for inhibiting said coupling means for a predetermined time,

supplemental data generator means with an output coupled to a data multiplexing means for generating supplemental data,

data multiplexing means responsive to video data read from said buffer and to supplemental data from said supplemental data generator means for outputting said video data and supplemental data in time sequence,

whereby detection of a sync signal inhibits reading of said buffer for said predetermined time allowing insertion of said supplemental data.

The standard T3 data rate and format constrains the availability of DPCM encoded video data which can be transmitted, and the asynchronous nature of the video sync and T3 data rate is a further complicating factor especially when it is desired to add horizontal line scan synchronous supplemental data which is therefore asynchronous with the frame rate. The T3 rate provides for 56 85-bit words wherein a first bit is control data and 84 bits are nominally data bits. The 56 words can be thought of as 28 odd and 28 even words. Of the 28 odd 84-bit data words, a 4-bit nibble of each is used for substantially only digitized audio. These 4-bit nibbles provide for two channels of digitized audio. Two of the 28 4-bit nibbles are unused (occupied by non-information bearing pulses), a single 4-bit nibble is used as a single stuff opportunity per frame, and another 4-bit nibble is used for encoded video. The remaining 84 data bits in each of the even words and the remaining 80 data bits in the odd words are used for encoded video with one exception. The supplemental data, in an embodiment of the invention, consists of three 4-bit nibbles representing horizontal line scan start, and a 4-bit parity nibble. Because of the asynchronous relation between the period between horizontal line scans and the T3 frame rate, the supplemental data does not appear in each frame. Rather, the supplemental data appears in at least one of each pair of frames, and in some instances occurs in both frames of a pair.

Therefore, in accordance with another aspect, the invention provides a method of transmitting DPCM encoded NTSC video in a T3 format comprising the steps of:

assembling a serial digital bit stream into sequential T3 frames, repeating at a T3 frame rate,

said serial digital bit stream consisting of a sequence of 56 85-bit digital words, 28 odd and 28 even words, each such word including a single control bit and 84 data bits, a majority of said 84 data bits comprising encoded video, each such odd digital word including a single 4-bit nibble of substantially only digitized audio data, said digital bit stream associated with at least every other frame including a plurality of 4-bit nibbles signalling a video scan line start.

The multi-frame also includes provision for a frequency code which relates video sampling clock (directly tied to the color subcarrier) and the T3 clock. Actually, one is divided by the other and a single bit of the remainder is transmitted. At the receiver, this bit can be used to meter the generation of the color subcarrier to maintain the color subcarrier at receiver and transmitter in alignment relative to the T3 clock. More particularly, each multi-frame, a delay (four-stage counter) is pulsed, and the counter is clocked by the video sampling clock. At the same time, a one bit counter is clocked by the video sampling clock. The one bit counter is reset a fixed number of video sampling clocks after multi-frame start. A latch is enabled to latch the state of the one bit counter a fixed number of video sampling clocks after multi-frame start. The condition of this latch is the frequency code. It should be apparent that over many multi-frames, a sequence formed by a sequence of the single bit frequency code will indicate the relation between the multi-frame start period (tied to the T3 clock) and the video sampling clock period (tied to color subcarrier). Thus, in another aspect, the invention provides an interface for a video encoder arranged to output asynchronously encoded video and supplemental data in a T3 data format and rate for interfacing with DPCM encoded video sampled at an integer multiple of a color subcarrier comprising:

means for inserting DPCM encoded video into a T3 frame,

first means for comparing a video sampling clock directly related to said color subcarrier and a T3 signal directly related to said T3 rate, and wherein said means for inserting includes,

means responsive to said first means for inserting data representative of said comparison into said T3 frame.

BRIEF DESCRIPTION OF THE DRAWINGS

The present invention will now be described in further detail in the following portions of this specification when taken in conjunction with the attached drawings in which like reference characters identify identical apparatus and in which:

FIG. **1** illustrates the format of the prior art T3 frame structure;

FIG. **2** illustrates the format of the frame as employed in accordance with the invention to transmit NTSC DPCM encoded video and auxiliary information;

FIG. **3** is a block diagram of the interface which accepts DPCM encoded video, horizontal and vertical sync, a sampling clock and audio data and produces the formatted data;

FIGS. **4–7** are detailed block diagrams illustrating the sync detection and parity generation circuit **40**, audio data latch **50** and timing control **90**;

FIG. **8** is a timing diagram illustrating the timing relation between the video sampling clock, the DPCM encoded video and the horizontal sync;

FIG. **9** is a timing diagram illustrating the timing relationship between the audio nibble clock, the audio sampling clock and representative audio data; and

FIG. **10** is a timing diagram illustrating the response to detection of horizontal sync, and the insertion of the horizontal sync and parity code words onto the tristate data bus.

DETAILED DESCRIPTION OF THE PREFERRED EMBODIMENTS

FIG. **1** illustrates the standard T3 format. FIG. **1** is organized on seven lines, thus each line represents a different frame, the seven lines shown in FIG. **1** represent a single seven-frame multiframe. Those skilled in the art will be aware that this illustration is for convenience, in practice the entire frame is a serial time sequence. In accordance with the representation in FIG. **1**, X represents an alarm service channel wherein in any multiframe the two X bits are identical. The P bits are parity information for parity taken over all information time slots in the preceding multiframe; conventionally, both P bits are identical. The M bits are multiframe alignment signals appearing, as shown in the fifth, sixth and seventh frames. Conventionally M_0 and M_1 are 0 and 1, respectively. The F bits are frame bits, which appear in a sequence as shown in FIG. **1** wherein F_0 and F_1 are 0 and 1, respectively. Each frame provides for stuffing indicator bits C_{xy}, where x identifies a frame and y is an integer in the range of 1–3. In any one frame, only one of two stuffing indicator words are allowed, either 000 or 111. The stuff time slot is the first data slot following F_1 after C_{i3} in the ith frame. Finally, the representation **84I** represents 84 data bits.

Accordingly, FIG. **1** can be said to represent a multiframe consisting of 56 85-bit digital words with each 85-bit digital word consisting of a single control bit and 84 data bits.

The particularized frame produced by the apparatus, and used in the method, of the present invention is shown in FIG. **2**. In FIG. **2**, for convenience, each different frame is shown on four lines, those skilled in the art will understand of course that each frame is actually a time sequence which begins at the upper left, proceeds horizontally across to the end of the line, then drops down to the next line and begins at the left, and so forth. Accordingly, FIG. **2** also shows the seven different frames of each multiframe. However, in order to show the particular use for each of the significant bit positions, the illustration in FIG. **2** is somewhat expanded relative to that of FIG. **1**. In FIG. **2**, the X, P, M_0, M_1, F_0 and F_1 positions have the same meaning as in FIG. **1**.

Referring now to the first 85-bit word, FIG. **2** indicates that it includes the X bit, a 4-bit audio nibble comprising bits A_{1A} through A_{1D}, and 80 bits consisting of 20 4-bit video samples (each sample including bits V_1–V_4). In contrast, the second 85-bit word includes a frame alignment bit F_1 and 21 4-bit video samples. It should be apparent from the format of FIG. **2** that every even word, that is the 28 85-bit words numbered 2–56 include 21 4-bit video samples and the first (control) bit is either F_1 or F_0. In contrast, the 28 odd 85-bit words each include 20 4-bit video samples, but for the most part the first 4-bit nibble is not a video sample. The bit positions A_{1x} and A_{2x} (where x is an alphabetic character from A–L, inclusive) represents two different audio channels. The 4-bit nibbles represented by the U bit positions are unused, the 4-bit nibble represented by the S bit positions are a single stuff opportunity per multiframe. The stuffing indicator bits are represented by some of the C bit positions, but in contrast to FIG. **1** which shows seven stuff opportunities, one for each frame, in FIG. **2** there is only one stuff opportunity per multiframe. The 4-bit nibble represented by the N bit positions is a further video sample. The unused bit positions carry time-consuming pulses, but do not represent meaningful information and are not employed, other than marking time, in the decoding process.

As shown in FIG. **2**, there are 1148 4-bit video samples in each multiframe.

Once per horizontal line scan start, three 4-bit nibbles of the 1148 4-bit nibbles in a multiframe used for video samples, carry a digital code indicating the beginning of the horizontal line scan. Associated with these three 4-bit nibbles is a fourth, 4-bit nibble containing parity information and an indication of the presence or absence of vertical sync. Accordingly, a total of 16 bits (or 4 4-bit nibbles) is devoted to a predetermined code word indicating beginning of horizontal line scan. Because the T3 frame rate and the horizontal line scan rate are asynchronous, there is no rational relation between the rate of horizontal line scan start and any T3 related rate. However, this code word is inserted into at least one of every pair of multiframes, and in some cases, both multiframes of a pair.

As was mentioned above, the 4-bit nibble identified by the S bit positions is used for stuffing, if needed. The stuffing indicator word corresponds to the bit positions C_{xy} (where x is 1–3 and y is 1–3). In any one multiframe, either all these bits are one, indicating the presence of stuffing, or all are zero, indicating the absence of stuffing. The bit positions identified by C_{xy} (where x is 4 or 5 and y is 1–3, inclusive) are not used. The bit positions identified by C_{xy} (where x is 6 or 7 and y is 1-3, inclusive) is a single bit frequency indicator appearing in six different locations in the multiframe. The frequency indicator indicates the measured value of f_{sc} at the encoder (relative to the T3 rate) so the decoder can properly track and recreate the NTSC waveform.

The apparatus to transmit the encoded video and auxiliary information in the T3 format is shown in FIG. 3.

As shown in FIG. 3, the 4-bit video samples which have been DPCM encoded via apparatus not shown, is input to one terminal of a FIFO 10 on a 4-bit wide basis. A second input to the FIFO 10 is a horizontal sync signal. Both inputs to the FIFO 10 are written with the aid of a sampling clock which is related to the color subcarrier frequency f_{sc}. In a particular embodiment of the invention, a sampling clock is three times this subcarrier, i.e. typically 10.7 Mhz. The data path taken by the video samples proceeds from an output of the FIFO 10 to one input of a multiplexer 20. Under control of a multiplexer control signal 21 from a sync detection and parity generation circuit 40, the multiplexer 20 accepts the 4-bit video samples, 4-bit sync and parity code words from the sync detection and parity generator 40 or 4-bit audio samples from a latch 50. The data multiplexer 20 produces a 4-bit wide XMIT data output, which is itself input to an overhead multiplexer 30. The other input to the overhead multiplexer 30 is frame and overhead bits from a frame and overhead generation circuit 70. The overhead multiplexer 30 accepts one or the other of its inputs in response to a multiplexer control signal 31 also derived from the frame and overhead generation circuit 70. The output of the multiplexer 30 is a serial stream at the T3 rate and in the T3 format which may, for example, be coupled through an amplifier 35 and output. In addition, in an embodiment of the invention, a B3ZS encoder 80 accepts the T3 data and produces at an output bipolar T3 data for transmission purposes.

The clock for reading the FIFO 10 is initially derived from a T3 clock which is input to a timing and control circuit 90. The timing and control circuit 90 produces an audio sampling clock and an audio nibble clock, both related to the T3 rate. These are used in the audio sampling equipment (not shown) which produces the audio data input to latch 50. Accordingly, the audio data is synchronous with the T3 clock. In addition, the timing and control circuit 90 produces a data clock 91, at ¼ the T3 rate, or 11.184 MHz. This data clock 91 is input to the sync detection and parity generation circuit 40 which, in response thereto, couples a read clock 41 at 11.184 MHz to the FIFO 10.

The T3 rate clock is also input to the frame and overhead generation circuit 70. This allows the production of the overhead multiplexer control signal 31 as well as the frame and overhead bits which provide one input to the overhead multiplexer 30. The frame and overhead generation circuit 70 also produces a multiframe signal 71 which is input to a frequency code generation circuit 60. The other input to the frequency code generation circuit is the sampling clock used to write the FIFO 10. The frequency code generation circuit 60 compares the periods of its two input signals. An output of the frequency code generation circuit 60 is a digital frequency code 61 which is input to the frame and overhead generation circuit 70 where it is used as a frequency indicator indicative of the frequency (f_{sc}), representing the sampling clock rate, and therefore the write clock rate for FIFO 10.

In response to the audio sampling clock and audio nibble clock, produced by the timing and control circuit 90, audio data samples are coupled as an input to the latch 50. At specified times in the frame, as represented in the format shown in FIG. 2, this audio data is accepted by the data multiplexer 20 for insertion into the serial 4-bit wide stream XMIT data.

The sync detection and parity generation circuit 40 produces the data multiplexer control signal 21 and a sync and parity code word as referred to above. It receives the data clock 91 from the timing and control circuit 90 and from that fashions the read clock 41 to read the FIFO 10. The sync detection and parity generation circuit 40 is also responsive to a vertical sync signal from the video output as well as an indication that the FIFO 10 has read a horizontal sync signal. In response to the horizontal sync, the sync detection and parity generation circuit 40 interrupts the read clock 41 and provides the sync and parity code words to the data multiplexer 20 for insertion in the frame as shown in FIG. 2. Depending on random relative timing between the T3 clock and the video scan line start, the sync and parity words can be inserted into any video nibble. At the conclusion of generating the necessary sync and parity bits, the read clock 41 is again enabled to continue reading FIFO 10.

Accordingly, as shown in FIG. 2, the T3 frame is composed of a time sequence of frame and other control bits dictated by the T3 format, video samples in the position specified in FIG. 2, a specified plurality of 4-bit nibbles representing the beginning of each horizontal line scan and the frequency code indicator produced by the frequency code generator 60.

FIGS. 4–7 show, in more detail, the sync detection and parity generation circuit 40 as well as the associated data multiplexer 20, audio data latch 50 and timing and control circuit 90. FIGS. 8–10 show relevant waveforms.

Referring now to FIG. 4, a latch 100 is subjected to 4-bit parallel video samples, and a horizontal sync pulse; the inputs are clocked via the video sampling clock. A timing diagram is shown in FIG. 8 relating the video sampling clock to the video difference code and horizontal sync. In a typical embodiment of the invention, the period T is 93.5 nanoseconds, the data setup time t_{SU} is 30 nanoseconds minimum and the data hold time t_{HD} is 30 nanoseconds minimum. Of course, FIG. 8, in indicating the timing of the horizontal sync pulse relative to the video sampling clock should not be taken as implying that a horizontal sync pulse is coincident with each video sampling clock, rather the horizontal sync pulse is produced at the beginning of each horizontal line scan.

After latching via the latch 100, the state of any video difference code and horizontal sync are shifted into the FIFO's 101 and 102 by the shift in clocks SIA and SIB. The latter two clocks are the video sampling clock divided in half via the flip-flop 103 (see FIG. 5). Since SIA and SIB are sequential, time sequential samples are stored in different FIFO's. It should be apparent that this use of two FIFO's is not essential to the invention.

The video difference samples and the horizontal sync pulse are shifted out of the FIFO's 101 and 102 by the shift out clocks SOA and SOB. Both these shift out clocks are 11.184 MHz burst rate derivatives of the T3 clock. The shift out clocks' derivation is shown in FIG.

5 and will be discussed hereinafter. Accordingly, the FIFO's **101**, **102** provide the function of a buffer.

The output conductors of FIFO's **101** and **102** carry the video difference codes to the input of parity generator **104**. The video difference codes, as read out from FIFO **101** or FIFO **102** are latched by the 11.184 MHz clock into a latch **105**. The output of the latch **105** comprises a 4-bit wide tristate data bus (DB0–DB3) and the latch **105** is controlled by the signal DISABLE VIDEO or $\overline{\text{ENABLE VIDEO}}$/Enable Sync & Parity (which is abbreviated $\overline{\text{EV}}$/ES & P in FIG. **4**). The production of both signals is described hereinafter.

Audio samples are latched into a latch **106** via a nibble clock also derived from the timing control **90**. As shown in FIG. **3**, the timing control **90** produces an audio sampling clock and an audio nibble clock; the timing is shown in FIG. **9**. The sampled audio produces audio data; and as is shown in FIG. **9**, the clocks provide for three 4-bit nibbles for each of two channels. The output of latch **106** is tristated onto the bus under control of the signal $\overline{\text{ENABLE AUDIO}}$. Since the signals DISABLE VIDEO and $\overline{\text{ENABLE AUDIO}}$ are mutually exclusive, only one of the latches **105** or **106** places an output on the tristate data bus at any one time.

Horizontal sync and parity is derived from the latch **107** which is clocked by the parity/sync clock, the production of which is shown in FIG. **5**. The timing for the operation is shown in FIG. **10**. The first two lines on FIG. **10** illustrate the 11.184 MHz derivative of the T3 clock and the second line illustrates one of the shift out clocks (SOA).

The third line of FIG. **10** shows horizontal sync read out of either FIFO **101** or FIFO **102**. When present, this pulse is latched, by the parity/sync clock, into the hex "D" flip-flop **108**. At the Q5 output, the rising edge of the signal clocks flip-flop **109** producing the SYNC DETECT pulse; which is shown in the fifth line in FIG. **10**. SYNC DETECT/LATCH PARITY is used to latch and then reset (via flip-flops **111**, **112**), the parity on the difference code.

The opposite phase $\overline{\text{LATCH PARITY}}$ (output of the inverter **113**) resets parity (flip-flop **111**) and is latched by the 11 MHz clock (see FIG. **5**) in flip-flop **116**. The output of flip-flop **116** is STOP VIDEO, which is gated with the 11 MHz clock (gate **117**) to stop the shift out clocks produced by flip-flop **118** is the absence of STOP VIDEO.

Returning to FIG. **4**, the SYNC DETECT (LATCH PARITY) pulse is reclocked (in the zero stage) by the PARITY/SYNC CLOCK in the flip-flop **108** and the result, $\overline{\text{EV}}$/ES & P $\overline{\text{ENABLE VIDEO}}$/ENABLE SYNC AND PARITY) at the Q0 output (shown in line **6** of FIG. **10**) is used to disable latch **105** and enable (via inverter **114**) the output of the parity/sync latch **107**. The output of latch **107** is tristated onto the common data bus (DB0–DB3).

Vertical sync is received at the multiplexer **110** to control its output. The output of multiplexer **110** is also controlled by the sync detect. When horizontal sync occurs, the outputs of the multiplexer **110** are all ones. Output YA remains high for three parity/sync clocks and then changes to the difference code parity.

As shown in FIG. **10**, line **10** the sync detect results in $\overline{\text{EV}}$/ES & P after a short delay. $\overline{\text{EV}}$/ES & P is present for a total of four nibble times. During the first three nibble times, three 4-bit nibbles, each 1's, are output from latch **107**. The multiplexer **110** output Y_A provides a single bit for each nibble and Y_B provides the other three bits for each nibble. At the last nibble time, another 4-bit nibble is again put on the bus, however the makeup of this depends on the condition of S_1. If S_1 is in a state indicating the presence of vertical sync, then the first three bits are all 1 and vice versa. The last bit is a parity bit from flip-flop **112**.

Part of the timing and control is shown in FIG. **6**. The T3 rate clock causes two counters **120** and **121** to count. The counters are arranged to divide the T3 rate clock by 85. One output produces the 11.184 MHz clock. The counters also produce a pulse, FRAME, which causes two additional counters **122** and **123** to count. These are arranged to divide the pulse FRAME by 56. The outputs of counters **122**, **123** are the address lines of two 256 by 4 ROMs **124** which store multiframe format data to generate necessary control signals. The output of the ROMs **124** is coupled to a hex flip-flop **125**, one output of which is latched (by FRAME) into a flip-flop **126**. The true output of flip-flop **126** is latched by an 11.184 MHz clock in a latch **127**. DISABLE VIDEO and $\overline{\text{ENABLE AUDIO}}$ are the outputs of latch **127**. DISABLE VIDEO disables the outputs of the video latch **105** and the parity/sync latch **107** when audio data is to be put on the data bus.

$\overline{\text{ENABLE AUDIO}}$ is used to enable the output of the audio latch **106** and stops the shift out clocks via gate **128** (FIG. **5**).

By storing data in ROM **124** identifying the locations of the audio nibbles in the frame of FIG. **2**, the output of flip-flop **127** can be controlled to disable the video and enable audio at the appropriate times in the frame.

The ROM **124** also stores other signals identifying other overhead locations in the frame. Specifically, the location of F, M, P, X, U and N (see FIG. **2**) are stored in ROM **124**. When read out of ROM **124**, these are used to set two positions of the hex flip-flop **125** which produce the MUX select signals, also shown in FIG. **6**.

By referring briefly to FIG. **5**, it should be apparent that in the presence of $\overline{\text{LATCH PARITY}}$ or $\overline{\text{ENABLE AUDIO}}$, clocking of flip-flop **118** can be prevented via AND gate **117**. Thus, under either of these circumstances, the shift out clocks SOA and SOB are inhibited, preventing the read out from FIFO **101** or **102**. This operation provides a "gap" in the frame for the insertion or either audio or the digital code word represen-ting beginning of a video line scan. As shown in FIG. **10**, line **5**, $\overline{\text{LATCH PARITY}}$, when present inhibits the buffer reading operation for four nibble times (via gate **117**).

FIG. **5** also illustrates generation of the signal FREQ CODE. This is a digital signal representing the video clock rate. The video sampling clock is used to clock a quad flip-flop **141** with its outputs of one stage connected to inputs of the succeeding stage; the video sampling clock is also used to clock flip-flop **142** via exclusive OR gate **145**. Accordingly, the flip-flop **142** provides for dividing the video sampling clock in half. At every multiframe, the output of flip-flop **142** is latched in flip-flop **143**; the multiframe start input at stage D_0 is output of stage Q1 to clock flip-flop **143**. Every multiframe, the flip-flop **142** is reset via NAND gate **144**. The frequency code output is a single bit indicating a relation between the video sampling clock and the multiframe start. By integrating the transmitted bit at the receiver, the video clock can be made to track the transmitter's video sampling clock since the T3 rate is identical at both transmitter and receiver. This single bit, from flip-flop **143** is actually inserted six times into each multiframe. Sending the same bit six times is not, of course, necessary.

The frequency code output of flip-flop **143** provides one input to the multiplexer **130** (FIG. 7), another input to the multiplexer **130** corresponds to parity information coupled from parity generator **133** through flip-flop **135**. The control signals for multiplexer **130** are provided by MUX select, i.e. from hex flip-flop **125** (FIG. 6). Periodically, in the frame a $\overline{\text{LOAD SR}}$ pulse is produced via flip-flop **146**. This is clocked with the T3 clock, and is set by the output of NAND gate **148**, one of whose inputs is provided by the 4-bit counter **120**, and the other of whose inputs is provided via NOR gate **147** from another output of the 4-bit counter **120**. The $\overline{\text{LOAD SR}}$ signal is used to load shift register **131** (see FIG. 7). The information for this parallel loading of shift register **131** is provided by a buffer **132** which receives its input in turn from the 4-bit wide tristate data bus DB0–DB3. The inputs to the buffer **131** are coupled in parallel to the parity generator **133** to generate the parity information referred to above.

The shift register **131** can also be serially loaded from the multiplexer **130** in a manner well-known to those skilled in the art. The MUX SELECT signals select frequency code parity information, or fixed data from the other multiplexer inputs, for shifting through the multiplexer **130** to the shift register **131** and out on the line labelled SERIAL DATA. It should be apparent how the ROM data stores a frame plan to generate appropriate MUX select signals to produce insertion of the appropriate inputs into the multiframe.

Although not illustrated in FIG. 6, the ROM **124** is also addressed by alarm and stuffing information. An alarm signal is generated (by equipment not illustrated), and its presence (or absence) produces the MUX SELECT signals to insert bits indicating the presence (or absence) of an alarm condition. Likewise, by monitoring the filled status of the FIFO's **101**, **102**, the necessity for stuffing is determined. If stuffing is required, the signal disable video is inhibited. If stuffing is not required, the signal is not inhibited and instead the MUX SELECT provides a fixed bit pattern in the unused stuff opportunity. The stuffing requirement is also used via ROM **124** to generate the appropriate stuffing indicator bits in the same fashion.

From the foregoing, it should be apparent that the interface described herein provides for a standard T3 format in which the information bearing signals consist of 4-bit nibbles of DPCM video and 4-bit audio nibbles. Coincident with each horizontal sync pulse and/or vertical sync pulse, the clock reading out the 4-bit video DPCM nibbles is inhibited and a predetermined sequence (depending on whether or not horizontal or vertical sync is occurred) is inserted onto the data bus in place of a video nibble. In addition, at selected times in the frame, parity, frequency code information or other data is inserted in a predetermined format.

The multiplexer **20** (FIG. 3) which multiplexes, video samples, audio samples, sync digital codes and parity corresonds to the latches **105–107** and the attached tri-state bus DB0–DB3. On the other hand the multiplexer **30** which multiplexes the output of multiplexer **20** with frame and overhead bits corresponds to the multiplexer **130**, shift register **131** and buffer **132** (all in FIG. 7). The use of the shift register allows concurrent multiplexing and serializes the input from the tri-state bus DB0–DB3.

We claim:

1. An interface for a video encoder arranged to output asynchronously DPCM encoded NTSC video and supplemental data in a T3 data format and rate for interfacing with encoded video sampled at an integer multiple of f_{sc} where f_{sc} is the color subcarrier frequency comprising:
   - a buffer with DPCM encoded video and video sync inputs, means to write video and video sync into said buffer in response to a sampling clock;
   - timing circuit control means responsive to a T3 rate clock for producing a data read clock at a submultiple of said T3 rate clock;
   - coupling means for coupling the data clock to said buffer for reading out said buffer memory at said data read clock rate to produce a data stream;
   - sync detection means responsive to a sync signal read from said buffer for inhibiting said coupling means for a predetermined time;
   - supplemental data generator means with an output coupled to a data multiplexing means for generating supplemental data;
   - data multiplexing means responsive to video data read from said buffer and to supplemental data from said supplemental data generator means for outputting said video data and supplemental data in time sequence;
   - whereby detection of a sync signal inhibits reading of said buffer for said predetermined time allowing insertion of said supplemental data.

2. The apparatus of claim 1 in which said data multiplexing means is also responsive to a vertical sync pulse for outputting supplemental data uniquely representative of vertical sync.

3. The apparatus of claim 2 in which said multiplexing means comprises:
   - a tristate data bus,
   - first and second latches coupled thereto,
   - a multiplexer with a first fixed input and coupled to said second latch, and
   - control means responsive to said sync detection means for disabling said first latch and enabling said second latch in response to a sync pulse whereby said first fixed input to said multiplexer is coupled to said tristate data bus through said second latch.

4. The apparatus of claim 3 wherein said multiplexer includes a second fixed input and a control terminal coupled to a vertical sync input,
   - whereby either said first or second fixed input is coupled to said tristate data bus depending on the presence or absence of said vertical sync pulse.

5. The apparatus of claim 1 which further includes:
   - comparator means with two inputs and an output for comparing a first input rate directly related to F_{sc} with a second input rate directly related to said T3 rate and for producing a frequency code signal, and wherein:
   - said data multiplexing means includes a frequency code multiplexer responsive to said comparator means and said time sequence of video and supplemental data for inserting said frequency code signal into said time sequence.

6. A method of transmitting DPCM encoded NTSC video in a T3 format comprising the steps of:
   - (a) creating a stream of said DPCM encoded NTSC video,
   - (b) creating a parallel stream of horizontal sync pulses,

(c) momentarily halting said stream of DPCM encoded NTSC video at the occurrence of each horizontal sync pulse,
(d) inserting a predetermined bit pattern into said DPCM stream once for each horizontal sync pulse during said momentary halt,
(e) inserting overhead bits into said stream corresponding to a standard T3 format, and
(f) transmitting the resulting stream in a bit serial fashion.

7. The method of claim 6 which includes the further steps of:
(i) periodically, with reference to a T3 clock, halting said streams of steps (a) and (b), and
(ii) inserting a digital word representing audio information into a stream produced as a result of step (c) for each halt of step (i).

8. The method of claim 6 which comprises the further steps of:
(i) continually comparing a rate directly related to a subcarrier of said NTSC video with a rate directly related to T3 rate to produce a frequency code signal, and
(ii) inserting said frequency code signal into said stream in said step (e).

9. A method of transmitting DPCM encoded NTSC video in a T3 format comprising the steps of:
assembling a serial digital bit stream into sequential T3 frames, repeating at a T3 frame rate,
said serial digital bit stream consisting of a sequence of 56 85-bit digital words, 28 odd and 28 even words, each such word including a single control bit and 84 data bits, a majority of said 84 data bits comprising encoded video, each such odd digital word including a single 4-bit nibble of substantially only digitized audio data, said digital bit stream associated with at least every other frame including a plurality of 4-bit nibbles signalling a video scan line start.

10. An interface for a video encoder arranged to output asynchronously encoded video and supplemental data in a T3 data format and rate for interfacing with DPCM encoded video sampled at an integer multiple of a color subcarrier comprising:
means for inserting DPCM encoded video into a T3 frame,
first means for comparing a video sampling clock directly related to said color subcarrier and a T3 signal directly related to said T3 rate, and wherein said means for inserting includes:
means responsive to said first means for inserting data representative of said comparison into said T3 frame.

11. The interface of claim 10 wherein said first means includes:
a delay means responsive to said T3 signal and clocked by said video sampling clock,
a counter clocked by said video sampling clock resetting means for resetting said counter in response to said delay means reaching a predetermined condition,
a latch responsive to said counter and clocked each time said delay means reaches a different predetermined condition for supplying said data representative of said comparison.

12. The apparatus of claim 10 wherein said means for inserting includes:
buffer means responsive to sequentially presented video differences and to periodically presented horizontal sync pulses for storing the same,
data read clock means responsive to said T3 rate for generating a data read clock,
coupling means for coupling said data read clock to said buffer means for reading said buffer means,
sync detection means responsive to said buffer means for inhibiting said coupling means for a predetermined time on detection of a sync pulse read from said buffer means,
supplemental data generating means for generating supplemental data in response to said sync pulse from said buffer means,
multiplexing means including a pair of latches, a first latch responsive to video differences read from said buffer means and a second responsive to said supplemental data for generating a time sequence of said video differences and said supplemental data,
second multiplexing means responsive to said time sequence and to said data representative of said comparison for generating a composite time sequence.

* * * * *

X	(841)	F_1	(841)	C_{11}	(841)	F_0	(841)	C_{12}	(841)	F_0	(841)	C_{13}	(841)	F_1	(841)	1st FRAME
X	(841)	F_1	(841)	C_{21}	(841)	F_0	(841)	C_{22}	(841)	F_0	(841)	C_{23}	(841)	F_1	(841)	2nd FRAME
P	(841)	F_1	(841)	C_{31}	(841)	F_0	(841)	C_{32}	(841)	F_0	(841)	C_{33}	(841)	F_1	(841)	3rd FRAME
P	(841)	F_1	(841)	C_{41}	(841)	F_0	(841)	C_{42}	(841)	F_0	(841)	C_{43}	(841)	F_1	(841)	4th FRAME
M_0	(841)	F_1	(841)	C_{51}	(841)	F_0	(841)	C_{52}	(841)	F_0	(841)	C_{53}	(841)	F_1	(841)	5th FRAME
M_1	(841)	F_1	(841)	C_{61}	(841)	F_0	(841)	C_{62}	(841)	F_0	(841)	C_{63}	(841)	F_1	(841)	6th FRAME
M_0	(841)	F_1	(841)	C_{71}	(841)	F_0	(841)	C_{72}	(841)	F_0	(841)	C_{73}	(841)	F_1	(841)	7th FRAME

FIG. 1 PRIOR ART

X	A_{1A}	A_{1B}	A_{1C}	A_{1D}	$[20(V_1\ V_2\ V_3\ V_4)]$	F_1	$[21(V_1\ V_2\ V_3\ V_4)]$	THE 1st FRAME
C_{11}	A_{1E}	A_{1F}	A_{1G}	A_{1H}	$[20(V_1\ V_2\ V_3\ V_4)]$	F_0	$[21(V_1\ V_2\ V_3\ V_4)]$	
C_{12}	A_{1I}	A_{1J}	A_{1K}	A_{1L}	$[20(V_1\ V_2\ V_3\ V_4)]$	F_0	$[21(V_1\ V_2\ V_3\ V_4)]$	
C_{13}	A_{2A}	A_{2B}	A_{2C}	A_{2D}	$[20(V_1\ V_2\ V_3\ V_4)]$	F_1	$[21(V_1\ V_2\ V_3\ V_4)]$	
X	A_{2E}	A_{2F}	A_{2G}	A_{2H}	$[20(V_1\ V_2\ V_3\ V_4)]$	F_1	$[21(V_1\ V_2\ V_3\ V_4)]$	THE 2nd FRAME
C_{21}	A_{2I}	A_{2J}	A_{2K}	A_{2L}	$[20(V_1\ V_2\ V_3\ V_4)]$	F_0	$[21(V_1\ V_2\ V_3\ V_4)]$	
C_{22}	N_1	N_2	N_3	N_4	$[20(V_1\ V_2\ V_3\ V_4)]$	F_0	$[21(V_1\ V_2\ V_3\ V_4)]$	
C_{23}	A_{1A}	A_{1B}	A_{1C}	A_{1D}	$[20(V_1\ V_2\ V_3\ V_4)]$	F_1	$[21(V_1\ V_2\ V_3\ V_4)]$	
P	A_{1E}	A_{1F}	A_{1G}	A_{1H}	$[20(V_1\ V_2\ V_3\ V_4)]$	F_1	$[21(V_1\ V_2\ V_3\ V_4)]$	THE 3rd FRAME
C_{31}	A_{1I}	A_{1J}	A_{1K}	A_{1L}	$[20(V_1\ V_2\ V_3\ V_4)]$	F_0	$[21(V_1\ V_2\ V_3\ V_4)]$	
C_{32}	A_{2A}	A_{2B}	A_{2C}	A_{2D}	$[20(V_1\ V_2\ V_3\ V_4)]$	F_0	$[21(V_1\ V_2\ V_3\ V_4)]$	
C_{33}	A_{2E}	A_{2F}	A_{2G}	A_{2H}	$[20(V_1\ V_2\ V_3\ V_4)]$	F_1	$[21(V_1\ V_2\ V_3\ V_4)]$	
P	A_{2I}	A_{2J}	A_{2K}	A_{2L}	$[20(V_1\ V_2\ V_3\ V_4)]$	F_1	$[21(V_1\ V_2\ V_3\ V_4)]$	THE 4th FRAME
C_{41}	S_1	S_2	S_3	S_4	$[20(V_1\ V_2\ V_3\ V_4)]$	F_0	$[21(V_1\ V_2\ V_3\ V_4)]$	
C_{42}	A_{1A}	A_{1B}	A_{1C}	A_{1D}	$[20(V_1\ V_2\ V_3\ V_4)]$	F_0	$[21(V_1\ V_2\ V_3\ V_4)]$	
C_{43}	A_{1E}	A_{1F}	A_{1G}	A_{1H}	$[20(V_1\ V_2\ V_3\ V_4)]$	F_1	$[21(V_1\ V_2\ V_3\ V_4)]$	
M_0	A_{1I}	A_{1J}	A_{1K}	A_{1L}	$[20(V_1\ V_2\ V_3\ V_4)]$	F_1	$[21(V_1\ V_2\ V_3\ V_4)]$	THE 5th FRAME
C_{51}	A_{2A}	A_{2B}	A_{2C}	A_{2D}	$[20(V_1\ V_2\ V_3\ V_4)]$	F_0	$[21(V_1\ V_2\ V_3\ V_4)]$	
C_{52}	A_{2E}	A_{2F}	A_{2G}	A_{2H}	$[20(V_1\ V_2\ V_3\ V_4)]$	F_0	$[21(V_1\ V_2\ V_3\ V_4)]$	
C_{53}	A_{2I}	A_{2J}	A_{2K}	A_{2L}	$[20(V_1\ V_2\ V_3\ V_4)]$	F_1	$[21(V_1\ V_2\ V_3\ V_4)]$	
M_1	U_1	U_2	U_3	U_4	$[20(V_1\ V_2\ V_3\ V_4)]$	F_1	$[21(V_1\ V_2\ V_3\ V_4)]$	THE 6th FRAME
C_{61}	A_{1A}	A_{1B}	A_{1C}	A_{1D}	$[20(V_1\ V_2\ V_3\ V_4)]$	F_0	$[21(V_1\ V_2\ V_3\ V_4)]$	
C_{62}	A_{1E}	A_{1F}	A_{1G}	A_{1H}	$[20(V_1\ V_2\ V_3\ V_4)]$	F_0	$[21(V_1\ V_2\ V_3\ V_4)]$	
C_{63}	A_{1I}	A_{1J}	A_{1K}	A_{1L}	$[20(V_1\ V_2\ V_3\ V_4)]$	F_1	$[21(V_1\ V_2\ V_3\ V_4)]$	
M_0	A_{2A}	A_{2B}	A_{2C}	A_{2D}	$[20(V_1\ V_2\ V_3\ V_4)]$	F_1	$[21(V_1\ V_2\ V_3\ V_4)]$	THE 7th FRAME
C_{71}	A_{2E}	A_{2F}	A_{2G}	A_{2H}	$[20(V_1\ V_2\ V_3\ V_4)]$	F_0	$[21(V_1\ V_2\ V_3\ V_4)]$	
C_{72}	A_{2I}	A_{2J}	A_{2K}	A_{2L}	$[20(V_1\ V_2\ V_3\ V_4)]$	F_0	$[21(V_1\ V_2\ V_3\ V_4)]$	
C_{73}	U_1	U_2	U_3	U_4	$[20(V_1\ V_2\ V_3\ V_4)]$	F_1	$[21(V_1\ V_2\ V_3\ V_4)]$	

FIG. 2

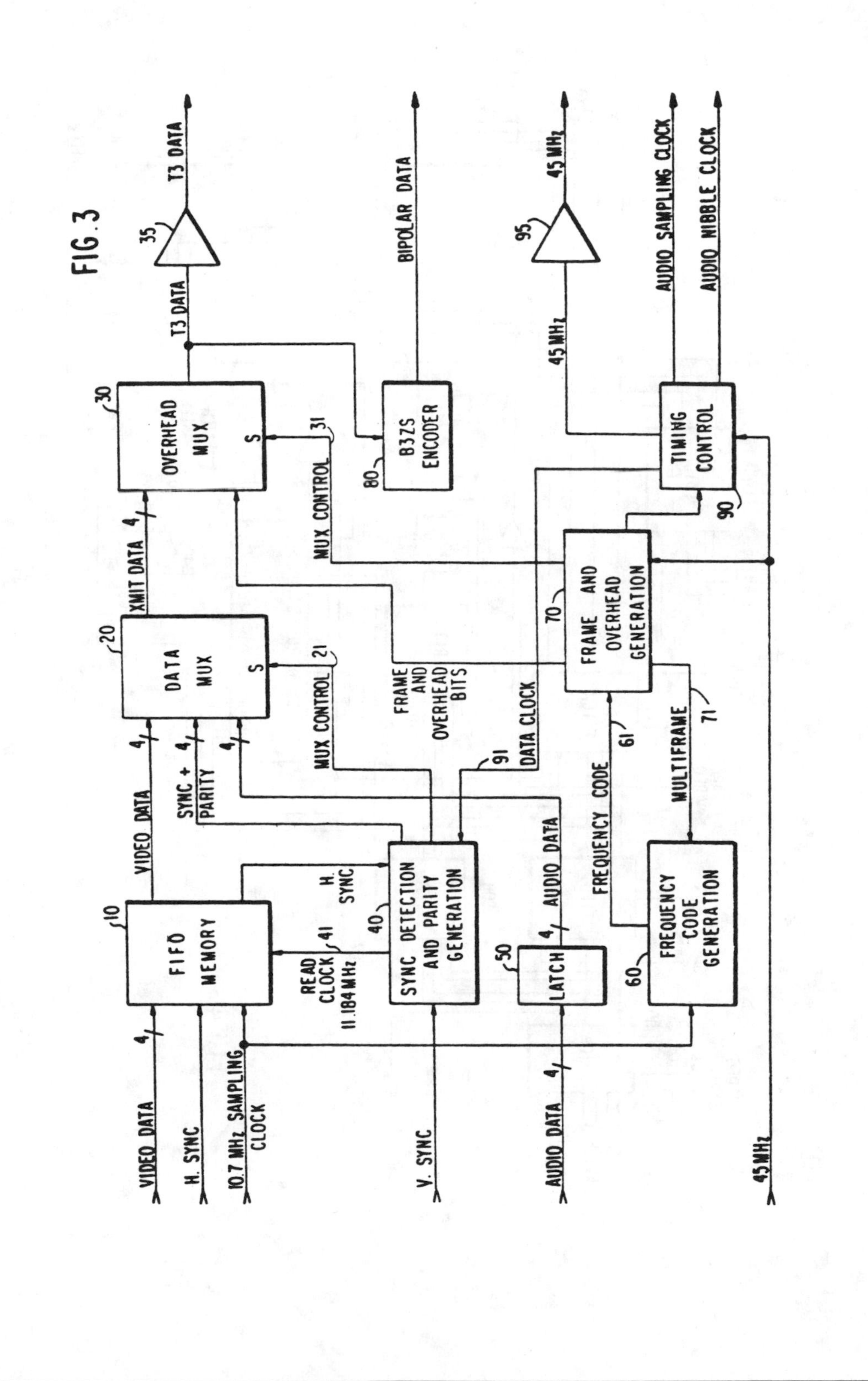
FIG. 3
VIDEO DATA
H. SYNC
10.7 MHz SAMPLING CLOCK
V. SYNC
AUDIO DATA
45MHz
FIFO MEMORY
10
READ CLOCK 11.184 MHz
41
H. SYNC
SYNC DETECTION AND PARITY GENERATION
40
LATCH
50
AUDIO DATA
FREQUENCY CODE GENERATION
60
FREQUENCY CODE
61
VIDEO DATA
SYNC + PARITY
DATA MUX
20
S
MUX CONTROL
21
FRAME AND OVERHEAD BITS
91
DATA CLOCK
FRAME AND OVERHEAD GENERATION
70
MULTIFRAME
71
XMIT DATA
OVERHEAD MUX
30
MUX CONTROL
31
B3ZS ENCODER
80
TIMING CONTROL
90
T3 DATA
35
BIPOLAR DATA
45MHz
95
AUDIO SAMPLING CLOCK
AUDIO NIBBLE CLOCK
4

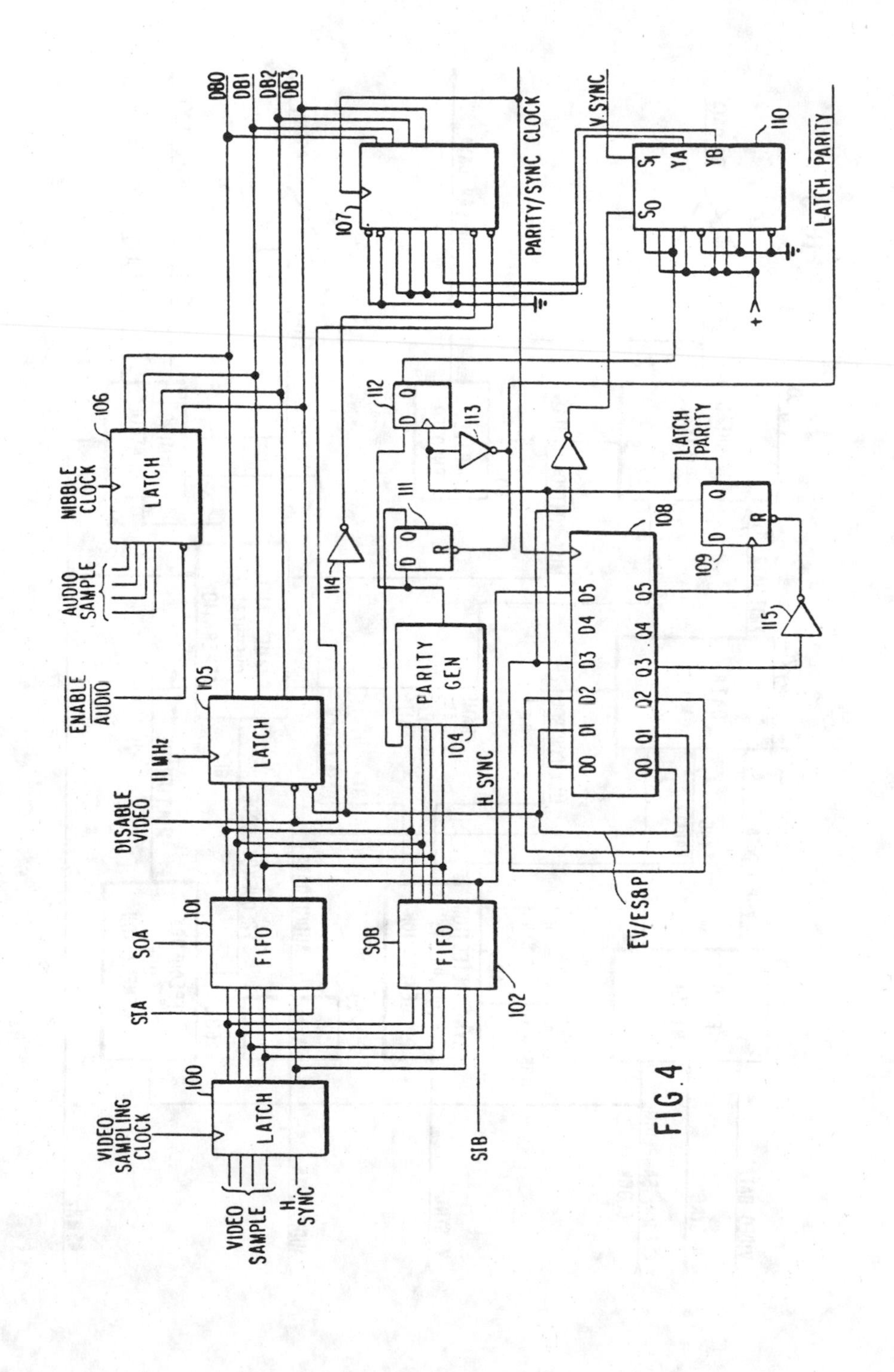
VIDEO SAMPLING CLOCK
LATCH
100
VIDEO SAMPLE
H. SYNC
SIA
SOA
FIFO
101
SOB
FIFO
102
SIB
DISABLE VIDEO
11 MHz
LATCH
105
ENABLE AUDIO
AUDIO SAMPLE
NIBBLE CLOCK
LATCH
106
DB0
DB1
DB2
DB3
107
PARITY/SYNC CLOCK
V.SYNC
110
S1
YA
YB
S0
LATCH PARITY
112
D
Q
113
114
111
D
Q
R
PARITY GEN
104
H. SYNC
108
D0 D1 D2 D3 D4 D5
Q0 Q1 Q2 Q3 Q4 Q5
LATCH PARITY
109
D
Q
R
115
EV/ES8P

FIG. 4

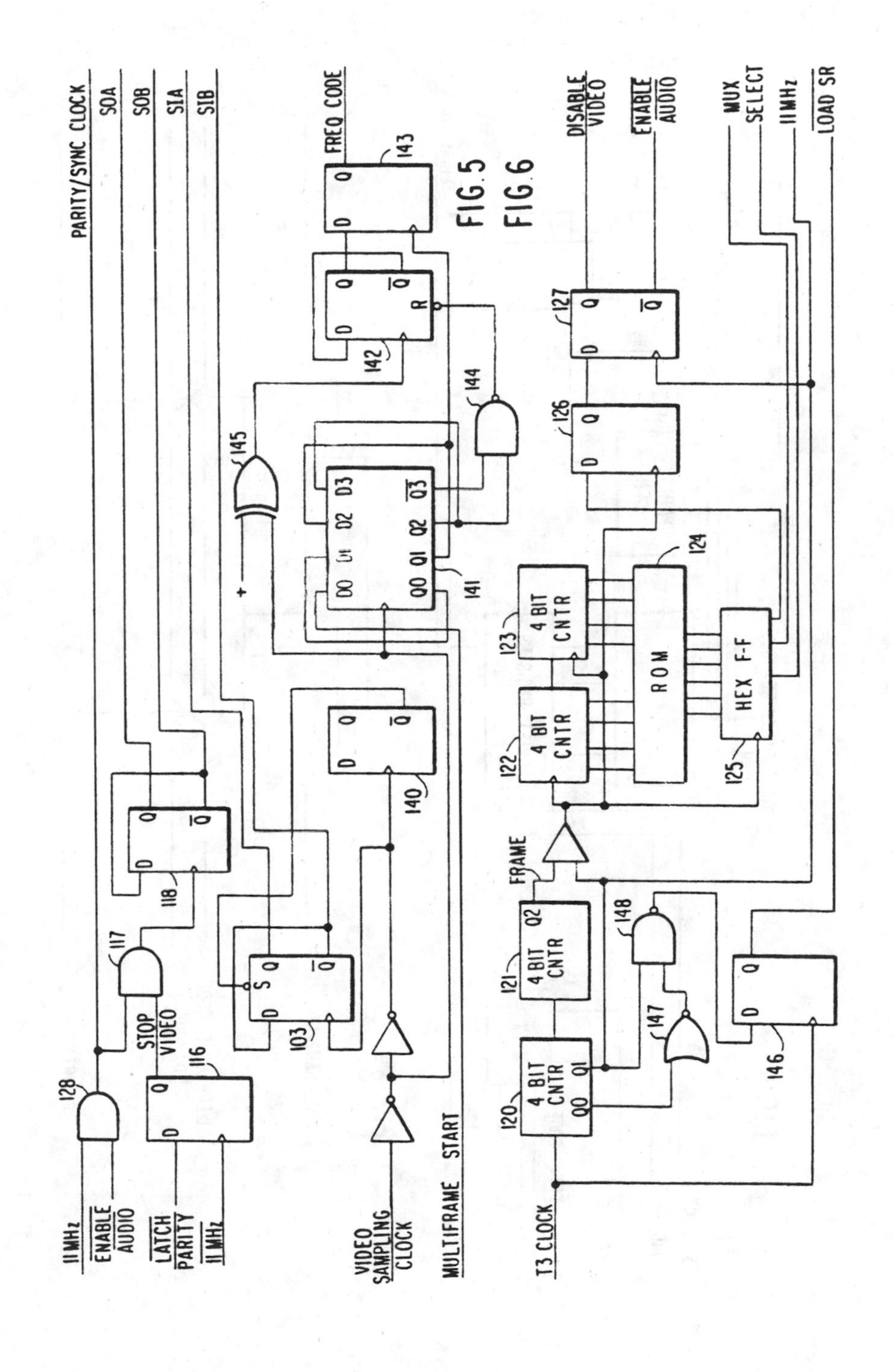
FIG. 5
FIG. 6
PARITY/SYNC CLOCK
SOA
SOB
SIA
SIB
FREQ CODE
DISABLE VIDEO
ENABLE AUDIO
MUX SELECT
11 MHz
LOAD SR
11 MHz
ENABLE AUDIO
LATCH PARITY
11 MHz
STOP VIDEO
VIDEO SAMPLING CLOCK
MULTIFRAME START
T3 CLOCK
FRAME
4 BIT CNTR
ROM
HEX F-F
103
116
117
118
120
121
122
123
124
125
126
127
128
140
141
142
143
144
145
146
147
148

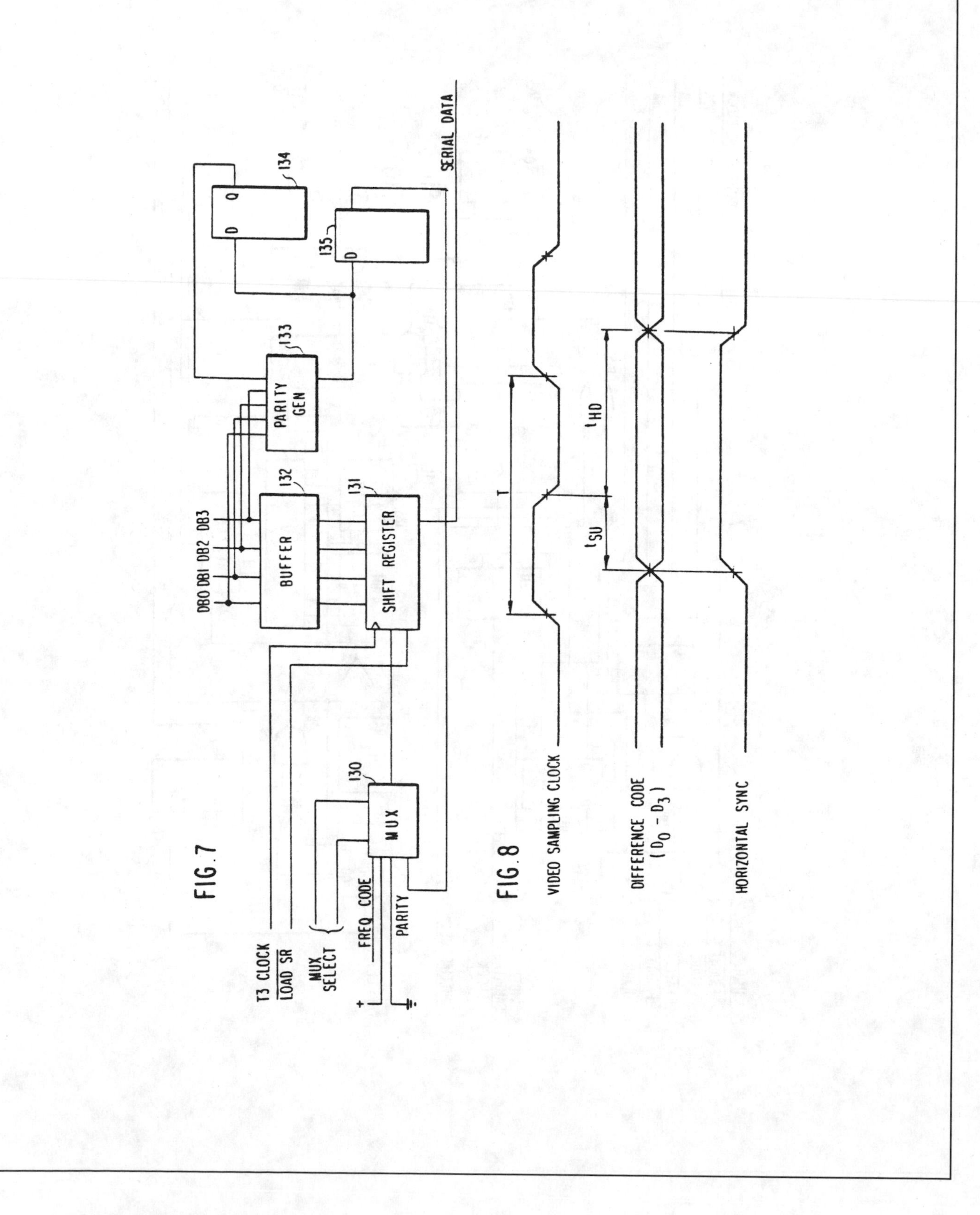
FIG. 7
T3 CLOCK
LOAD SR
MUX SELECT
FREQ CODE
PARITY
MUX
130
BUFFER
132
SHIFT REGISTER
131
PARITY GEN
133
DB0 DB1 DB2 DB3
D
Q
134
135
SERIAL DATA
FIG. 8
VIDEO SAMPLING CLOCK
DIFFERENCE CODE
(D_0 - D_3)
HORIZONTAL SYNC
T
t_{HD}
t_{SU}

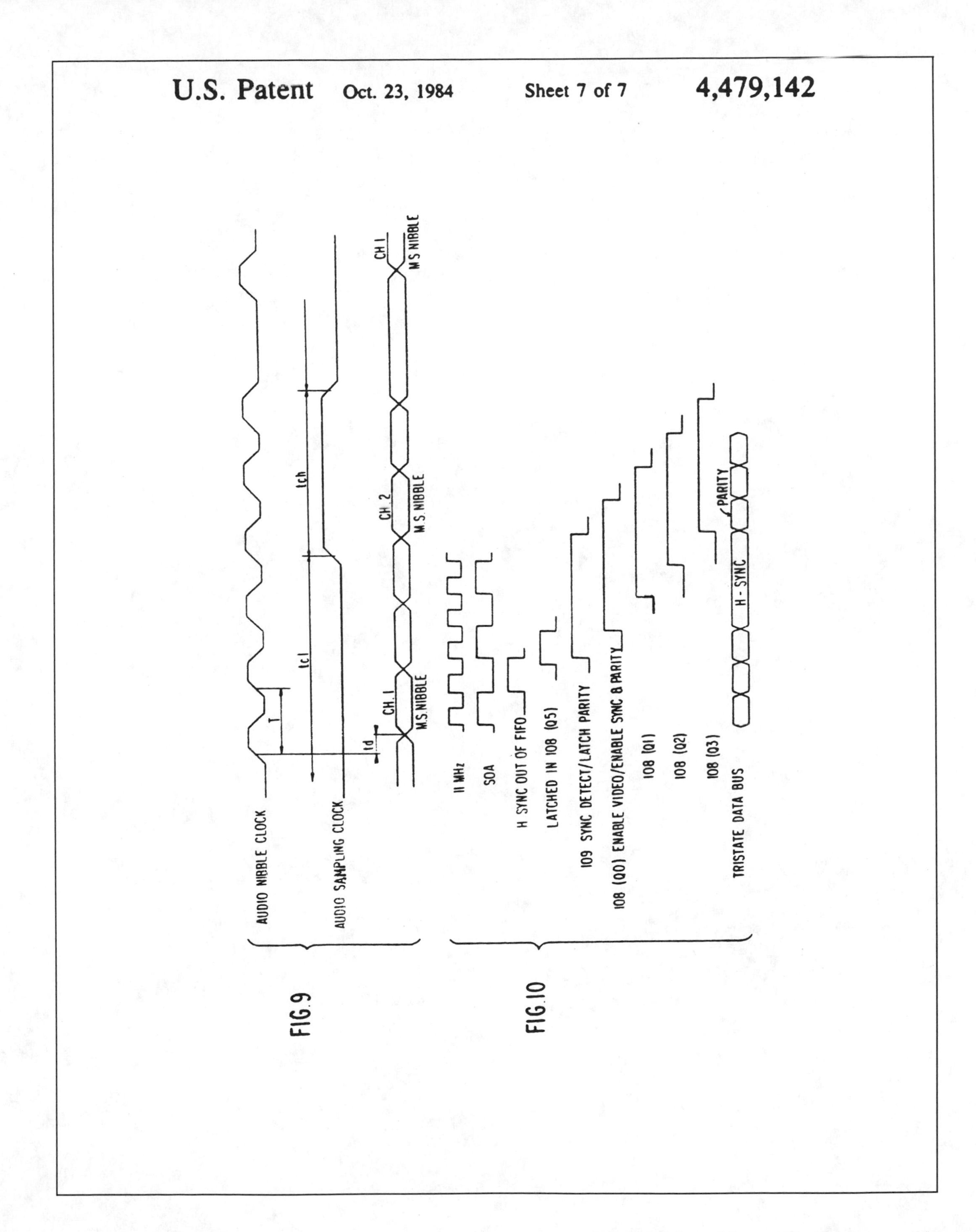
U.S. Patent
Oct. 23, 1984
Sheet 7 of 7
4,479,142
FIG.9
AUDIO NIBBLE CLOCK
AUDIO SAMPLING CLOCK
T
tcl
tch
td
CH. 1
CH. 2
CH. 1
M.S. NIBBLE
M.S. NIBBLE
M.S. NIBBLE
FIG.10
11 MHz
SOA
H SYNC OUT OF FIFO
LATCHED IN 108 (Q5)
109 SYNC DETECT/LATCH PARITY
108 (Q0) ENABLE VIDEO/ENABLE SYNC & PARITY
108 (Q1)
108 (Q2)
108 (Q3)
TRISTATE DATA BUS
H - SYNC
PARITY

15

Technical Data

This chapter is for the reader who wishes to know more about video levels, video frequencies, and the worldwide television standards. Some documents which might interest such a reader are the EIA Standards. These standards which are designed to help eliminate misunderstandings between the manufacturer and the purchaser and to assist the purchaser in selecting the proper product for a particular need.

EIA Standard RS-240, Electrical Performance Standards for Television Broadcast Transmitters, which contains definitions and other pertinent information on transmitters for the TV broadcast channels (Channels 2 through 83), will be of interest to technicians and electronic hobbyists. However, it is not included in this book as it is quite lengthy. Interested readers can obtain copies from:

Electronics Industries Association
Engineering Department
2001 Eye Street, N.W.
Washington, DC 20006

TV Channels

The TV channel allocations for various countries worldwide are given in Tables 15-1 through 15-10. This information should be of particular interest to technicians, hobbyists, and other radio/TV listeners, because, in addition to the channel numbers and their bandwidth frequencies, the color, sound, and picture frequencies are also given.

Table 15-1. US & Canada

CHANNEL	BW(MHz)	PIX	COLOR	SOUND
		SUB-BAND		
T-7	5.75-11.75	7.00	10.58	11.50
T-8	11.75-17.75	13.00	16.58	17.50
T-9	17.75-23.75	19.00	22.58	23.50
T-10	23.75-29.75	25.00	28.58	29.50
T-11	29.75-35.75	31.00	34.58	35.50
T-12	35.75-41.75	37.00	40.58	41.50
T-13	41.75-47.75	43.00	46.58	47.50
		VHF-LOW BAND		
TVIF	40-46	41.25	44.83	45.75
2	54-60	55.25	58.83	59.75
3	60-66	61.25	64.83	65.75
4	66-72	67.25	70.83	71.75
5	76-82	77.25	80.83	81.75
6	82-88	83.25	86.83	87.75
		*FM (PSEUDO)**		
FM-1	88-94	89.25	92.83	93.75
FM-2	94-100	95.75	98.83	99.75
FM-3	100-106	101.25	104.83	105.75
		VHF-MID BAND		
A-2	108-114	109.25	112.83	113.75
A-1	114-120	115.25	118.83	119.75
A	120-126	121.25	124.83	125.75
B	126-132	127.25	130.83	131.75
C	132-138	133.25	136.83	137.75
D	138-144	139.25	142.83	143.75
E	144-150	145.25	148.83	149.75
F	150-156	151.25	154.83	155.75
G	156-162	157.25	160.83	161.75
H	162-168	163.25	166.83	167.75
I	168-174	169.25	172.83	173.75
		VHF-HIGH BAND		
7	174-180	175.25	178.83	179.75
8	180-186	181.25	184.83	185.75
9	186-192	187.25	190.83	191.75
10	192-198	193.25	196.83	197.75
11	198-204	199.25	202.83	203.75
12	204-210	205.25	208.83	209.75
13	210-216	211.25	214.83	215.75

(continued)

Table 15-1. (cont.)

CHANNEL	BW(MHz)	PIX	COLOR	SOUND
		VHF-SUPER BAND		
J	216-222	217.25	220.83	221.75
K	222-228	223.25	226.83	227.75
L	228-234	229.25	232.83	233.75
M	234-240	235.25	238.83	239.75
N	240-246	241.25	244.83	245.75
O	246-252	247.25	250.83	251.75
P	252-258	253.25	256.83	257.75
Q	258-264	259.25	262.83	263.75
R	264-270	265.25	268.83	269.75
S	270-276	271.25	274.83	275.75
T	276-282	277.25	280.83	281.75
U	282-288	283.25	286.83	287.75
V	288-294	289.25	292.83	293.75
W	294-300	295.25	298.83	299.75
		UHF		
14	470-476	471.25	474.83	475.75
15	476-482	477.25	480.83	481.75
16	482-288	483.25	486.83	487.75
17	488-494	489.25	492.83	493.75
18	494-500	495.25	498.83	499.75
19	500-506	501.25	504.83	505.75
20	506-512	507.25	510.83	511.75
21	512-518	513.25	516.83	517.75
22	518-524	519.25	522.83	523.75
23	524-530	525.25	528.83	529.75
24	530-536	531.25	534.83	535.75
25	536-542	537.25	540.83	541.75
26	542-548	543.25	546.83	547.75
27	548-554	549.25	552.83	553.75
28	554-560	555.25	558.83	559.75
29	560-566	561.25	564.83	565.75
30	566-572	567.25	570.83	571.75
31	572-578	573.25	576.83	577.75
32	578-584	579.25	582.83	583.75
33	584-590	585.25	588.83	589.75
34	590-596	591.25	594.83	595.75
35	596-602	597.25	600.83	601.75
36	602-608	603.25	606.83	607.75
37	608-614	609.25	612.83	613.75
38	614-620	615.25	618.83	619.75
39	620-626	621.25	624.83	625.75
40	626-632	627.25	630.83	631.75
41	632-638	633.25	636.83	637.75
42	638-644	639.25	642.83	643.75
43	644-650	645.25	648.83	649.75
44	650-656	651.25	654.83	655.75
45	656-662	657.25	660.83	661.75
46	662-668	663.25	666.83	667.75
47	668-674	669.25	672.83	673.75
48	674-680	675.25	678.83	679.75
49	680-686	681.25	684.83	685.75
50	686-692	687.25	690.83	691.75
51	692-698	693.25	696.83	697.75
52	698-704	699.25	702.83	703.75
53	704-710	705.25	708.83	709.75
54	710-716	711.25	714.83	715.75
55	716-722	717.25	720.83	721.75
56	722-728	723.25	726.83	727.75
57	728-734	729.25	732.83	733.75
58	734-740	735.25	738.83	739.75
59	740-746	741.25	744.83	745.75
60	746-752	747.25	750.83	751.75
61	752-758	753.25	756.83	757.75
62	758-764	759.25	762.83	763.75
63	764-770	765.25	768.83	769.75
64	770-776	771.25	774.83	775.75
65	776-782	777.25	780.83	781.75
66	782-788	783.25	786.83	787.75
67	788-794	789.25	792.83	793.75
68	794-800	795.25	798.83	799.75
69	800-806	801.25	804.83	805.75
70**	806-812	807.25	810.83	811.75
71	812-818	813.25	816.83	817.75
72	818-824	819.25	822.83	823.75
73	824-830	825.25	828.83	829.75
74	830-836	831.25	834.83	835.75
75	836-842	837.25	840.83	841.75
76	842-848	843.25	846.83	847.75
77	848-854	849.25	852.83	853.75
78	854-860	855.25	858.83	859.75
79	860-866	861.25	864.83	865.75
80	866-872	867.25	870.83	871.75
81	872-878	873.25	876.83	877.75
82	878-884	879.25	882.83	883.75
83	884-890	885.25	888.83	889.75

*Official status unknown. Some CATV operators are using these as extra converter positions.

**The frequencies between 806 and 890 MHz, formerly allocated to television broadcasting, are now allocated to the land mobile services. Operation, on a secondary basis, of some television translators may continue on these frequencies.

Table 15-2. Western European (E) Channels

CHANNEL	BW(MHz)	PIX	COLOR	SOUND
2	47–54	48.25	52.68	53.75
3	54-61	55.25	59.68	60.75
4	61-68	62.25	66.68	67.75
S-3	118-125	119.25	123.68	124.75
S-4	125-132	126.25	130.68	131.75
S-5	132-139	133.25	137.68	138.75
S-6	139-146	140.25	144.68	145.75
S-7	146-153	147.25	151.68	152.75
S-8	153-160	154.25	158.68	159.75
S-9	160-167	161.25	165.68	166.75
S-10	167-174	168.25	172.68	173.75
5	174-181	175.25	179.68	180.75
6	182-188	182.25	186.68	187.75
7	188-195	189.25	193.68	194.78
8	195-202	196.25	200.68	201.75
9	202-209	203.25	207.68	208.75
10	209-216	210.25	214.68	215.75
11	216-223	217.25	221.68	222.75
12	223-230	224.25	228.68	229.75
S-11	230-237	231.25	235.68	236.75
S-12	237-244	238.25	242.68	243.75
S-13	244-251	245.25	249.68	250.75
S-14	251-258	252.25	256.68	257.75
S-15	258-265	259.25	263.68	264.75
S-16	265-272	266.25	270.68	271.75
S-17	272-279	273.25	277.68	278.75
21	470-478	471.25	475.68	476.75
22	478-486	479.25	483.68	484.75
23	486-494	487.25	491.68	492.75
24	494-502	495.25	499.68	500.75
25	502-510	503.25	507.68	508.75
26	510-518	511.25	515.68	516.75
27	518-526	519.25	523.68	524.75
28	526-534	527.25	531.68	532.75
29	534-542	535.25	539.68	540.75
30	542-550	543.25	547.68	548.75
31	550-558	551.25	555.68	556.75
32	558-566	559.25	563.68	564.75

Table 15-2. (cont.)

CHANNEL	BW(MHz)	PIX	COLOR	SOUND
33	566-574	567.25	571.68	572.75
34	574-582	575.25	579.68	580.75
35	582-590	583.25	587.68	588.75
36	590-598	591.25	595.68	596.75
37	598-606	599.25	603.68	604.75
38	606-614	607.25	611.68	612.75
39	614-622	615.25	619.68	620.75
40	622-630	623.25	627.68	628.75
41	630-638	631.25	635.68	636.75
42	638-646	639.25	643.68	644.75
43	646-654	647.25	651.68	652.75
44	654-662	655.25	659.68	660.75
45	662-670	663.25	667.68	668.75
46	670-678	671.25	675.68	676.75
47	678-686	679.25	683.68	684.75
48	686-694	687.25	691.68	692.75
49	694-702	695.25	699.68	700.75
50	702-710	703.25	707.68	708.75
51	710-718	711.25	715.68	716.75
52	718-726	719.25	723.68	724.75
53	726-734	727.25	731.68	732.75
54	734-742	735.25	739.68	740.75
55	742-750	743.25	747.68	748.75
56	750-758	751.25	755.68	756.75
57	758-766	759.25	763.68	764.75
58	766-774	767.25	771.68	772.75
59	774-782	775.25	779.68	780.75
60	782-790	783.25	787.68	788.75
61	790-798	791.25	795.68	796.75
62	798-806	799.25	803.68	804.75
63	806-814	807.25	811.68	812.75
64	814-822	815.25	819.68	820.75
65	822-830	823.25	827.68	828.75
66	830-838	831.25	835.68	836.75
67	838-846	839.25	843.68	844.75
68	846-854	847.25	851.68	852.75
69	854-862	855.25	859.68	860.75

Table 15-3. French

CHANNEL	PIX	SOUND
F-2	52.40	41.25
F-4	65.55	54.40
F-5	164.00	175.15
F-6	173.40	162.25
F-7	177.15	188.30
F-8A	185.25	174.10
F-8	186.55	175.40
F-9	190.30	201.45
F-10	199.70	188.55
F-11	203.45	214.60
F-12	212.85	201.70

Table 15-4. French Overseas

CHANNEL	PIX	SOUND
F-4	175.25	161.75
F-5	183.25	180.75
F-6	191.25	197.75
F-7	199.25	205.75
F-8	207.25	213.75
F-9	215.25	221.75

Table 15-5. Eastern European

CHANNEL	PIX	SOUND
OIRT-1	49.75	56.25
OIRT-2	59.25	65.75
OIRT-3	77.25	83.75
OIRT-4	85.25	91.75
OIRT-5	93.25	99.75
OIRT-6	175.25	181.75
OIRT-7	183.25	189.75
OIRT-8	191.25	197.75
OIRT-9	199.25	205.75
OIRT-10	207.25	213.75
OIRT-11	215.25	221.75
OIRT-12	223.25	229.75

Table 15-6. Japanese

CHANNEL	PIX	SOUND
J-1	91.25	95.75
J-2	97.25	101.75
J-3	103.25	107.75
J-4	171.25	175.75
J-5	177.25	181.75
J-6	183.25	187.75
J-7	189.25	193.75
J-8	193.25	197.75
J-9	199.25	203.75
J-10	205.25	209.75
J-11	211.25	215.75
J-12	217.25	221.75
J-45	663.25	667.75
J-46	669.25	673.75
J-47	675.25	679.75
J-48	681.25	685.75
J-49	687.25	691.75
J-50	693.25	697.75
J-51	699.25	703.75
J-52	705.25	709.75
J-53	711.25	715.75
J-54	717.25	721.75
J-55	723.25	727.75
J-56	729.25	733.75
J-57	735.25	739.75
J-58	741.25	745.75
J-59	747.25	751.75
J-60	753.25	757.75
J-61	759.25	763.75
J-62	765.25	769.75

Table 15-7. Italian

CHANNEL	PIX	SOUND
A	53.75	59.25
B	62.25	67.75
C	82.25	87.75
D	175.25	180.75
E	183.75	189.25
F	192.25	197.75
G	201.25	206.75
H	210.25	215.75
H-1	217.25	222.75

Table 15-8. Australian

CHANNEL	PIX	SOUND
A-0	46.25	51.75
A-1	57.25	62.75
A-2	64.25	69.75
A-3	86.25	91.75
A-4	95.25	100.75
A-5	102.25	107.75
A-5A	138.25	143.75
A-6	175.25	180.75
A-7	182.25	187.75
A-8	189.25	194.75
A-9	196.25	201.75
A-10	209.25	214.75
A-11	215.00	222.00

Table 15-9. British

CHANNEL	PIX	SOUND
B-1	45.00	41.50
B-2	51.75	48.25
B-3	56.75	53.25
B-4	61.75	58.25
B-5	66.75	63.25
B-6	179.75	176.25
B-7	184.75	181.25
B-8	189.75	186.25
B-9	194.75	191.25
B-10	199.75	196.25
B-11	204.75	201.25
B-12	209.75	206.25
B-13	214.75	211.25
B-14	219.75	216.25

Table 15-10. South African Channels

CHANNEL	BW(MHz)	PIX	SOUND
4	174-182	175.25	181.25
5	182-190	183.25	189.25
6	190-198	191.25	197.25
7	198-206	199.25	205.25
8	206-214	207.25	213.25
9	214-222	215.25	221.25
10	222-230	223.25	229.25
11	230-238	231.25	237.25
13		247.43	253.43
21	470-478	471.25	477.25
22	478-486	479.25	485.25
23	486-494	487.25	493.25
24	494-502	495.25	501.25
25	502-510	503.25	509.25
26	510-518	511.25	517.25
27	518-526	519.25	525.25
28	526-534	527.25	533.25
29	534-542	535.25	541.25
30	542-550	543.25	549.25
31	550-558	551.25	557.25
32	558-566	559.25	565.25
33	566-574	567.25	573.25
34	574-582	575.25	581.25
35	582-590	583.25	589.25
36	590-598	591.25	597.25
37	598-606	599.25	605.25
39	614-622	615.25	621.25
40	622-630	623.25	629.25
41	630-638	631.25	637.25
42	638-646	639.25	645.25
43	646-654	647.25	653.25
44	654-662	655.25	661.25
45	662-670	663.25	669.25
46	670-678	671.25	677.25
47	678-686	679.25	685.25
48	686-694	687.25	693.25
49	694-702	695.25	701.25
50	702-710	703.25	709.25
51	710-718	711.25	717.25
52	718-726	719.25	725.25
53	726-734	727.25	733.25
54	734-742	735.25	741.25
55	742-750	743.25	749.25
56	750-758	751.25	757.25
57	758-766	759.25	765.25
58	766-774	767.25	773.25
59	774-782	775.25	781.25
60	782-790	783.25	789.25
61	790-798	791.25	797.25
62	798-806	799.25	805.25
63	806-814	807.25	813.25
64	814-822	815.25	821.25
65	822-830	823.25	829.25
66	830-838	831.25	837.25
67	838-846	839.25	845.25
68	846-854	847.25	853.25

Anatomy of a TV Channel

Amplitude modulation of a carrier results in an upper and a lower sideband. These sidebands are spaced from the carrier with a frequency difference that is equal to the modulating frequency. (For example, if a 54-MHz carrier is amplitude-modulated with a 4-MHz video signal, sidebands will occur at 54 $\pm$4 MHz, or 50 and 58 MHz, respectively.)

Each sideband contains the same intelligence. With standard AM detector circuits, both sidebands are necessary for proper detection. In practice, one sideband (the lower) is reduced in amplitude in order to conserve spectrum space (see FCC standards). This sideband is called the *vestigial* sideband. The NTSC signal spaces the picture and sound carriers by 4.5 MHz, with the color-signal subcarrier set 3.58 MHz from the picture frequency and 920 kHz from the sound frequency. The sound carrier is frequency-modulated with program audio and occupies a total bandwidth of about 60 kHz (mono audio).

Comparison of Video Levels

There are several international standards for specifying the amplitude and video components of a waveform for use with television and video monitors. The *National Television System Committee* (NTSC) was introduced in 1948, with *Sequential a Memoire* (SECAM) being introduced in 1957, and *Phase Alternating Line* (PAL) being introduced in 1961. In North America, the NTSC format is the standard, while the PAL or SECAM video format is used in Europe and Far East.

A comparison of the various video levels is shown in Table 15-11 for reference, and Fig. 15-1 is a drawing of a composite video waveform. The international standards specify a voltage amplitude of 0.714 volts between the blanking level and reference white for the video portion of the waveform and is defined as having 100 IRE units. The reference black to blank level (typically referred to as the setup) is used to shut off the beam during the retrace time and varies between the three formats. The total amplitude is 140 IRE units from synch tips to reference white.

To assist those who need to determine the output current when terminated into 75-ohm (and doubly-terminated 75-ohm) loads, a comparison chart is shown in Table 15-12.

Other items of interest, when comparing the various video formats for number of lines per frame, field frequency, and nominal video bandwidth are summarized in Table 15-13. Figs. 15-2 and 15-3 show the NTSC Field Scanning and Horizontal Scan methods.

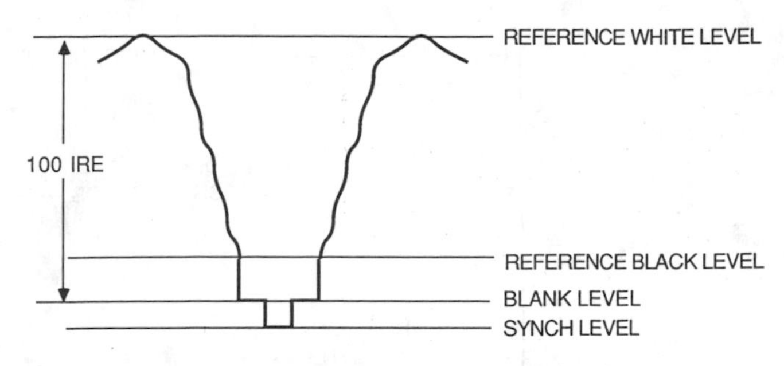

Fig. 15-1. Composite waveform.

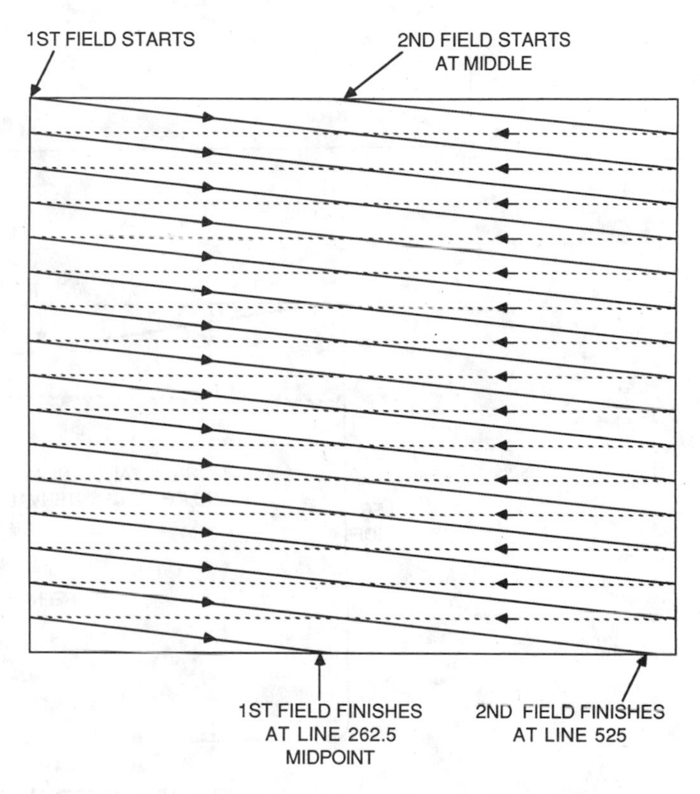

Fig. 15-2. NTSC field scanning and interlacing.

Table 15-11. IRE Levels

Video Levels	NTSC	PAL	SECAM
Blank Level	0	0	0
Reference White	100	100	100
Synch Level	−40	−43	−43
Reference Black Level To Blank Level	5–10	0	0 color 0–7 monochrome

Table 15-12. Comparison Between Video Formats

	Video Output Levels	Voltage	Current in mA (75 ohm)	Current in mA (37.5 ohm)
NTSC (RS-343-A)	Reference White	0.714	9.52	19.1
	Reference Black (10 IRE)	0.071	0.952	1.89
	Blank Level	0.00	0.00	0.00
	Synch Level	−2.86	−3.81	−7.63
PAL	Reference White	0.714	9.52	19.1
	Reference Black (10 IRE)	0.00	0.00	0.00
	Blank Level	0.00	0.00	0.00
	Synch Level	−0.307	−3.81	−7.63
SECAM	Reference White	0.714	9.52	19.1
	Reference Black (10 IRE)	0.049	0.653	1.30
	Blank Level	0.00	0.00	0.00
	Synch Level	−0.307	−4.09	−8.19

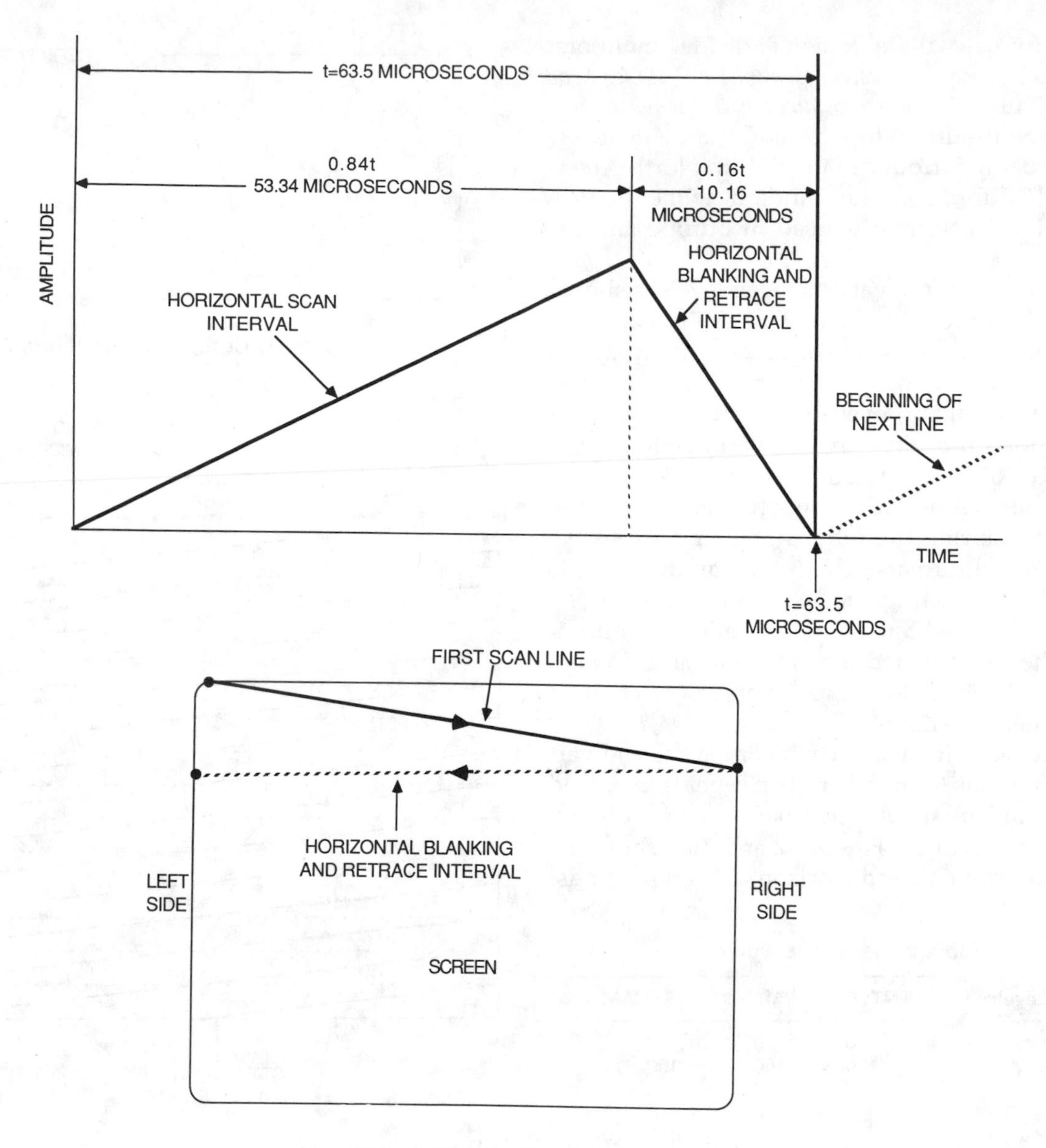

Fig. 15-3. NTSC Horizontal scan cycle.

Table 15-13. Comparison of Various Video Formats

	NTSC	PAL	SECAM
Number of lines per frame	525	625	625
Field frequency	60	50	50
Video bandwidth (MHz)	4.2	5	6

An older standard known as RS-170-A is also used in the United States and it differs from the RS-343-A level for the voltage level from BLANK to REFERENCE WHITE. The RS-170-A white level is 1 volt as compared to the RS-343-A level of 0.714 volt. The RS-170-A levels are listed in Table 15-14.

Table 15-14. RS-170-A Levels

	Video Output Levels	Voltage	Current in mA (75 ohm)	Current in mA (37.5 ohm)
NTSC (RS-170-A)	Reference White	1.00	13.3	26.66
	Reference Black (7.3 IRE)	0.075	1.00	2.00
	Blank Level	0.00	0.00	0.00
	Synch Level	−0.40	−5.33	−10.66

Picture Transmission Characteristics

Fig. 15-4 illustrates the idealized picture transmission amplitude characteristic *before* vestigial-sideband suppression, while Fig. 15-5 shows the amplitude characteristic *after* vestigial-sideband suppression.

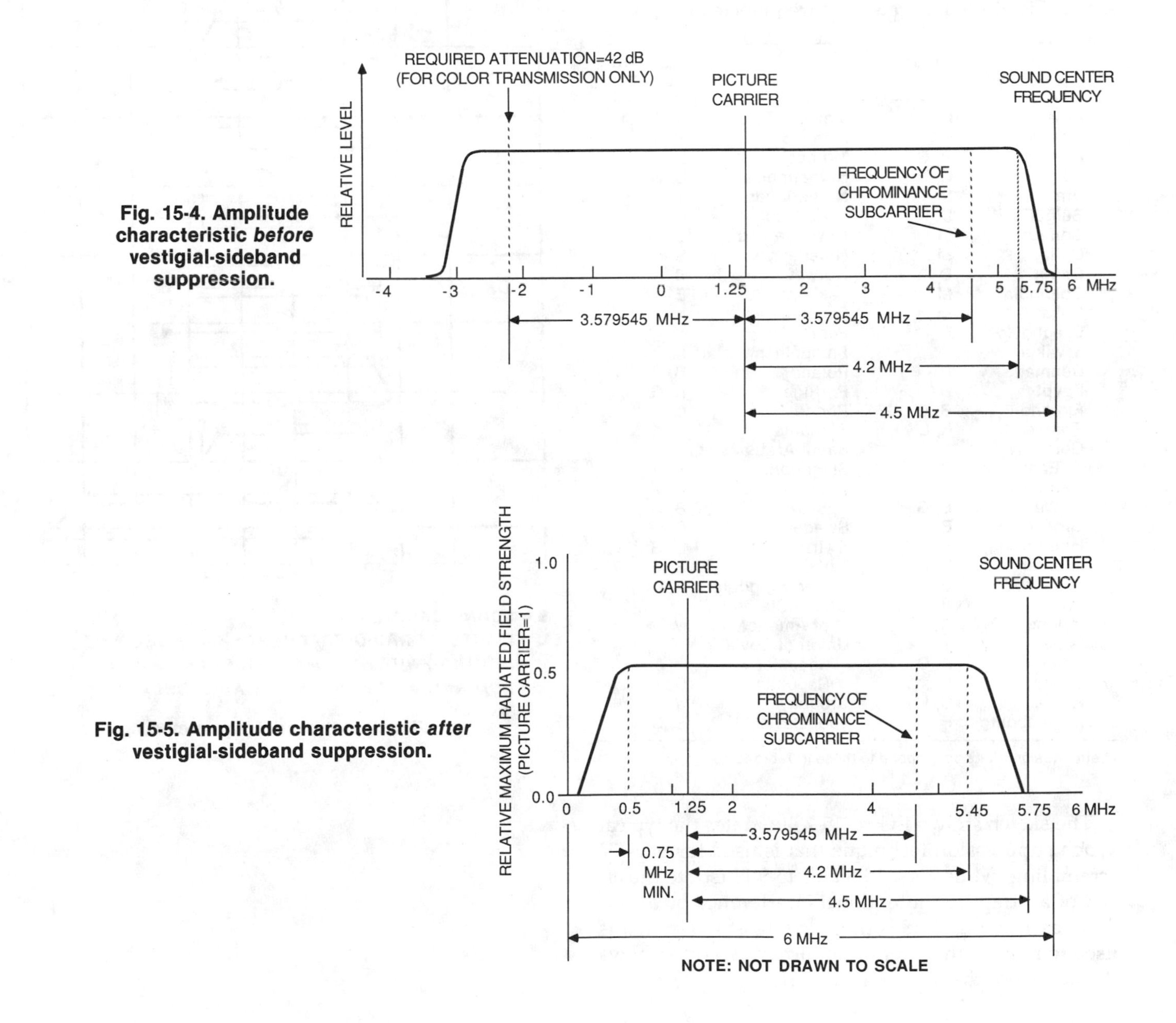

Fig. 15-4. Amplitude characteristic *before* vestigial-sideband suppression.

Fig. 15-5. Amplitude characteristic *after* vestigial-sideband suppression.

Worldwide Television Standards

The following listing outlines pertinent characteristics of the current TV standards used throughout the world. The video frequency-channel arrangements are also shown (Table 15-15). The systems have also been designated by a letter, which shows the systems that are in use or proposed for use in the countries listed. Fig. 15-6 illustrates the frequencies used.

Country	Standard Used*	Country	Standard Used*
Argentina	N	Mexico	M
Australia	B	Monaco	E, G
Austria	B, G	Morocco	B
Belgium	C, H	Netherlands	B, G
Brazil	M	Netherlands Antilles	M
Bulgaria	D, K	New Zealand	B
Canada	M	Nigeria	B
Chile	M	Norway	B
China	D	Pakistan	B
Columbia	M	Panama	M
Cuba	M	Peru	M
Czechoslovakia	D	Philippines	M
Denmark	B	Poland	D
Egypt	B	Portugal	B, G
Finland	B, G	Rhodesia	B
France	E, L	Romania	K
Germany (East)	B	Saudi Arabia	B
		Singapore	B
Germany (West)	B, G	South Africa	I
		Spain	B, G
Greece	B	Sweden	B, G
Hong Kong	B, I	Switzerland	B, G
Hungary	D, K	Turkey	B
India	B	United Kingdom	A, I
Iran	B	United States of America	M
Ireland	A		
Israel	B	Union of Soviet Socialist Republic	D
Italy	B, G		
Japan	M		
Korea	C, L	Uruguay	N
Luxembourg	F	Yugoslavia	B, G

*Letter designations correspond to those in Table 15-15.

The sketch shown in Fig. 15-7 illustrates the typical synch suppression technique that is used for pay-TV scrambling, while Figs. 15-8 and 15-9 illustrate the effect of a "trap" on intentional interference by a jamming carrier. Fig. 15-8 is the drawing of a trap that is used to remove the interfering carrier. Fig. 15-9 shows the curves of the jamming carrier(s).

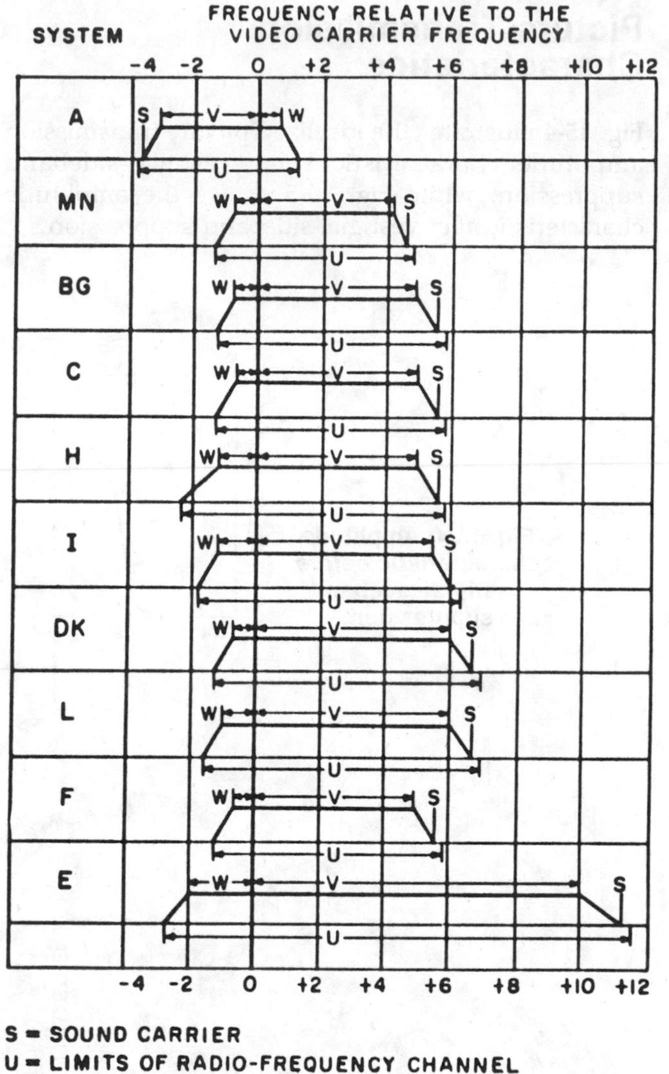

Fig. 15-6. System frequencies.

Table 15-15. Current TV Standards in Use Worldwide*

	A	M	N	B	C	G	H	I	D,K	L	F	E
Lines/frame	405	525	625	625	625	625	625	625	625	625	819	819
Fields/sec	50	60	50	50	50	50	50	50	50	50	50	50
Interlace	2/1	2/1	2/1	2/1	2/1	2/1	2/1	2/1	2/1	2/1	2/1	2/1
Frames/sec	25	30	—	25	25	25	25	25	25	25	25	25
Lines/sec	10,125	15,750	—	15,625	15,625	15,625	15,625	15,625	15,625	15/625	20,475	20,475
Aspect ratio[1]	4/3	4/3	—	4/3	4/3	4/3	4/3	4/3	4/3	4/3	4/3	4/3
Video band (MHz)	3	4.2	4.2	5	5	5	5	5.5	6	6	5	10
RF band (MHz)	5	6	6	7	7	8	8	8	8	8	7	14
Visual polarity[2]	+	—	—	—	+	—	—	—	—	+	+	+
Sound modulation	A3	F3	—	F3	A3	F3	F3	F3	F3	F3	A3	A3
Pre-emphasis in microseconds	—	75	—	50	50	50	50	50	50	—	50	—
Deviation (kHz)	—	25	—	50	—	50	50	50	50	—	—	—
Gamma of picture signal	0.45	0.45	—	0.5	0.5	0.5	0.5	0.5	0.5	0.5	0.5	0.6

Notes:
[1]In all systems, the scanning sequence is from left to right and top to bottom.
[2]All visual carriers are amplitude modulated. Positive polarity indicates that an increase in light intensity causes an increase in radiated power. Negative polarity (as used in the U.S.—Standard *M*) means that a decrease in light intensity causes an increase in radiated power.
*From Electronic Databook, 3rd Edition, by Rudolf F. Graf.

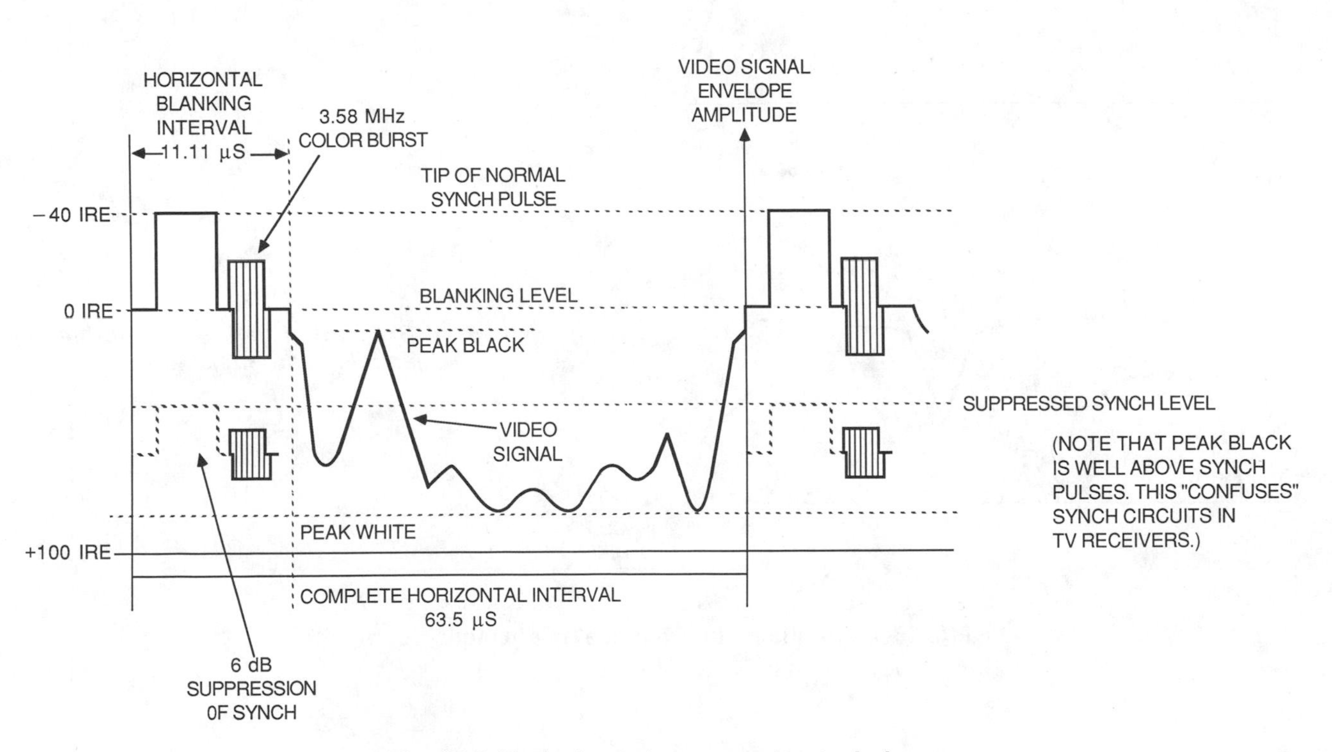

Fig. 15-7. Typical synch suppression technique.

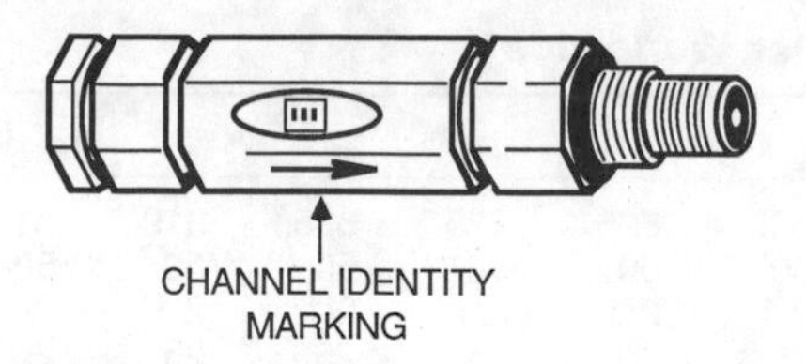

Fig. 15-8. Trap used to remove interfering carrier.

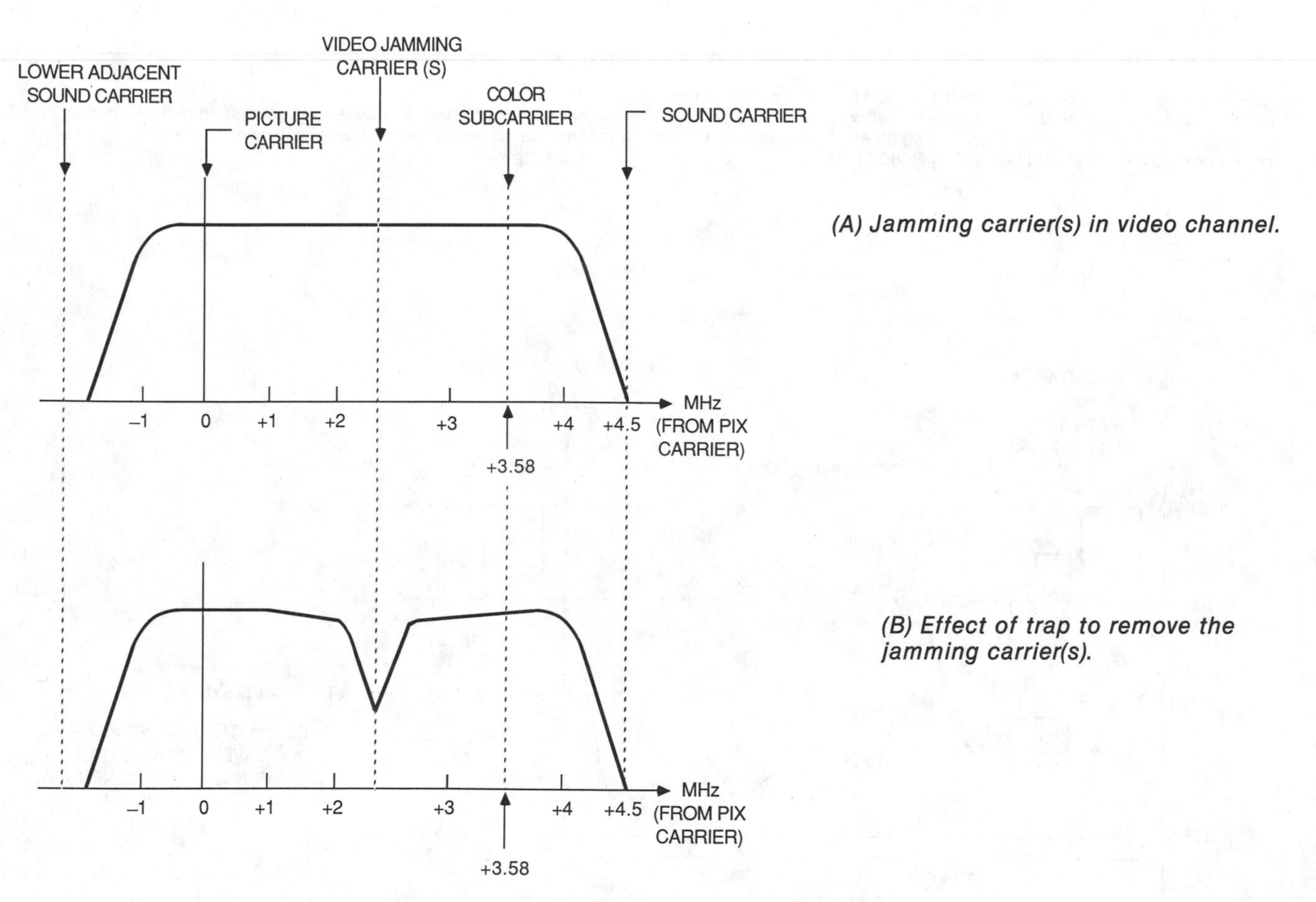

Fig. 15-9. Intentional interference by a jamming carrier.

Frequencies and Related Harmonics

Table 15-16 is useful for the determination of system clock frequencies and subcarrier frequencies for various encoding or decoding schemes.

Table 15-16. NTSC Scan Frequencies, Periods, and Related Harmonics

Horizontal Sweep			Vertical Sweep		
Harmonic	Hz	Period (Microseconds)	Harmonic	Hz	Period (Milliseconds)
0.5	7867.13	127.111	0.5	29.97002476	33.36667
1	15734.26	63.556	1	59.94004952	16.683336
2	31468.52	31.778	2	119.880	8.341
2.5	39335.65	25.422	3	179.820	5.5611
3	47202.79	21.185	4	239.760	4.1708
4	62937.05	15.888	5	299.700	3.3367
5	78671.32	12.7112	6	359.640	2.7806
6	94405.56	10.5926	7	419.580	2.383
7	110139.84	9.07936	8	479.520	2.085
8	125874.10	7.9444	9	539.46	1.8537
9	141608.37	7.0617	10	599.40	1.6683
10	157342.63	6.3556	12	719.280	1.3903
12	188811.16	5.2963	16	959.040	1.0427
16	251748.21	3.9722	24	1438.560	0.6951
20	314685.26	3.1778	32	1918.08	0.52135
24	377622.31	2.6482	64	3836.16	0.26067
32	503496.42	1.98611			
64	1006992.83	0.99305			

COLOR BURST = 3.579545 MHz

Harmonic	MHz
2	7.15909
3	10.738635
4	14.31818

Some Scrambling Systems in Use

Chart 15-1. Scrambling Systems

Trade Name and Type of Scrambling	Scrambing Methods Used	Relative Security Level
COMSAT SATGUARD Analog Audio Analog Video (Satellite)	Suppresses horizontal synch audio, combined with complex digital 15.734-kHz waveform. Video line inversion, with individual lines specifically.	Moderately high video and audio.
TELEASE Analog Audio Analog Video (Satellite)	Sine wave (94 kHz) added to video. Audio is modulated on 15.6-kHz (approximately) subcarrier.	Moderate.
OAK SYSTEMS	Video may be inverted; synch missing.	Audio high.
1. *Oak Orion* Digital Audio Analog Video (Satellite)	Color burst moved to nonstandard location; random video polarity sequences. Audio is digital, with data pulses in horizontal blanking interval. Audio may be further encrypted.	Moderate video.
2. *Oak Polaris* Digital Audio Digital Video (Satellite)	Video lines shuffled, DES encrypted. Audio in horizontal blanking interval is digital, DES encrypted.	Very high audio. Very high video.
3. *Oak (sine-wave cable)* Analog Audio Analog Video	Video modulated with 15.7-kHz or 31.5-kHz sine wave; audio is on 62.5-kHz subcarrrier.	Low to moderate.
4. *Oak SIGMA (cable)* Analog Video Digital Audio	Video randomly inverted on scene changes. Synch omitted; vertical and horizontal synch intervals have digitized encrypted audio. Decryption keys are on separate carrier (cable).	Moderate video. Very high audio.
GI STAR-LOK Digital Video Digital Audio (Satellite)	Synch and color burst are removed, and scanning lines diced and rotated according to DES algorithm. Audio is digital and DES encrypted.	Very high video. Very high audio.
MA-COM LINKABIT, INC.	Video lines diced in variable size segments and DES encrypted.	Very high video. Very high audio.
1. *VideoCipher I™* Digital Video Digital Audio (Satellite)	Audio is digital and DES encrypted.	
2. *VideoCipher II™* Analog Video Digital Audio (Satellite)	Video inverted synch is removed; color burst is relocated. Audio is digital, in the horizontal blanking intervals, and DES encrypted.	Moderate video. Very high audio.
JERROLD STARCOM 1. (Analog Video Cable)	Synch suppression is pseudo random (6 to 10 dB); audio is clear.	Low-to-moderate video. Audio is clear.
TOCOM Analog Video Clear Audio (Cable)	Video randomly inverted; random synch suppression and video may be compressed. Audio is clear.	Low to moderate video. Audio is clear.

Chart 15-1. (cont.) Scrambling Systems

Trade Name and Type of Scrambling	Scrambing Methods Used	Relative Security Level
ZENITH SSAVI, z TAC Analog Audio Analog Video (Cable, over the air)	Video randomly inverted. Audio on subcarrier is synch suppressed.	Moderate video.
HAMLIN Analog Video Analog Audio (Cable)	Random synch suppression similar to SSAVI.	Moderate video and audio.
SCIENTIFIC ATLANTA B-MAC Digital Video Digital Audio (Satellite)	Video time compressed, luminance and chroma sent separately. Synch and data sent in random-width packets of digitized audio. Non-NTSC format.	High video. High audio. Very high IF DES encryption is used.
SINE WAVE (Analog) From various manufacturers. (Cable, over the air UHF)	Video has interfering sine wave superimposed. Audio is on 62-kHz subcarrier.	Moderately low video and audio.
GATED SYNCH (Analog) From various manufacturers. (Cable, over the air UHF)	Synch pulses suppressed by fixed amount. Audio is on a 15-kHz subcarrier.	Low video and audio. Decoder is simple to build.
OUTBAND (Cable) Analog Video and Audio Various manufacturers (Cable)	Synch pulses on separate video carrier that is usually in FM broadcast band. Audio may be in clear. Main program video has synch pulses removed.	Low video. No audio. Decoder is simple to build.
"NASA" SYSTEM Analog Video (Satellite)	Horizontal lines shifted in groups. Video polarity inverted in various lines. Audio may or may not be encrypted.	Moderate video. No audio unless encrypted.

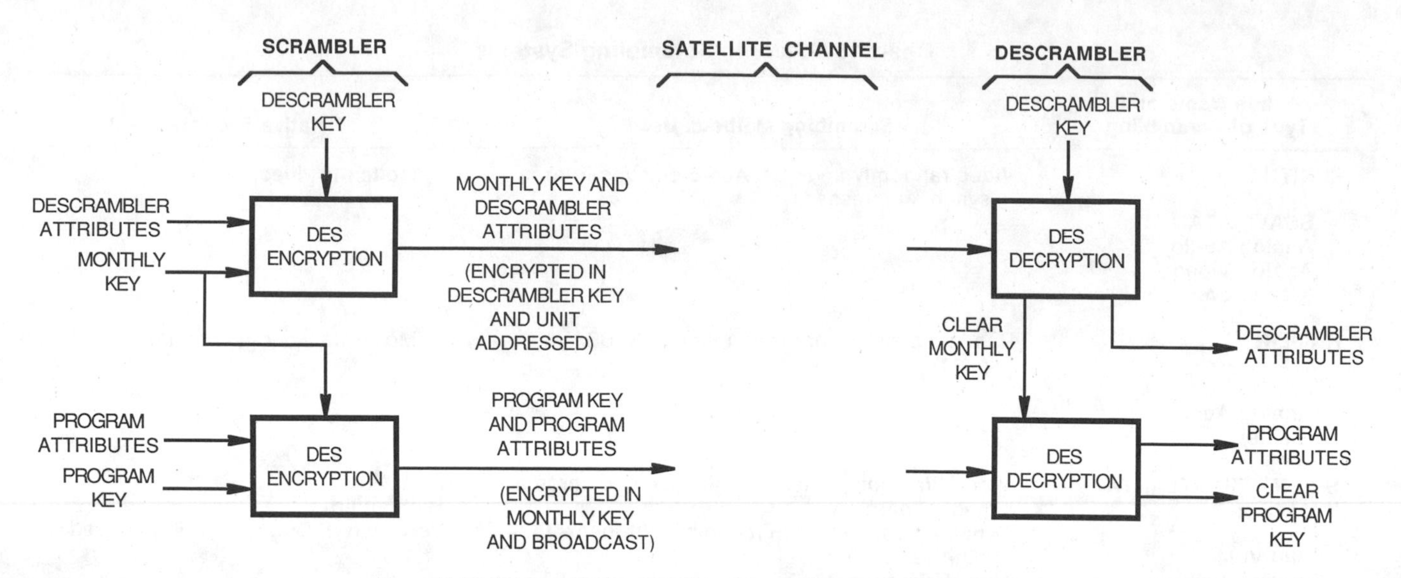

Fig. 15-10. VideoCipher II™ Key hierarchy and distribution.

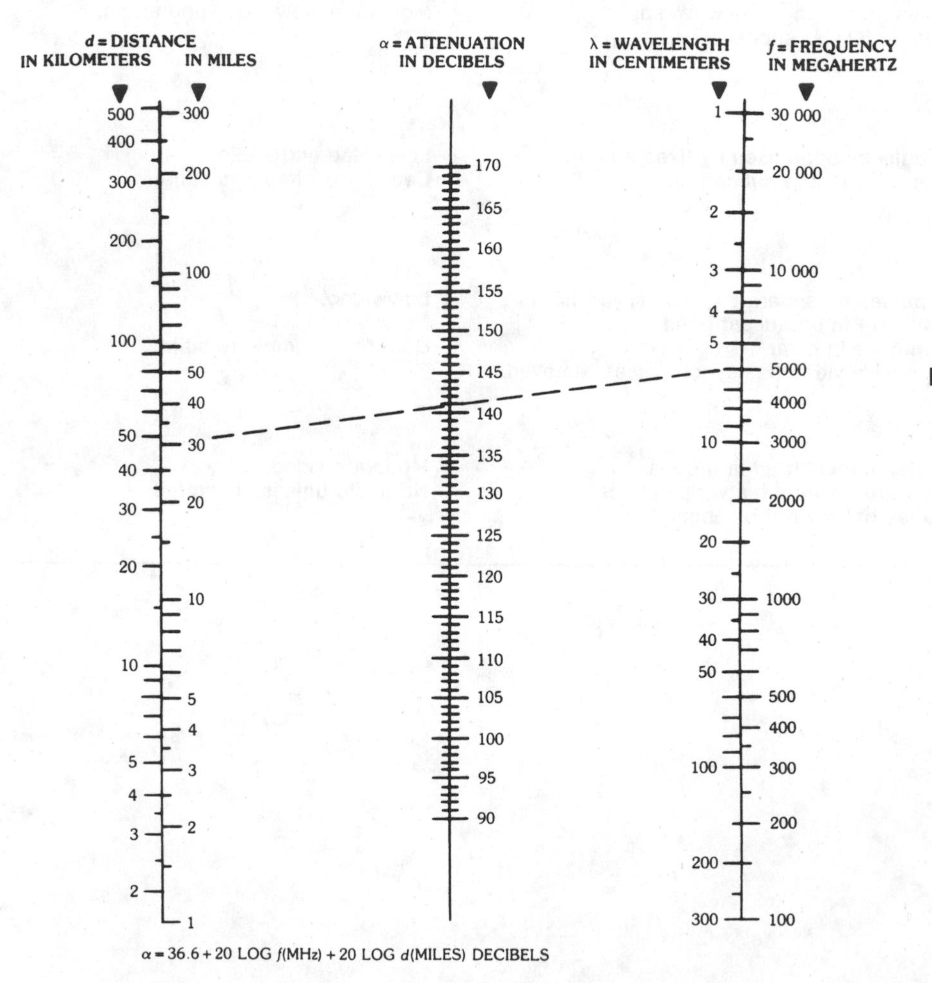

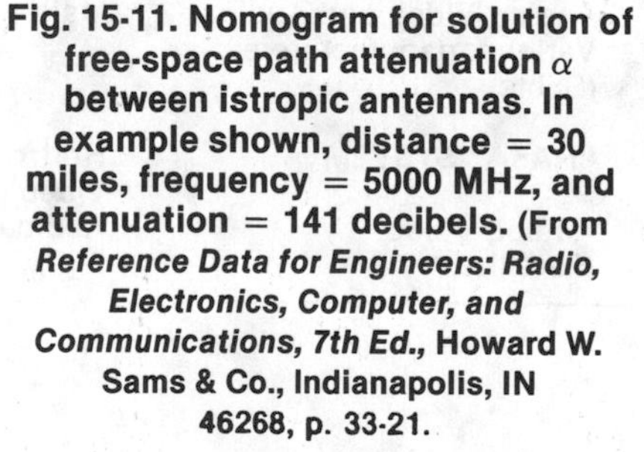

Fig. 15-11. Nomogram for solution of free-space path attenuation α between istropic antennas. In example shown, distance = 30 miles, frequency = 5000 MHz, and attenuation = 141 decibels. (From *Reference Data for Engineers: Radio, Electronics, Computer, and Communications, 7th Ed.*, Howard W. Sams & Co., Indianapolis, IN 46268, p. 33-21.

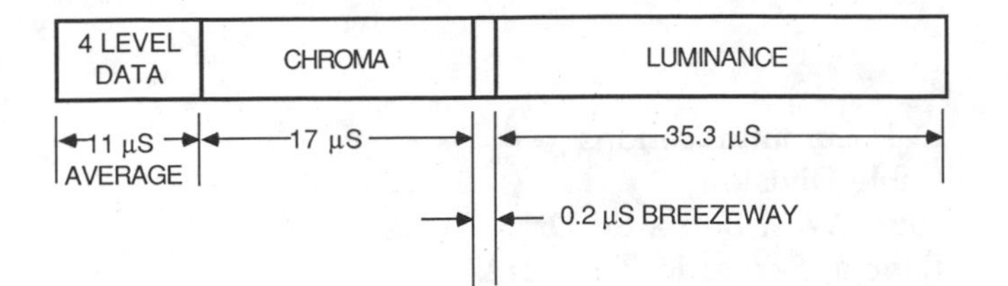

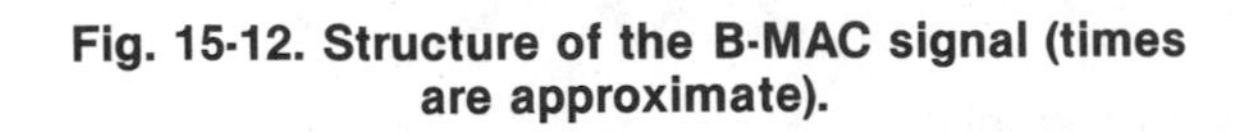

Fig. 15-12. Structure of the B-MAC signal (times are approximate).

Cable Networks

The following is a list of the cable networks that are fed from satellites.

Program	Approximate Number of Affiliate Stations	Estimated Number of Subscribers
AP Cable News	450	4,700,000
Black Entertainment Television (BET)	180	3,700,000
Bravo	1,600	2,000,000
Cable Health Network (CHN)	1,000	10,300,000
Cable News Network (CNN)	3,400	19,200,000
Cinemax	435	100,000
CNN Headline News (CNN2)	500	3,700,000
C-SPAN America's Network	1,000	13,000,000
CBN Cable Network	3,830	20,770,000
Daytime (Hearst/ABC)	675	9,200,000
Disney Channel	23	200,000
Dow Jones Cable News	110	1,115,000
Entertainment and Sports Programming Network (ESPN)	5,800	22,350,000
EROS	160	125,000
Eternal Word Television Network (EWTN)	72	1,124,000
Financial News Network (FNN)	820	8,200,000
Home Box Office (HBO)	340	160,000
Home Theater Network Plus (HTN)	2,500	2,350,000
Mens World	68	83,000
Modern Satellite Network (MSN)	520	7,805,020
The Movie Channel (TMC)	310	500,000
Music Television (MTV)	1,500	13,000,000
Nashville Network (TNN)	725	7,500,000
National Christian Network (NCN)	100	1,500,000
National Jewish Television (NJTV)	110	2,550,000
Nickelodeon	2,200	11,400,000
ON TV	240	420,000
Playboy Channel	35	225,000
PTL Satellite Network	725	7,500,000
Reuters Monitor Service	15	5,000
Reuters News View	375	3,500,000
Satellite News Channel (SNC)	600	5,700,000
Satellite Programming Network (SPN)	375	5,600,000
SelecTv™	2,400	4,000,000
Showtime	240	750,000
Trinity Broadcasting Network	245	2,874,000

Equipment Manufacturers

Conifer
1400 North Roosevelt Avenue
Burlington, IA 52601
(319) 752-3607

General Instruments Corporation
2200 Byberry Road
Hatboro, PA 19040
(215) 674-4800

Hamlin
13610 First Avenue South
Seattle, WA 98168
(206) 246-9330

Microwave Filter Company
6743 Kinne Street
East Syracuse, NY 13057
(800) 525-5571

M/A-COM LINKABIT, Inc.
Cable Com Group
P.O. Box 1729
Hickory, NC 28603
(800) 438-3331

Oak Communications
Satellite Systems Division
P.O. Box 517
Crystal Lake, IL 60014
(815) 459-5000

Oak Communications
Cable Division
16935 West Bernardo Drive
Rancho Bernardo, CA 92127
(619) 485-9880

Pico Products
103 Commerce Blvd.
Liverpool, NY 13088
(800) 822-7420

Scientific Atlanta
4356 Communications Drive
Norcross, GA 30093
(404) 925-5778

(Scientific Atlanta mailing address)
P.O. Box 105027
Atlanta, GA 30348

Tekscan
1440 Goodyear Drive
El Paso, TX 79936
(915) 594-3555

Zenith Electronics Corporation
1000 Milwaukee Avenue
Glenview, IL 60025
(312) 391-8338

Glossary

Access Time—The time it takes a computer to produce a bit of information from its memory section (also called "read time,") or, the time it takes a computer to store information in its memory section (also called "write time.")

Active Component—A device whose output is dependent on a source of power other than the input signal.

A/D—Abbreviation for analog-to-digital.

A/D Converter—A circuit that converts an analog signal sample to a digital representation that is suitable for digital processing and switching.

Address—A series of bits that denotes the location of information in a computer memory. (Also called *Address Code.*)

AFC—Abbreviation for automatic frequency control. An electronic circuit common in radios, television sets, satellite receivers, etc., which locks onto the desired frequency and keeps the receiver from drifting off frequency (channel).

AGC—Abbreviation for automatic gain control. A control input to an amplifier that is used to control its gain, usually to keep the amplifier output constant as the input-signal amplitude varies.

Algorithm—1. A set of rules, or steps, prescribed for solving a problem. Usually the problem is a mathematical one and the algorithm prescribes its solution in a finite number of steps. 2. A procedure for implementing a formula.

Aliasing Noise—A distortion component that will be created if a sampled signal bandwidth is effectively greater than one-half the sampling rate.

Alignment—The process of fine-tuning a circuit or an antenna to maximize its sensitivity and its signal-receiving capability.

AM—Abbreviation for amplitude modulation. A technique for sending information as patterns of amplitude variations of a carrier sinusoid.

Amateur Radio Operator—A private citizen who operates electronic communications equipment as a hobby.

Amplifier—An electronic device used to increase signal power or amplitude.

Amplitude Modulation—A method of transmission where the signal voltage is impressed on a carrier wave of higher frequency such that the amplitude of the carrier wave is varied proportionately with the amplitude of the signal. (Also called *AM.*)

Analog—Information represented by a continuous and smoothly varying signal amplitude or frequency over a certain range, such as in human speech or music.

Analog-to-Digital Conversion—Conversion of an analog signal to a digital signal.

ANIK—The Canadian domestic satellite system used to transmit the network TV feeds of the Canadian Broadcasting Corporation. All ANIK satellites are operated by TeleSat Canada of Ottawa. ANIK satellites have both 4-GHz C-band and 12-GHz Ku-band transponders. ANIK means "brother" in Inuit (Eskimo).

Antenna—A structure for radiating and receiving electromagnetic waves.

Anti-Aliasing Filter—A filter (normally low pass) which band-limits the input signal before sampling to less than half the sampling rate—to prevent aliasing noise.

Aperture—The diameter of an antenna dish.

Apogee—The point in an elliptical satellite orbit that is farthest from the Earth's surface.

ARO (Audio Receive Only)—Small dish antennas used by radio networks for music and news programming distribution from TV satellites (mostly WESTARS). Dishes that are 2 meters and smaller have been considered by radio broadcast stations.

ASCII—An acronym for American Standard Code of Information Interchange.

Aspect Radio—1. The width-to-height ratio of a two-dimensional image, such as a television picture. 2. In the

NTSC system for America, the aspect ratio for any picture tube is four units wide versus three units high, and said to equal 4:3. In high-definition TV, however, this ratio will probably become 5:3.

Astable—Having no stable state.

Asynchronous—Refers to circuitry and operations without common timing (clock) signals.

Asynchronous Transmission—A transmission method in which each character of information is individually synchronized, usually by the use of "start" and "stop" elements. (Compare with "synchronous transmission.")

Attenuation—1. Signal or energy reduction experienced by analog or digital information travelling through transmission lines, waveguides, or through a space medium. Deliberate attenuation is normally induced by specially designed attenuators, with results expressed in decibels (dB). 2. The natural lessening in strength of an electrical current or an electromagnetic wave as it moves through a transmission medium, such as a wire pair, coaxial cable, or air. If the wave must travel a long distance, it will require repeaters at regular intervals to prevent a weakening of the signal to the point where the information is lost.

Audio Carrier—Always 4.5 MHz above the video carrier, it extends to within 0.25 MHz of the assigned channel frequency. For Channel 2, its active frequency is 59.75 MHz.

Audio Frequencies—The frequencies that can be heard by the human ear (usually between 30 and 15,000 hertz) when transmitted as sound waves.

Audio Modulation—Formerly ±25 kHz, it has now been broadened to ±50 kHz for multichannel stereo sound. It remains an FM carrier and must be detected accordingly.

Audio Subcarrier—Subcarriers of satellite video signals that are modulated by audio. The frequency range of these subcarriers can be from 5 to 8 MHz, but is usually 6.2 or 6.8 MHz.

Automatic Frequency Control (AFC)—A circuit which uses the frequency of an input signal as a reference for tracking another oscillator to the input frequency.

Automatic Gain Control (AGC)—A circuit which maintains a constant signal amplitude at a reference point, within a system that has varying input-signal levels.

AVC—Abbreviation for automatic volume control. A term used in radio for AGC.

Az/El Mount—A dish mount whose position is changed by two separate adjustments (in azimuth and elevation).

Azimuth—the angular displacement of a dish in a horizontal plane with respect to true north (measured in a clockwise direction).

Bandpass Amplifier—An amplifier which has a gain characteristic such that gain is relatively constant for frequencies that fall between an upper and lower limit.

Bandpass Filter—A device or circuit that only allows selected frequencies to pass through.

Bandwidth—1. The range of signal frequencies that a circuit or network will respond to or pass. 2. The difference between the high and low frequencies of a transmission band, expressed in hertz (cycles per second). 3. The range of frequencies, from DC to gigahertz (GHz), that are passing through some transmission medium. Bandpass filters are often used to limit or protect such transmissions as the signals progress through the transmitting or receiving equipment. In television, RF channels are 6 MHz wide, with the baseband limited to 4.5 MHz.

Baseband—1. As applied to television, this means pure audio and video with no RF modulation. 2. The frequency band occupied by the modulating signal. For example, baseband video signals occupy a band from zero to 4.3 MHz. 3. The frequency band occupied by information-bearing signals before they are combined with a carrier in the modulation process.

Baud—1. A unit of signaling speed. Speed as expressed in bauds is equal to the number of signaling elements per second. 2. Measure of signaling speed for transmitting data that is roughly equivalent to "bits per second." 3. A unit of signaling speed equal to the number of signal events per second.

Binary Code—A pattern of binary digits (0 and 1) used to represent information, such as instructions or numbers.

Bistable—Having two stable states.

Bit—1. The smallest possible piece of binary information. The specification of one of two possible alternatives. 2. A contraction of *bi*nary dig*it*, the smallest unit of information in a binary system of notation. Data bits are used in combination to form characters; framing bits are used for parity, transmission synchronization, and so on. 3. A unit of data in the binary numbering system.

Bit-Mapped Display—A method of CRT display that uses a separate area of computer memory to specify the locations of individual pixels, resulting in high-quality images. As yet, few computers use bit-mapped displays.

Bit Rate—Speed at which bits are transmitted; usually expressed in bits per second.

Block Downconverter—1. A type of downconverter which changes the microwave signal into an IF frequency which contains all the transponder frequencies (channels) of the satellite. The block downconverter allows inexpensive multiple receivers to tune all the channels simultaneously using one central downconverter—an advantage when using multiple receivers with a single antenna. 2. A device that converts an entire band (for example, the 3.7- to 4.2-GHz C-band) down to a lower band of frequencies.

BNC Connector—A weatherproof twist-lock coaxial cable connector used on some satellite TV receivers and standard on commercial video equipment.

Boresight Point—The area of maximum signal strength of a down-link signal. The center of the transponder "footprint."

BPS—Abbreviation for bits per second (also expressed as

b/s). A measure of speed in a serial transmission. Also used to describe hardware capabilities, as in specifying a 9600-bps modem.

Broadcast Satellite Service (BSS)—International designation for direct-to-home satellite transmissions. In the Western Hemisphere, BSS operates in the 12.2- to 12.7-GHz frequency band.

BSE—Abbreviation for Broadcasting Satellite for Experimental Purposes.

Bucket Brigade—1. Term for a delay line or shift register in which data moves from one element to the next, at a rate determined by an external clock, for storage or delay functions. 2. A shift register that transfers information from stage to stage in response to timing signals.

Burden Test—Test to determine whether the offering of new service or the continued provision of an existing service will cause customers (or other services) to pay prices that are no higher than those prices which would otherwise be required if the service were not offered.

Burst—A pulsed RF envelope, generally 2.3 μs (8 cycles) in duration, that appears on the back porch of the horizontal synch pulse. This 3.579545-MHz frequency is used to synchronize the receiver's regenerated subcarrier color oscillator.

Byte—1. A generic term to indicate a measurable portion of consecutive binary digits; e.g., an 8-bit or 16-bit byte. 2. A group of eight binary bits operated on as a unit.

Cable Television Service—Reception and retransmission of broadcast television programming, and the origination of television programming, which is supplied to subscribers for a fee and transmitted over coaxial cable. This process differs from standard television which is broadcast through the air without charge to any receiver.

Cable TV—A system for distributing television programming by a cable network rather than by electromagnetic radiation.

Carrier—A signal which is modulated with the information to be transmitted.

Carrier-to-Noise Ratio (C/N)—The ratio of the received carrier power to the noise power in a given bandwidth. The C/N is an indicator of how well an earth station will perform in a particular location and is calculated from satellite power levels, antenna gain, and the combined antenna and LNA noise temperature.

Cassegrain Antenna—An antenna that uses a sub-reflector at the focal point to reflect the received signals to a feed located at the dish's apex (center).

Cathode-Ray Tube—A device made of glass, with either dot or vertical strip phosphors, which is excited through a grill shadow mask. CRTs are the display devices for most visual images now available. Flat-panel devices already coming out of the laboratories, however, are expected to replace CRTs in the not-too-distant future, removing the CRT high-voltage problem and permitting a tremendous reduction in the depth of TV receivers.

CATV—Abbreviation for Community Antenna TeleVision. Now generally used to mean Cable TV.

Cavity—A structure that may be rectangular, circular, spherical, or any geometric shape. It is used as an electrically resonant element, with its physical dimensions determining the resonant frequency.

C Band—A band of downlink frequencies having a range from 3.7 GHz to 4.2 GHz. This band also includes uplink signal frequencies from 5.92 to 6.42 GHz. It also designates frequencies between these limits.

CCD—Abbreviation for charge-coupled device. Microelectronic circuit elements that move quantities of electrical charge serially across the surface of a semiconductor. CCDs provide a small, reliable, volatile memory with very low power dissipation per bit. The primary use of CCDs is in signal processing.

CCIR—Abbreviation for Comite Consultant International des Radiocommunications (International Radio Consultative Committee). Part of the International Telecommunication Union (ITU), CCIR is a forum for coordination of international radio communication services.

Channel—1. In communications, a path for transmission (usually one way) between two or more points. Through multiplexing, several channels may share common equipment. 2. A band of frequencies used for carrying video and/or audio signals, data, news, etc. Its bandwidth depends on the amount of information to be transmitted. Channel frequencies are specified in the United States by the FCC.

Character—Any coded representation of an alphabet letter, numerical digit, or special symbol.

Checksum—An error-detecting method for block of information loaded into computer memory from magnetic media. It is the sum of the numerical values of the bytes in the block. As the block loads, the checksum is computed. Once loaded, the computed value is compared with the checksum value that was placed on the media when it was generated. If they are equal, the block has loaded error free.

Chroma—1. An abbreviation for chrominance, and refers to the color signals. A reference to a combination of hue, purity, and saturation. 2. Color produced by I and Q sidebands having bandwidths of 1.5 and 0.5 MHz, respectively.

Chrominance—That portion of a television signal that contains the color information of the picture.

Chrominance Subcarrier—See *color subcarrier.*

Circuit—An interconnect of electrical components or devices that will perform some electronic function.

Circular Polarization—A form of polarization of electromagnetic signals (such as from a satellite) in which the electric field contains two quadrature components (90°) in space and time phase. Form can be either clockwise or counterclockwise.

Clark Belt—The name given in honor of Arthur C. Clarke,

the noted science-fiction writer, to the orbit belt that is 22,300 miles directly above the equator. Satellites placed in this orbit travel at the same rate as the earth's rotation, thereby maintaining a stationary position in relation to the earth. (Also called the geostationary or geosynchronous orbit.)

Clock—In data communications, a device that generates precisely spaced timing pulses (or the pulses themselves) used for synchronizing transmissions and recording elapsed times.

Close-Captioned TV—A text service for the hard-of-hearing TV audience which decodes a text subcarrier and displays it at the bottom of the TV frame on the accompanying video picture. It does not interfere with the standard audio FM subcarrier.

Coaxial Cable—A cable that consists of an outer conductor concentric with an inner conductor with the two being separated from each other by an insulating material. (Also called Coax Cable or simply Coax.)

Code—A specific way of using symbols and rules to represent information.

Color—That characteristic perceived by the eye, which is produced by a specific spectral distribution of light reflected or transmitted through or by a medium. The human eye recognizes approximately 30,000 different colors, all within a spectrum of 400 to 700 nanometers. The primary colors of red, blue, and green are seen at 680, 470, and 540 nanometers.

Color Edging—Extraneous colors that appear along the edges of video pictures, but don't have any color relationship to those areas.

Color Killer—A stage designed to prevent signals in a color receiver from passing through the chrominance channel during monochrome telecasts.

Color Modulation—This consists of 500 kHz of Q information and 1.5 MHz of information, modulated on 90° out-of-phase suppressed subcarriers. Still AM, it is then "folded" into video at odd multiples of one half the line-scanning frequencies, but is decoded (detected) separately from luminance.

Color Purity—An area (or field) of pure color with no contaminations, resulting from the absence of stray magnetic fields and the presence of correct color-beam landings on cathode-ray tube phosphors.

Color Subcarrier—The 3.569545-MHz color subcarrier oscillator that is always suppressed in transmission but reconstituted in the receiver to synchronously time the color synch for correct color demodulation. (Also called the chrominance subcarrier.)

Companding—The process of the dynamic range compression of a signal and its subsequent expansion in accordance with a given transfer characteristic (companding law), which is usually logarithmic. The purpose of companding is to allow transmission of signals.

Comparator—1. A circuit that compares two signal levels and provides a digital output indicating which signal is larger. 2. In an FM demodulator using a phase-lock loop (PLL) circuit, this is the electronic component that compares the phrase relationship of the input signal with the signal from the tracking local oscillator.

Composite TV Signal—A combination of video picture, color, audio, and synchronization information.

Composite Video—A signal for monitors (and television sets) which use a single video signal. Composite color monitors, though usually cheaper than RGB displays, tend to be slow and fuzzy in producing images.

Compression—The reduction of a signal's dynamic range in such a way that the small-signal characteristics are maintained; usually a logarithmic-type conversion is used (see *Companding*).

Computer—The combination of a central processor, input/output, and memory for the storing and processing of data.

COMSAT—Abbreviation for Communication Satellite Corporation. Established by the 1982 Satellite Communication Act to represent U.S. interests in the development of a global commercial-satellite-communications network (INTELSAT).

Conjugate Impedance—An impedance, the value of which is the conjugate of a given impedance. For an impedance associated with an electric network, the conjugate is an impedance with the same resistance component and an opposite reactive component of the original.

Conjugate Matching—A condition of source- and load-impedance matching in which the source impedance and the load impedance have equal resistive parts and equal reactance values with opposite signs. This results in maximum power transfer.

Convergence—The colors, Red, Blue, and Green, blended in every given area to produce white. Crosshatch, dot-hatch, and dots are used to ensure correct color-beam phosphor landings, as adjusted by the various types of convergence devices and magnets as an integral part of the deflection system. Satisfactory convergence on the CRTs of today should be within at least 95% of the screen area. Both dynamic (operational control) and static (fixed) beam-convergence systems continue to be available in modern TV receivers, but with vastly more effective and simpler methods.

C/N—The ratio of the satellite signal-carrier strength to the power of the received noise, measured in decibels (see *Carrier-to-Noise ratio*).

CPU—Abbreviation for Central Processing Unit, the part of the computer that includes the circuitry for interpreting and executing instructions. It contains the main storage, arithmetic unit, and special register groups. (Synonymous with "mainframe.")

Cross Modulation—A condition that exists when interfering signals from one circuit or system modulate another.

Crosstalk—Interference or an unwanted signal from one transmission circuit that is detected on another (usually parallel) circuit.

CRT—Abbreviation of cathode-ray tube. An electron tube—essentially a funnel-shaped glass bottle—in which electron beams (cathode rays) trace fine horizontal patterns, called a raster scan, on a phosphor-coated screen. Selectively activated phosphors form the images you see.

CRT Display—1. An electronic device similar to a TV screen that provides a visual-display output of stored or transmitted information. 2. An electronic vacuum tube, such as a television picture tube, that can be used to display images.

C.R.T.C.—Abbreviation of Canadian Radio-Television and Telecommunications Commission. A regulatory body of the Canadian Government which controls TV and radio broadcasting in Canada.

Cutoff Frequency—1. The frequency at which signals are handled by a circuit or network in a different manner from those below or above that frequency. 2. The frequency at which the gain of an amplifier falls below 0.707 times the maximum gain.

D/A Converter—A circuit that converts signals from digital form to analog form.

Data—1. A representation of facts, concepts, or instructions in a formalized manner suitable for communication, interpretation, or processing; any representations, such as characters, to which meaning may be assigned. 2. Data connotes basic elements of information which can be processed or produced by a computer. 3. Information about the physical state or properties of a system.

Data Compression—The technique which provides for the transmission of fewer data bits without the loss of information. The receiving location expands the received dated bits into the original bit sequence.

Data Stream—Generally, the flow of information being transmitted in a communications system, or path along which it flows.

dB—Abbreviation for decibels. dB is the standard unit for expressing relative power, voltage, or current.

$$dB = 10 \log^{10} \frac{P_1}{P_2}$$

dBm—A measure of power in communications; decibels referenced to 1 milliwatt. 0 dBm = 1 milliwatt, with a logarithmic relationship as the values increase or decrease. In a 50-ohm system, 0 dBm = 0.223 volt.

dBmV—Abbreviation for decibel/millivolt. The level at any point in a system, expressed in decibels above or below a 1-millivolt/75-ohm standard, is said to be the level in decibel/millivolts, or dBmV. 0 dBmV is equal to 1 millivolt across 75 ohms.

DBS—Abbreviation for Direct Broadcast Satellites, a term commonly used when referring to the 12- to 14-GHz satellites. Also called the Ku-band. DBS satellites can be received by 3- to 5-foot antennas, although numerous problems must be overcome before widespread use is available.

dBw—Decibels referred to 1 watt.

DC Amplifier—1. A circuit that amplies signals with frequencies from zero to high cut-off limit. 2. An amplifier capable of boosting DC voltages.

Decibel—Listed as one-tenth of a Bel and used to define the ratio of two powers, voltage or current, in terms of gains or losses. It is 10 times the log of the power ratio and 20 times that of either voltage or current.

Decoder—1. Any device which modifies transmitted information to a form that can be understood by a receiver. 2. A device for unscrambling video signals.

Decoding—A process in which one of a set of reconstructed samples is generated from another related signal representing that sample.

Decoupling—The prevention of the transfer of energy from one circuit or subsystem to another. Capacitors, for instance, are often used as RF decouplers.

Deemphasis—Also called post-emphasis or post-equalization. The introduction of a frequency-response characteristic which is complementary to that introduced in preemphasis.

Deemphasis Network—A network inserted into a system to restore the preemphasized frequency spectrum to its original form.

Demodulation—The process of retrieving an original signal from a modulated carrier wave. The technique used in data sets to make communication signals compatible with business machine signals.

Demultiplexer—A circuit that distributes an input signal to a selected output line (with more than one output line available).

DES—Abbreviation for Data Encryption Standard. A standard used for encrypting data (such as computer data or digital audio), using an algorithm described by the National Bureau of Standards.

Descrambler—A device that is used to unscramble intentionally scrambled satellite transmissions so that a television picture can be viewed. Satellite signals are generally scrambled by companies offering subscription-supported services (for pay) to prevent their unauthorized reception.

Desensitization—The saturation of one component (an amplifier, for instance) by another so that the first component cannot perform its proper function.

Digital—1. Information in discrete or quantized form; i.e., not continuous. 2. Of or relating to the technology of computers and data communications, wherein all information is encoded as binary bits (1s or 0s) that represent *on* or *off* states. 3. In data communications, the description of the binary (off/on) output of a computer or terminal. Modems can convert the pulsating digital signals into analog waves for transmission over conventional telephone lines.

Digital Television—Analog signals following the video detector are digitized for both audio and video by A/D converters, and, then, returned to baseband once again by D/A converters for sound and the cathode-ray tube. With the introduction of LCD or other flat panel displays, completely digitized receivers should appear and eventually become less expensive and more reliable than analog types.

Digitize—To convert an analog quantity into digital format.

Diode—Simple two-terminal devices, with a cathode and anode, devoted principally to AC rectification in power supplies. Also, a device that permits current to flow in one direction through the device, but not in the opposite direction. May be either a vacuum-tube type device or a solid-state device.

Direct Broadcast Satellites (DBS)—Service that uses satellites to broadcast multiple channels of television programming directly to small and, thus, widely affordable dish-type antennas.

Directional Radiation—Radiation that is controlled or concentrated by an antenna or focusing device such that it is received only by certain receivers.

Discriminator—A device in which output signals are produced in response to frequency or phase variations of the input signal.

Dish—A device that reflects received satellite microwave signals to a feedhorn. Sometimes incorrectly referred to as an antenna. Also, the popular name for a spherical or parabolic reflector for UHF or microwave antennas.

Dish Illumination—The area of a dish as seen by the feedhorn.

Distortion—The unwanted change in waveform that occurs between two points in a transmission system. "Amplitude vs. frequency distortion" is caused by the nonuniform gain or attenuation of the system with respect to frequency. "Delay vs. frequency distortion" is caused by differences in the transit time of frequencies within a given bandwidth (under specified conditions). "Nonlinear distortion" is a deviation from the normal linear relationship between the input and output of a system or component.

Dithering—The process of spreading the energy of a signal. The 6-MHz satellite signal is shifted up and down the 36-MHz satellite transponder spectrum to distribute the energy of the video signal. The purpose is to reduce the interference that any terrestrial microwave transmitter could cause to the satellite transmission.

Doppler Effect—The change in the observed frequency of an acoustic or electromagnetic wave caused by the relative motion of a source (such as a beacon) and an observer (such as a polar-orbiting satellite). The frequency rises as the observer approaches the source and falls as the observer retreats from the source.

Dot Pitch—A measure of picture quality or resolution in RGB color monitors, which is the more common pixel resolution. Dot pitch is the distance between screen dots (pixels) measured in millimeters. The shorter the distance, the better the resolution. It is specified in pixels/mm.

Downconverter—1. The circuit associated with a satellite receiver that lowers the high-frequency signal to a lower intermediate range. There are three distinct types of downconversion: single downconversion, dual downconversion, and block downconversion. 2. The part of a satellite receiving system that converts the downlink signals to the intermediate frequency that is used by the receiver. Although it is sometimes part of the receiver, the downconverter is more often externally mounted directly at the LNA so that inexpensive coaxial cable can be used to bring the signal to the receiver.

Downlink—The carrier used by satellites to transmit information to Earth stations.

Downlink Signals—The signals that are transmitted from a satellite transponder to Earth.

Dual Feedhorn—A waveguide feed system designed for both vertically and horizontally polarized signals.

Dual Orthomode Coupler—A dish-mounted device that allows reception of both vertically and horizontally polarized signals.

Duplexer—A device that combines audio and video signals, or separates signals on a single-transmission path.

Dynamic Range—The weakest through the strongest signals that a receiver will accept as input. Signals which are too weak will cause excess noise and signals which are too strong cause overloading and possible modulation distortion.

Earth Station—A station equipped with transmitting equipment for the production of uplink signals, *and*, also, a complete receiving system for picking up downlink signals. Sometimes used synonymously, but incorrectly, with TVRO.

ECL—Abbreviation for emitter-coupled logic. A type of unsaturated logic performed by emitter-coupled transistors. Higher speeds may be achieved with ECL than are obtainable with standard logic circuits.

EHF (Extremely High Frequency)—The portion of the electromagnetic spectrum from 30,000 megahertz to 300,000 megahertz.

EIA—Abbreviation for Electronics Industries Association. The U.S. national organization of electronic manufacturers. It is responsible for the development and maintenance of industry standards.

EIRP—Abbreviation for effective isotropic radiated power. 1. A measure of the relative strength of the satellite TV signal expressed in dBW. These EIRP figures usually range from 30 dBW to 37 dBW. 2. The energy level of a transmitted signal expressed in dBW.

Electromagnetic Spectrum—The entire available wavelength range of electromagnetic radiation. In electronics, those frequencies from DC (zero) to the microwaves (> 3,000 GHz), or higher.

Electronic Mail—Messages sent electronically between computers or terminals (e.g., remote printer).

Elevation—The vertical angular displacement of a dish, with the earth considered as a horizontal frame of reference.

Encoder—Any device which modifies information into the desired pattern or form for a specific method of transmission.

Encoder (PCM)—A device which performs repeated sampling, compression, and A/D conversion in order to convert an analog signal to a serial stream of PCM samples representing the analog signal.

Encoding—1. The generation of digital character signals to represent quantized analog samples. 2. The "scrambling" of a signal to prevent viewing of a program by nonsubscribers.

Encryption—A process of electronically manipulating a satellite-transmitted television signal so that it cannot be received without special equipment. This technique is employed to reduce and/or eliminate unauthorized reception of subscription-supported services.

Equalizing Pulses—A group of pulses (part of the composite TV signal) that permit synch circuits to recognize and lock on the vertical synch. Their durations are 2.54 microsecond each. Sequences are in sets of six before and after the 6 vertical pulses; each sequence occupies 3H, or a period of three horizontal lines.

Error-Detecting and Feedback System—A system wherein any signal detected as an error automatically initiates a request for retransmission of the data in error. (Also called a "decision feedback system," "requests repeat system," and an "ARQ system.")

Error-Detecting Code—A code in which each data signal conforms to specific rules of construction so that departures from the norm—errors—are automatically detected. (Synonymous with "self-checking code.") Such codes require more signal elements than are necessary for conveying the fundamental information.

Error-Detecting System—A system which employs an error-detecting code and is so designed that the errors detected are either automatically deleted from the delivered data (with or without an indication that such a deletion has taken place), or delivered together with an indication that an error has been detected.

Error Detection—Code in which each data signal conforms to specific rules of construction so that departures from this construction in the received signals can be automatically detected. Any data detected as being in error is either deleted from the data delivered to the destination, with or without an indication that such deletion has taken place, or delivered to the destination together with an indication that it has been detected as being in error.

Error Rate—The ratio of incorrectly received data (bits, elements, characters, or blocks) to the total amount of data transmitted.

Exlusive-OR—A logical function whose output is "O" if either of the two input variables is a "1" but whose output is a "1" if both inputs are "1" or both are "0." Also called XOR.

Expendable Launch Vehicle (ELV)—An unmanned, nonreusable booster used to launch satellites, space capsules, and other payloads into orbit.

"Fantasy" Decoder—A type of descrambler used for satellite descrambling. It uses a 94-kHz sine-wave scrambling plus video inversion. Audio is encoded on a 15.7-kHz carrier that is one sixth that of the scrambling sine wave. The 94-kHz sine wave is *not* an exact multiple of the scan rate, in general.

FCC (Federal Communications Commission)—A government agency that regulates and monitors the domestic use of the electromagnetic spectrum for communications.

FDM (Frequency-Division Multiplexing)—A method by which the available transmission frequency range is divided into narrower bands, each used for a separate channel. As utilized by broadband technology, the frequency spectrum is divided up among discrete channels, to allow one user, or a set of users, access to single channels.

FDMA (Frequency-Division Multiple Access)—The technique of allocating frequency-multiplexed communication channels to satisfy user demands.

F/D Ratio—The ratio of the focal length of a dish to its diameter. A method for indicating the depth of a dish.

Feedback—The return of part of the output of a machine, process, or system to the input, especially for self-correcting or control purposes.

Feedhorn—The entrance to the waveguide that is used to channel signals from the focal point of the dish to the LNA. This is actually the antenna, not the dish. (See also *Horn.*)

FET—Abbreviation for junction field-effect transistors. A single-channel device with source, gate, and drain electrodes. A reverse bias applied to the gate depletes the electron flow, reducing drain-to-source current.

Fiber Optics—A method for the transmission of information (sound, picture, data) over high-purity, hair-thin, glass fibers. The capacity of fiber-optic cable is much greater than that of conventional cable or copper wire because of the higher carrier frequencies.

Field—Every other scan line in a frame of 525 scan lines which make up a transmitted or received television picture. There are two fields per frame; 262½ lines per field.

Figure of Merit—A quality factor used to compare dish gain to system noise and written as G/T, where T is the system noise in Kelvins and G is the gain of the dish in dB. Also, a figure used in electronics that compares specific quantities to that of a reference value, or is used as a comparison between various circuits (i.e., gain-bandwidth).

Flat-Square—Refers to cathode-ray tubes that are both full-square and have relatively flat screen surfaces also.

Flip-Flop—A type of digital circuit that can be either of two stable states, depending both on the input received and on which state it was in when the input was received. Used as a memory element or as a frequency divider.

Flyback Pulse—The pulse in a television (or facsimile) signal that causes the scanning mechanism to return rapidly to the left side of the picture.

FM—Abbreviation for frequency modulation. A technique for sending information as patterns of frequency variations of a carrier signal.

FM Discriminator—A circuit that converts signal frequency variations to corresponding amplitude variations; i.e., a device that demodulates FM signals.

FM Improvement—The potential noise reduction in an FM signal due to the demodulation process in a satellite TV receiver. This figure is at most 38.6 dB, and is attained above the FM threshold. Below this point, it rapidly drops from 38.6 dB. Above threshold: S/N = CN +38.6 dB.

FM Threshold—An input signal level which is just enough to enable the demodulator circuits to extract a good picture from the carrier. With test equipment, the static threshold is the point at which S/N drops more than 1 dB from the straight graph line; S/N = CN +38.6 dB. Typically, the FM threshold is 8 dB in a satellite TV receiver.

Focal Length—The distance from the center of the dish to its focal point.

Focal Point—The point at which all the signals reflected by a dish join or cross.

Footprint—The area covered by downlink signals transmitted from a satellite.

Fourier Analysis—1. The process of analyzing a complex wave by separating it into a plurality of component waves, each of a particular frequency, amplitude, and phase displacement. 2. The representation of arbitrary functions as the superposition of sinusoidal functions, whereby the representations themselves are referred to as *Fourier series* or *Fourier integers.*

Fourier Series—A mathematical analysis that permits any complex waveform to be resolved into a fundamental, plus a finite number of terms involving its harmonics.

Fourier Transform—A mathematical relationship that provides a connection between information in the frequency domain and the time domain. It is derived from *Fourier series,* in which the frequency spacing between component harmonic approaches zero. It can be used for nonperiodic waveform analysis.

Frame—A set of 525 electron scan lines that completes the entire image of an electronically transmitted or received television picture.

Frequency—The rate, in hertz, at which a signal pattern is repeated.

Frequency Coordinator—A test procedure that is part of a site survey to determine interference signal levels.

Frequency-Division Multiplex—A system of transmission in which the available frequency transmission range is divided into narrower bands, so that separate messages may be transmitted simultaneously on a single circuit.

Frequency Multiplexed System—A communication system serving several users.

Frequency Reuse—A method that allows two different TV channels to be broadcast simultaneously on the same transponder by vertically polarizing one channel and horizontally polarizing the other. Another method of frequency reuse is to space satellites about 4° apart. A TVRO pointed at one satellite will not detect any signal from the other satellite, even if it is operating at the same frequency.

FSK—Abbreviation for frequency-shift keying. The most common form of frequency modulation, in which the two possible states (1/0, off/on, yes/no, etc.) are transmitted as two separate frequencies.

Full Duplex—Term used to describe a communications system or component capable of transmitting data simultaneously in two directions.

Full-Square—This refers to the new sharply-rectangular cathode-ray tubes; they come in sizes of 14, 20, 26, and 27 inches.

Gain—A term used to indicate the amount of signal amplification; expressed in decibels. The gain of an antenna or of an LNA is usually a manufacturer specification, and is referred to a reference antenna (usually a dipole or isotropic radiator).

Gain Insurance—The idea of having a little more signal than the minimal acceptable level. A slightly larger antenna or a lower-noise LNA gives some gain insurance.

Gallium Arsenide (GaAs)—A semiconductor material well-suited for low-noise microwave applications.

Gamma—A number that indicates the degree of contrast in a photograph, facsimile reproduction, or received television picture, as referred to the original scene.

Geodetic North—A reference to the nonmagnetic poles. (Also called *true north.*)

Geostationary—A fixed position relative to the Earth. For example, a satellite in the Clarke belt is geostationary. (See also *geosynchronous.*)

Geostationary Satellite—A satellite that orbits the Earth from west to east at a speed and at an altitude that keeps it fixed over the Earth's equator, making one rotation every 24 hours in synchronism with the Earth's rotation. It is also known as a *geosynchronous satellite.*

Geosynchronous—See *geostationary.*

Geosynchronous Orbit—1. An area that is 22,300 miles above Earth's equator in which satellites circle the Earth at same speed as the Earth's rotation. 2. The orbit of a satellite that matches the Earth's rotation so that any spot in the coverage area will always remain in the same relative position with respect to the satellite.

Gigahertz (GHz)—Unit of frequency equal to one billion cycles per second.

Global Beam—A broad pattern of signal radiation from a satellite that covers one third of the Earth's surface. (Type beam used by INTELSAT satellites.)

Ground Noise—The unwanted microwave signals generated from the warm ground and detected by a dish.

Ground Station—An installation specially constructed for communications with a satellite.

G/T—A figure of merit that describes the capability of a TVRO system to receive a signal from a satellite. The ratio (in decibels) of the gain of a receiving system to the noise temperature of the system.

Hacker—A computer enthusiast. Term is sometimes used in reference to a hobbyist or enthusiast in other fields as well, such as a "video" hacker.

Halation—In a cathode-ray tube, the glow surrounding a bright spot that appears on the fluorescent screen as the result of the screen's light being reflected back by the front and rear surfaces of the tube's face.

Half Duplex—Term used to describe a communications system or component capable of transmitting data alternately, but not simultaneously, in two directions.

Hamming Code—A code that checks and corrects errors in binary information by comparing the parity of received information with transmitted parity information.

Handshaking—An exchange of predetermined signals for the purpose of control when a connection is established between two data sets.

Hard Scrambling—An encryption method that uses proprietary, highly secure technology (i.e., digital), such as that used by VideoCipher II™.

Harmonic Distortion—The production of harmonic frequencies at the output by the nonlinearity of a transducer when a sinusoidal voltage is applied to the input.

Hertz (Hz)—A term that is synonymous with cycles per second; one hertz is equal to 1 cycle per second.

High Frequency (HF)—That portion of the electromagnetic spectrum from 3 MHz to 30 MHz.

High-Pass Amplifier—A circuit that increases the amplitude of signals whose frequencies are above some cutoff frequency.

Horizontal Blanking Interval—This nominally occupies 11 microseconds of the composite video signal and contains both the horizontal blanking pulse and color burst.

Horizontal Synch—A pulse that appears during each horizontal scan line and lasts for 4.76 microseconds. On its "back porch" is the color burst information, consisting of 8 or 9 cycles that are 2.24 microseconds in duration.

Horn—A type of waveguide having a flared end. Its shape is selected so as to provide a better impedance match between the feed and open air. It is used as an antenna. (See also *Feedhorn.*)

Hyperboloidal Subreflector—The secondary reflector used in a Cassegrain antenna system. The surface has a hyperbolic shape.

IF Amplifier—Refers to bandpass amplifiers in superheterodyne receivers. They normally provide high gain and stable operation over a fixed range of frequencies, with fix-tuned (generally) filtering. They usually have variable signal gain, that is controlled by AGC or AVC systems.

Image Noise—When a signal is downconverted using a mixer and a local oscillator, noise can be passed through the system that is on the mirror-image frequency of the selected channel, with the local oscillator frequency as the point of symmetry. Subsequent bandpass filters can remove this noise in double-conversion downconverters. A preselector filter in single-conversion receivers does the same thing.

Impedance—The algebraic sum of resistance and reactance; represented by the letter Z. (Generally a complex quantity, such as $Z = R \pm jX$.)

Information—In communications, a physical pattern that has been assigned a unique and commonly understood meaning.

Information Bit—A bit used as part of a data character within a code group (as opposed to a framing bit).

Infrared—Abbreviated IR, and pertaining to or designating those radiations, such as are emitted by a hot body, with wavelengths between the microwaves and visible light. The longest IR wavelengths are those at the upper end of the microwave spectrum (300 GHz, or 0.1 cm). The shortest are just below red light (0.7 micron).

Instruction Code—Digital information that represents an instruction to be performed by a computer.

Integrated Circuit—A circuit whose connections and components are fabricated into one integrated structure on a certain material, such as silicon.

INTELSAT—Abbreviation for International Telecommunications Satellite Consortium. A series of commercial communications satellites designed to relay telephone and television signals among the member INTELSAT nations. Established in 1964, in part through the efforts of COMSAT, the INTELSAT network of commercial satellites now services more than 100 member nations.

Interface—A shared connection or boundary between two devices or systems. The point at which two devices or systems are linked. Some common interface standards include the EIA Standard RS-232B/C, adopted by the Electronic Industries Association to ensure uniformity among most manufacturers; the MIL STD 188B, a mandatory standard established by the Department of Defense; and the CCITT, the world recommendation for interface, that is mandatory in Europe and closely resembles the American EIA standard.

Interlaced Scanning—The sequential display of odd and even fields to form a flicker-free linear video picture. Normally, the better the interlace, the better the picture, because there are few video dropouts, resulting in less information loss.

Intermediate Frequency (IF)—A frequency to which a signal wave is shifted locally as an intermediate step in transmission or reception.

Intermodulation Distortion—A nonlinearity characterized by the appearance of frequencies in the output that are equal to the sums and differences of integral multiples of the component frequencies that are in the input signal.

Ionosphere—A layer of electrically charged particles at the top of the Earth's atmosphere.

IRAS—Abbreviation for Infrared Astronomical Satellite. Launched in January 1983, the IRAS carried the cryogenically cooled, space-based telescope to record long-term observations in the 8- to 120-μm wavelength range. A joint venture by NASA, the Netherlands Agency for Aerospace Programs, and the United Kingdom's Science and Engineering Research Council, the instrument operated for 300 days, at which time it exhausted its cryogen supply.

IRE Unit—An arbitrary unit used to describe the amplitude characteristics of a video signal. Pure white is defined as +100, pure black is 0 (zero) and synch tips are defined as −40. The intermediate shades of gray lie between 0 and 100. Medium gray is 50.

Isolator—1. A device which acts as a one-way valve for microwave signals to prevent stray receiver signals from leaking out past the LNA onto the antenna. It also facilitates the design of the LNA by impedance matching the feed probe to the first LNA amplifier stage. Most LNAs have an isolator attached between the CPR-229 feed flange and the main amplifier box. They are generally ferrite devices. 2. A device that allows the transmission of signals in one direction while blocking or attenuating them in the other.

ITU—Abbreviation for the International Telecommunication Union. A specialized United Nations agency with broad responsibility for coordinating the use of the electromagnetic frequency spectrum and for intergovernmental coordination of technical standards for radio and wired communication systems.

Jitter—1. Short-term variations of the significant instants of a digital signal from their ideal positions in time. 2. Instability of a signal in either its amplitude or its phase, or both, due to mechanical disturbances, changes in supply voltage, or component characteristics, etc.

Kelvin (K)—A temperature scale on which the unit of measurement equals the Celsius degree. A scale where absolute zero, the temperature at which all molecular motion stops, is 0 Kelvin. Absolute zero also equals −273°C (−459°F).

Kilobyte—Denotes 1000 bytes.

Ku Band—Also called the K band. A band of frequencies that extends from 11.7 to 12.7 GHz (for broadcasting satellite service or DBS).

LANDSAT—A system of earth-sensing satellites used to take inventory of the natural resources, monitor environmental quality, and map natural and man-made surface features. Originally called Earth Resources Technology Satellite (ERTS-1), the first satellite in the series was launched by NASA in 1972. It was then followed by three additional satellites in 1975, 1978, and 1982. LANDSAT is now managed by the U.S. National Oceanic and Atmospheric Administration.

Laser—An acronym for *L*ight *A*mplification by *S*timulated *E*mission of *R*adiation. A device which transmits an extremely narrow and coherent beam of electromagnetic energy in the visible light spectrum.

LCD—Abbreviation for liquid crystal display. LCD screens, made up of liquid crystals sandwiched between two glass plates, are often used to replace bulky CRTs where portability is a major concern. LCDs are typically small and flat, and require very little power for operation. The quality of the display is often poor and of low contrast.

Line Amplifier—An amplifier inserted in any part of the transmission line, following the downconverter, to compensate for signal losses caused by long lengths of coaxial cable or the insertion of passive devices, such as splitters. Line amplifiers are also used when the signal must drive a number of television receivers.

Line Scan—The forward scan time of 52.4 microseconds required to trace one of the 525 horizontal scan lines from left to right on the cathode-ray tube.

LNA—See *low-noise amplifier.*

LNC—See *low-noise converter.*

Log-Periodic Antenna—A type of directional antenna that achieves its wideband properties by geometric iteration.

Look Angle—The positioning angle of a mount which permits a dish to focus on (to "see") a satellite.

Low-Noise Amplifier (LNA)—A wideband nontunable amplifier that accepts satellite signals from an antenna probe and delivers them to a downconverter.

Low-Noise Converter (LNC)—A combination LNA and remote downconverter in a single housing.

Low-Pass Amplifier—A circuit that increases the amplitude of signals whose frequencies are below a certain cut-off frequency.

Luminance—1. The sum of all the color signal voltages; it is the algebraic addition of reds, blues, and greens. Y (luminance) = 30% Red, 59% Green, and 11% Blue. 2. That portion of a television signal that contains the gray level information of the picture (the black and white component).

Luminance Signal—The monochrome portion of a color television signal. Also known as a brightness signal.

MAC-B—An acronym for *m*ixed *a*nalog *c*omponent signals. It refers to placing TV sound into the horizontal line-blanking period and, then, separating the color and luminance for periods of 20 to 40 microseconds each during horizontal scan. In the process, luminance and chroma are compressed during transmission and expanded during reception, enlarging their bandwidths considerably. Transmitted as FM, this system, when used in satellite transmission, provides considerably better TV definition and resolution. Its present parameters are within the existing NTSC format.

MATV—Abbreviation for master antenna television system. A small, less-expensive, cable system that is usually restricted to one or two buildings, such as hospitals, apartments, libraries, hotels, office buildings, etc.

Microprocessor—A miniaturized device that is the control and processing portion of a small computer or microcomputer, which can usually be built on a single chip.

Microsecond (μsec)—One-millionth of a second.

Microwave—Generally, the electromagnetic waves in the frequency spectrum above 890 MHz. A generic term that refers to the very short waves used for the transmission of voice, data, and television signals.

Millisecond (msec)—One-thousandth of a second.

Mode—A method of operation (as in binary mode, alphameric mode, etc.).

Modem—An acronym for *MO*dulator *DEM*odulator. A device that converts data from a form which is compatible with data-processing equipment to a form that is compatible with the transmission facilities, and vice versa.

Modulation—1. The process of controlling the properties of a carrier signal so that it contains the information patterns to be transmitted. 2. The process by which a characteristic of one wave is varied in accordance with another wave or signal. 3. Types of modulation include: *amplitude modulation* (AM), in which the amplitude of the carrier is varied in accordance with the instantaneous value of the modulating signal; *differential modulation,* in which the choice of the significant condition for any given signal element is dependent upon the choice of the previous signal element; *frequency modulation* (FM), in which the instantaneous frequency of a sine-wave carrier departs from the carrier frequency by an amount proportional to the instantaneous value of the modulating signal; *phase modulation,* in which the angle relative to the unmodulated carrier angle is varied in accordance with the instantaneous value of the amplitude of the modulating signal; *pulse amplitude modulation,* in which the amplitude of the pulse carrier is varied in accordance with successive samples of the modulating signal; and *pulse code modulation,* in which the modulating signal is sampled, quantitized, and coded so that each element of the information consists of different kinds and/or numbers of pulses and spaces.

Modulo-*n* Counter—A counter with *n* unique states. (Also called a programmable counter.)

Moire—A wavy or satiny effect produced by the convergence of lines. Usually appears as a curving of the lines in the horizontal wedges of the test pattern.

Monochrome Monitor—A monitor having only one chromaticity—usually achromatic, or black and white and all shades of gray. Monochrome monitors usually offer a much sharper display than do the best multicolor monitors.

Monostable—A device having only one stable state.

Morse Code—A code developed by Samuel Morse which uses patterns of long and short pulses (dashes and dots) to represent numbers and letters of the alphabet.

MOSFET—An insulated-gate FET with two modes of operation: depletion and enhancement. Depletion-mode FETs are normally "on" with zero gate bias. Enhancement types are "off" under zero-gate bias conditions.

Multichannel Sound—A system of stereo sound transmission for TV applications. Approved in early 1984 by the FCC, it is AM double-sideband for stereo L-R and operates on a 15,734-Hz pilot carrier, which is doubled to 31,468 Hz for stereo. Multichannel sound also contains higher-frequency carriers for SAP (second audio program) and professional channel(s).

Multiple LNAs—A pair of LNAs, with one being used for the amplification of vertically polarized signals and the other for horizontally polarized signals.

Multiplex—To interleave, or simultaneously transmit, two or more messages on a single channel.

Multiplexing—1. The process of dividing a transmission facility into two or more channels. 2. Used in the sense of time-division multiplexing, a process where samples of information are transmitted during separate time intervals. (Multiplexing is used in the MAC system for companding the luminance and chroma.)

Multipoint Distribution Service (MDS)—One-way domestic public radio service rendered on the microwave frequencies (usually 2.1–2.8 GHz). The service is from a fixed station transmitting (usually in an omnidirectional pattern) to multiple-receiving facilities which are located at fixed points that are determined by the subscriber's location.

Multistage LNA—Three or more transistor amplifier stages placed end to end (cascaded) so that the gain contribution of each will add up to total gain of approximately 50 dB. In most LNAs, the first stage (closest to the antenna feed probe) has the best noise characteristics needed to minimize the noise propagated along through the remaining stages.

NAND—A Boolean logic expression used to identify the logic operation wherein two or more variables must all be a logic "1" for the result to be a logic "0."

NAND Gate—A binary digital building block whose output is a logic "1" if any of its inputs are a logic "0." All inputs must be "1" for the output to be "0."

Nanosecond—One-billionth of a second (10^{-9} second).

Noise—1. Any unwanted signal not present in the original transmitted information. 2. Generally, any disturbance that tends to interfere with the normal operation of a communication device or system. 3. Random electrical signals, introduced by circuit components or natural disturbances, which degrade the performance of a communications channel.

Noise Factor—1. The ratio of the available signal-to-noise power ratio at the input to the available signal-to-noise power ratio at the output of a semiconductor device. 2. The ratio of input S/N to output S/N.

Noise Figure—The common logarithm of the ratio of the input signal-to-noise ratio to the output signal-to-noise ratio.

Noise Filter—A filter which attenuates a very narrow portion of the frequency spectrum, but will pass signals on either side.

Noise Temperature—1. Noise of a component expressed in Kelvins, and ordinarily used with reference to an LNA. 2. The thermal noise that the LNA adds to the TVRO system. 3. The temperature at which the thermal noise available from a device equals the total fluctuation noise actually available from the device.

Nonuniform Quantizing—Quantizing in which the intervals are not all equal.

NOR—A Boolean logic operation which yields a logic "0" output with one or more logic "1" input signals.

NTSC—Acronym for the *N*ational *T*elevision *S*ystem *C*ommittee, the committee that worked with the FCC in formulating standards for the present-day North American color-television system. Now used to describe the American system of color telecasting.

Operational Amplifier—1. An amplifier that performs various mathematical operations. Application of negative feedback around a high-gain DC amplifier produces a circuit with a precise gain characteristic that depends on the feedback used. By the proper selection of feedback components, operational amplifier circuits can be used to add, subtract, average, integrate, and differentiate. An operational amplifier can have a single-input/ single-output, differential-input/single-output, or differential-input/differential-output configuration. 2. Generally, any high-gain amplifier whose gain and response characteristics are determined by external components.

OR—A Boolean logic operation wherein one or more true inputs add up to one true output.

ORION—An acronym for *O*ak *R*estricted *I*nformation and *O*perational *N*etwork.

Oscillator—An electronic device used to produce periodic signals of a given frequency and amplitude.

PAL—An acronym for *P*hase *A*lternation *L*ine. So-called because the chroma phase is alternated 180° from one line to the next to cancel differential phase errors. PAL is the German system of colorcasting and was developed by Walter Brunch in 1961.

Parabolic Antenna—1. An antenna with a radiating element and a parabolic reflector that concentrates the radiated power into a beam. 2. A highly directional microwave antenna that uses a parabolic reflector.

Parallel—The internal handling of data in groups, with all elements of a group being handled simultaneously.

Parallel Data—The transfer of data bits simultaneously over two or more wires or transmission links.

Parity—1. A binary bit that indicates the number of "1s" in a given binary code. 2. In ASCII, a check of the total number of "1" bits in a character. A final eighth bit is then set so that the count, when transmitted, is always even or always odd. This even or odd state can easily be checked at the receiving end.

Parity Bit—A "1" or "0" added to a group of bits to identify the sum of the bits as either odd or even.

Parity Check—A checking system that tests to ensure that the 1s and 0s in any array of binary digits is consistently odd or even. Parity checking detects characters, blocks, or any other bit groupings that contain single errors.

Parity Error—An error which occurs in a particular entity of data, where an extra or redundant quantity of bits is sent with the data, as based on a specific calculation made at the transmit end. The same calculation is also performed at the receive end. If the results of both calculations do not agree, then a parity error has occurred.

Path Attenuation—The power loss between transmitter and receiver which results from all causes.

Path Loss—Also known as space loss. The attenuation of a signal as it travels through space.

Pay Television—A service that provides a TV viewer, for an extra fee, special extra programming, such as first-run movies, sporting events, news, and features. Term generally refers to an additional channel service—over and above the regular TV/network cable service—in exchange for an extra monthly charge.

PCM—An abbreviation for pulse code modulation. A communications systems technique of coding signals with binary digital codes to carry the information.

PDM (Pulse Duration Modulation)—See PWM.

Perigee—The point in an elliptical orbit closest to the Earth.

Period—The time between successive similar points of a repetitive signal.

Petalized Dish—A receiving dish that is shipped in sections or "petals," and then assembled at the installation site.

Phase—The time a signal is delayed with respect to some reference position. For sine-wave signals, the instantaneous angular difference between the signals, in degrees or radians.

Phosphor—A material that emits visible light when excited by an electron beam.

Photodetector—A device that converts light-intensity variations into corresponding electrical current variations.

Picosecond—One-trillionth of a second (10^{-12} second).

Picture Element—Abbreviated Pixel or PEL. 1. The smallest area of a picture whose light characteristics are converted to an equivalent electrical current or voltage. 2. A single point (or dot) in the grid of dots that forms the image on a CRT display. The more pixels to a screen unit, the higher the picture resolution.

Picture Frames—Two interlaced video fields make a frame. There are 30 frames per second in the United States' NTSC television transmit and receive systems.

Picture Tube—An electronic vacuum tube containing the elements necessary to convert transmitted electronic signals into visual images.

Pixel—See Picture Element.

Polarization—1. The process of making light or other radiation vibrate perpendicular to the ray. The vibrations are straight lines, circles, or ellipses—giving plane, circular, or elliptical polarization, respectively. 2. The property of a radiated electromagnetic wave that describes the direction of its electric field with respect to a reference plane.

Polar Mount—Support for a dish that permits simultaneous movements in azimuth and elevation.

Port—The entry channel to which a data set is attached. For example, in a central computer, each user is assigned one port.

Primary Colors—A set of colors that can be combined to produce any desired set of intermediate colors, within a certain limitation called the ''gamut.'' For color TV and photography, the primary colors are red, blue, and green.

Prime Focus Feed—An arrangement in which the entrance to a section of a waveguide is positioned at the signal focal point.

Processor—The central control element of a computer that obtains and executes the instructions contained in the program memory.

Professional Channel(s)—Subcarrier channels in FM broadcasting. They are usually 6.5 × the pilot carrier, or they may be interspersed between the stereo position and 102 kHz, if there is no SAP conflict.

Program—The sequence of instructions stored in the computer memory that state the execution of a particular operation.

Pulse Code Modulation—Abbreviated PCM. A process in which an analog signal is sampled, and the magnitude of each sample, with respect to a fixed reference, is quantized and converted (by coding) into a digital signal.

PWM—An abbreviation for pulse width modulation. A technique for sending information as patterns of the width of carrier pulses. Also called *pulse duration modulation* or *PDM*.

Quadrature—The state or condition of two related periodic functions or two related points, which are separated by a quarter of a cycle, or 90 electrical degrees.

Quadrature Modulation—The modulation of two carrier components, which are 90° apart in phase, by separate modulating functions.

Quantization—The process of converting a continuous analog input into a set of discrete output levels.

Quantization Distortion—Also called *quantization noise.* The inherent uncertainty or distortion introduced during quantization.

Quantizing—A process in which samples are classified into a number of adjacent intervals, with each interval being represented by a single value called the quantized value.

Radiation Patterns—The distribution in space of the electromagnetic energy produced by an antenna.

Random-Access Memory (RAM)—A storage arrangement from which information can be retrieved with a speed that is (ideally) independent of the location of the information in the storage.

Random-Noise Generator—A generator of a succession of random signals which are distributed over a wide frequency spectrum.

Random Numbers—A sequence of integers or group of numbers (often in the form of a table) which show absolutely no mathematical relationship to each other anywhere in the sequence.

Raster—The scanning pattern used in reproducing television or facsimile images.

Rate of Information Transfer—The speed at which information can be communicated from sender to receiver in a given length of time without significant error.

Ratio Detector—An FM detector which inherently discriminates against amplitude modulation. Two diode detectors and tuned circuits are used so that the output is proportional to the ratio of the applied voltages to those circuits.

Real Time—Generally, an operating mode under which receiving the data, processing it, and returning the results takes place so quickly that it actually affects the functioning of the environment, guides the physical processes in question, or interacts instantaneously with the human user(s). Examples include a process-control system used in manufacturing, or a computer-assisted instruction system used in an educational institution.

Receiver—A person or device to which information is sent over a communication link.

Regeneration—The process of recognizing and reconstructing a signal (usually digital) so that the amplitude, waveform, and timing are constrained within state limits.

Register—A series of identical circuits placed side by side so that they are able to store digital information.

Resolution—A measure of the smallest detail that can be seen in a reproduced picture.

RF—Abbreviation for radio frequency. Radiation having a frequency between zero and the lowest infrared frequency (≈ 300 GHz).

RGB—An abbreviation for red-green-blue, the three primary colors to which the retina of the human eye responds. RGB color monitors use three video signals (one for each primary color) to produce a variety of color shades. RGB monitors produce the sharpest color pictures, but are generally of a lesser resolution than most monochrome monitors.

RG-11/U Coaxial Cable—A coaxial cable that is electrically similar to RG-59/U coax, but with larger conductors for less signal loss over long distances.

RG-59/U Coaxial Cable—A type of coaxial cable characterized by an impedance of 75 ohms, which is commonly used for the transmission of video signals.

Rise Time—The time required for the leading edge of a pulse to rise from 10 percent to 90 percent of its final value.

ROM—An abbreviation for read-only memory. A storage arrangement where information is stored permanently or semipermanently and is read out, but not altered.

RS-232 Interface—The interface between a modem and the associated data terminal equipment, as standardized by EIA Standard RS-232.

R-Y Signal—In color television, the red-minus-luminance color-difference signal. When combined with the luminance (Y) signal, it produces the red primary signal.

Sample—1. To obtain sample values of a complex wave at periodic intervals. 2. The value of a particular characteristic of a signal at a chosen instant.

Sample and Hold—A circuit that holds or ''freezes'' a changing analog-input signal voltage. Usually, the voltage thus frozen is then converted into another form, either by a voltage-controlled oscillator, an analog-to-digital converter, or some other device.

Sampling—The process of taking samples, usually at equal time intervals.

Sampling therorem—A theorem (developed by Nyquist in 1928) which states that two samples per cycle will completely characterize a band-limited signal; that is, the sampling rate must be twice the highest-frequency component. (In practice, the sampling rate is ordinarily three to ten times the highest signal frequency.)

SAP—An acronym for *S*econd *A*udio *P*rogram, which is transmitted in FM. Generally used for dual-language purposes or other audio at a bandwidth of only 10 kHz.

SATCOM—The name of satellites built by RCA Astro-Electronics and operated by RCA Americom. They distribute programming to cable-TV systems and provide network radio transmissions, as well as commercial and government voice, data, and video services.

SATCOM F1—American TV satellite (operated by RCA) used to supply most of the cable-TV programming on 24 transponders (12 are vertically and 12 are horizontally polarized). It is located at 135° west longitude. Also referred to as F1.

SATCOM F2—American TV satellite (operated by RCA) used to supply assorted video and data programming to Alaska and other points in the United States. Like its sister, F1, it too has 24 transponders. It is located at 119° west longitude. Also referred to as F2.

Satellite Communication Systems—A remote communications technique using a satellite in orbit to receive signals from one location and, then, retransmit them to another location.

Satellite Receiver—A component used for tuning in a selected satellite transponder. It may contain one or two downconverters, or none. The receiver recovers the original baseband signals and delivers them to a remodulator. The receiver can also supply the DC operating voltages for an external LNA and downconverter.

Satellites—1. Geosynchronous space vehicles launched from Earth and positioned at specific slots in an arc directly above the equator and 22,300 miles above the earth. Being synchronous with earth movements, they remain in fixed positions, with only minor station-keeping thruster propulsion required during their (typical) 8- to 10-year lifetimes. 2. TV stations licensed to rebroadcast the programming of parent stations.

Saturation—A term applied to a color which is ''pure'' to the extent that it is not mixed with white light. The less white light, the more saturated the color is said to be. This definition is suggestive of ''signal-to-noise'' ratio, where the ''noise'' in this case is white light.

SAW—Abbreviation for Surface Acoustic-Wave (device). A technology for broad-bandwidth signal delay, custom-designed filters, and complex generation and correlation at IF frequencies. SAW devices make use of a single-crystal planar substrate, with aluminum or gold electrode patterns fabricated by photolithography, of some piezoelectric material. The electrode patterns are used to excite and detect minute acoustic waves that travel over the surface of the substrate, much like earthquake waves travel over the crust of the earth.

Sawtooth—A waveform that increases approximately linearly as a function of time for a fixed interval, returning to its original state sharply, and then repeating the process periodically.

Scalar Feed—A series of concentric rings positioned at the mouth of the feedhorn which act as an aid in picking up the reflected signals from the dish.

Scan Line—One pass of an electron beam across the face of a cathode-ray tube or the target of a CCD or vidicon.

Scan Pattern—The path of an electron beam which converts an image into electronic signals.

Schmitt Trigger—A regenerative circuit which changes stage abruptly when the input signal crosses specified (DC) triggering levels. (Also called a Schmitt limiter.)

Schottky TTL—A TTL circuit that incorporates Schottky diodes to greatly speed up TTL circuit operation.

SCR—An abbreviation for silicon-controlled rectifier. A four-layer PNPN semiconductor with cathode, anode, and gate; actually a reverse-blocking thyristor.

Scramble—To transpose and/or invert bands of frequencies, or otherwise modify the form of the intelligence at the transmitting end of a communication system, according to a prearranged scheme, to obtain secrecy.

Scrambler—A device that changes data so it appears to be in an unrecognizable pattern. A descrambling device can change (unscramble) this data back to its original pattern.

Scrambling—Technique to encipher a signal to prevent unauthorized reception without a descrambler device.

SECAM—An acronym for *S*equentiel à *M*emoire, which is the French method of broadcasting color that was invented by Henri de France in 1957. It transmits R-Y and B-Y in sequence, depending on frequency modulation for the video and amplitude modulation for the audio—just the reverse of the United States' NTSC method.

Serial Data—The transfer of data over a single wire in a sequential pattern.

Serial Transmission—A mode of transmission in which each bit of a character is sent sequentially on a single circuit or channel, rather than simultaneously as in a parallel transmission.

Shapiro Network—A class of passive networks having an asymmetrical matrix that allows for proper N-order sequence decryption of Crossmeyer functions of the first kind.

SHF—An abbreviation for super-high frequency. The portion of the electromagnetic spectrum from 3,000 megahertz to 30,000 megahertz.

S&H (Sample and Hold)—A circuit which samples a signal and holds the sampled value until the next sample is taken. In A/D conversion, S/H usually is an analog function. In D/A conversion, the S/H may be performed digitally, either making continuous use of the D/A converter or letting the D/A converter be shared by other functions, with its output signal held by an analog sample and hold circuit.

Signal Converter—A communications circuit that converts one form of information-signal input into another form of signal output.

Signal Threshold Level—Measurement in decibels of a satellite receiver's ability to produce noise-free pictures. With 7 or less indicating excellent, 8 is good, 9 is fair, and 10 or more is poor.

Signal-to-Noise Ratio—Ratio of the magnitude of a signal to that of the noise (often expressed in decibels). (Also called signal-noise ratio.)

Simplex—A communications system or device capable of transmission in one direction only.

Skew—The time difference of corresponding digital information on separate lines, measured at the rising or falling edges of digital data.

Slot—The longitudinal position of a satellite in its geosynchronous orbit.

Slow-Scan Television—A television system that employs a slow rate of horizontal scanning that is suitable for the transmission of printed matter, photographs, or drawings.

Smoothing Filter—A generally low-pass filter that removes the sampling rate high-frequency components; used in digital audio systems.

Soft Scrambling—An encryption method using analog techniques, which is not as difficult to defeat as hard scrambling. Examples of soft scrambling are video inversion or synch suppression technology.

Software Package—Various computer programs or sets of programs used in a particular applications package.

Sound Spectrum—The range of frequencies in the electromagnetic spectrum that can be heard by the human ear; usually from about 30 hertz to 15,000 hertz.

Space Station—A platform in earth orbit. The space station comprises four modules: the habitability module with life-support systems, the service and repair module, the microgravity research module, and the life-science module. In actuality, the station consists of two complexes: an unmanned structure in polar orbit (i.e., circling the earth from one pole to the other) and a manned complex of structures in low-inclination orbit near the equator.

Sparklies—Small black and/or white blips in a satellite-TV picture which indicates insufficient signal.

Spherical Dish—A dish whose surface is a section of a sphere.

Splitter—A device with one input that provides two or more outputs to allow multiple receiver hookups to one antenna. A splitter can be passive (i.e., an antenna coupler) or active (providing gain).

Stagger-Tuned Amplifier—An amplifier consisting of two or more stages, with each tuned to a different frequency.

Staircase—A video test signal containing several steps at increasing luminance levels. The staircase signal usually is amplitude-modulated by the subcarrier frequency and is useful for checking amplitude and phase linearities in video systems.

Subcarrier—1. A carrier used to generate a modulated wave which is applied, in turn, in a modulated form to modulate another carrier. 2. A second signal that is "piggybacked" onto another (main) electronic signal. In satellite-TV applications, the video is transmitted via the transponder's main carrier while the audio goes out over an FM frequency subcarrier. Some satellite transponders carry up to four audio or data subcarriers, which carry information that can be related or unrelated to the main signal's programming.

Subcarrier Oscillator—In a color-television receiver, the crystal oscillator operating at the chrominance subcarrier frequency of 3.48 MHz.

Subscription Television—Abbreviated STV. An over-the-air or cable pay-TV system in which scrambled signals are sent to decoder boxes connected to TV sets. Viewers pay a fee to receive the programming sent via STV.

Synchronous—Refers to two or more events that happen in a system at the same time.

Synchronous Transmission—A transmission method where the synchronizing of characters is controlled by timing signals generated at the sending and receiving stations (as opposed to start/stop communications). Both stations operate continuously at the same frequency and are maintained in a desired phase relationship. Any of several data codes may be used for the transmission, as long as the code utilizes the required line-control characters. (Also called "bi-synch," or "binary synchronous." Compare with "asynchronous transmission."

Tangential Sensitivity—A term generally used to indicate quality in a receiving system. The term can be used to define the minimum signal level that can be detected above the background noise. However, it is usually expressed as that signal power level which causes a 3-dB rise above the noise-level reading.

Tchebychev Filters—Filter networks that are designed to exhibit a predetermined ripple in the passband (ripple amplitudes from 0.01 dB to 3 dB are common) in exchange for which they provide a more rapid attenuation above cutoff—which, unlike their passband response, is monotonic.

Tearing—A form of weak signal interference which causes ragged streaks on the TV picture, in vertical lines that join light to dark transitions. If this occurs in a satellite-TV picture, it is usually an indication that the receiver is operating well below the FM threshold.

Television—The transmission or reception of a televised image, accompanied usually by vertical and horizontal synch, sound, and color (chroma), all delivered within a channel which is 6 MHz wide.

Terminal—An input/output device designed to send or receive data.

Thermal Noise—Noise that occurs in all transmission media and communications equipment as a result of random electron motion which is a function of temperature. Thermal noise sets the lower limit for the sensitivity of a receiving system.

Theshold Extension—A circuit technique, sometimes located in the loop filter of a phase-lock loop demodulator, which improves the low-signal performance of a receiver by lowering the FM threshold.

Time-Division Multiplexer—A device which permits the simultaneous transmission of many independent channels into a single high-speed data stream, by dividing the signal into successive alternate bits.

Time-Division Multiplexing—A communications system technique that separates information from channel inputs and places them on a carrier frequency in specific positions in time.

Time-Multiplexed System—A communications system serving several users, whereby each communicator takes his turn during a specific period of time.

Translation Frequency—The 2200-MHz (2.2-GHz) frequency difference between an uplink and downlink signal.

Transponder—1. A receiver-transmitter combination that receives (uplink) signals, and retransmits (downlink) the signals on a different carrier frequency. 2. Part of satellite that receives signals from earth, alters their frequency, amplifies the signals, and sends them back to earth.

Transverse Electromagnetic Wave—Abbreviated TEM wave. In a homogeneous isotropic medium, an electromagnetic wave in which the electric and magnetic field vectors are both perpendicular everywhere to the direction of propagation. This is the normal mode of propagation in coaxial line, open-wire line, and stripline.

TV Channels—A total of 12 VHF and 56 UHF television channels, beginning with Channel 2 and ending with Channel 69. Former UHF Channels 70 through 83 are now either used by repeaters or are directly assigned to the Land Mobile Service. Frequency ranges begin at 54 MHz and extend to 806 MHz.

Tweaking—The fine tuning of a circuit with overly critical adjustments, repeated several times until no further improvement in performance is obtained. This usually requires considerable skill and is not recommended as a requirement in circuits for production or construction by a hobbyist with limited technical resources.

Unguided Wave—A broadcast wave.

UHF—Abbreviation for ultrahigh frequency. The portion of the electromagnetic spectrum from 300 to 3000 megahertz.

Uniform Quantizing—Quantizing in which all the intervals are equal, i.e, a linear A/D converter.

Unscramble—A method used to restore a scrambled signal to its original format.

Uplink—1. Path of a transmission from an earth station to a satellite. 2. The earth station which transmits TV programs to a satellite for relay back to the ground. 3. The communications path from Earth to one or more of the geosynchronous satellites.

Up-Link—The carrier used by earth stations to transmit information to a satellite.

Vertical Blanking Pulse—In television, a pulse transmitted at the end of each field to cut off the cathode-ray beam while it returns to the start of the next field.

Vertical-Field Blanking Interval—This interval occupies 16,667 milliseconds and occurs at the end of each 262.5 field scan lines. There are 21 usable lines in this interval that are devoted to 6 vertical synch pulses and 12 equalizing pulses, as well as additional horizontal pulse intervals.

Very-Large-Scale Integration (VLSI)—A concept whereby a complete system function is fabricated as a single microcircuit. In this context, a system, whether digital or linear, is considered to be one that contains 1000 or more elements.

Video—Composite video contains color and luminance (brightness information) as well as horizontal and vertical synch pulses during the horizontal and vertical blanking intervals. One line of video, for instance, amounts to approximately 52 microseconds in visible horizontal scan time, and 63.5 microseconds for a complete line, including the color burst and horizontal synch.

Video Carrier—A frequency that is 1.25 MHz above the lower edge of the assigned 6-MHz frequency band of a TV channel. At Channel 2, for instance, with a bandwidth of 54–60 MHz, the video carrier rests at 55.25 MHz.

VideoCipher I, II™—A trademark for a video/audio encryption system developed by M/A-COM LINKABIT, Inc. VCI uses digital video and audio encryption; VCII uses analog video and digital audio encryption.

Video Inversion—A type of encoding or scrambling in which the transmitted downlink video signals are inverted.

Video Modulation—That component of a TV RF signal that includes both the color and monochrome information within a bandwidth of 4.5 MHz.

Video Signal—The electrical signal which contains the picture-information content in a television or facsimile system.

VHF—Abbreviation for very high frequency. The portion of the electromagnetic spectrum from 30 to 300 megahertz.

Voice-Grade Channel—A channel suitable for the transmission of speech, digital or analog data, or facsimile; generally having a frequency range of about 300 to 3000 hertz.

VCO—Abbreviation for voltage-controlled oscillator. An oscillator whose frequency is determined by a control voltage.

VTO—Abbreviation for voltage-tuned oscillator. In a TVRO, the local oscillator in an externally positioned downconverter whose operating frequency is changed by a DC voltage that is delivered from the in-home satellite receiver.

Waveguide—A transmission line that is comprised of a hollow conducting tube within which electromagnetic waves are propagated. It may, in some cases, be a solid dielectric or dielectric-filled conductor.

Wavelength—The distance between successive "peaks" of an energy sinusoid traveling at the speed of light (300,000,000 meters per second).

White Level—The carrier-signal level which corresponds to maximum picture brightness in television.

Wind Load Survival—The amount of wind pressure tolerated by a dish or other structure.

X-Band—A radio-frequency band of 5200 to 11,000 MHz, having wavelengths of 5.77 to 2.75 centimeters.

XSO—Abbreviation for crystal-stabilized oscillator.

Index

D

E

F

G

H

I